T0328265

Introduction to Addiction

Volumes in the series:

Volume 1: Introduction to Addiction
Volume 2: Psychostimulants
Volume 3: Alcohol
Volume 4: Opioids
Volume 5: Nicotine and Marijuana

Neurobiology of Addiction

Introduction to Addiction

Addiction, Animal Models, and Theories

VOLUME 1

GEORGE F. KOOB

National Institute on Alcohol Abuse and Alcoholism, National Institutes of Health, Bethesda, Maryland, United States
National Institute on Drug Abuse, National Institutes of Health, Bethesda, Maryland, United States

MICHAEL A. ARENDS

The Scripps Research Institute, La Jolla, California, United States

MANDY MCCRACKEN

National Institute on Drug Abuse, National Institutes of Health, Bethesda, Maryland, United States

MICHEL LE MOAL

University of Bordeaux, Neurocentre Magendie, Inserm, Bordeaux, France

Academic Press is an imprint of Elsevier
125 London Wall, London EC2Y 5AS, United Kingdom
525 B Street, Suite 1650, San Diego, CA 92101, United States
50 Hampshire Street, 5th Floor, Cambridge, MA 02139, United States
The Boulevard, Langford Lane, Kidlington, Oxford OX5 1GB, United Kingdom

Library of Congress Cataloging-in-Publication Data
A catalog record for this book is available from the Library of Congress

British Library Cataloguing-in-Publication Data
A catalogue record for this book is available from the British Library

ISBN: 978-0-12-816863-9

For information on all Academic Press publications visit our website at
https://www.elsevier.com/books-and-journals

Publisher: Nikki Levy
Acquisition Editor: Joslyn Chaiprasert-Paguio
Editorial Project Manager: Sam W. Young
Production Project Manager: Bharatwaj Varatharajan
Cover Designer: Matthew Limbert

Typeset by TNQ Technologies

Contents

Preface

The present series of volumes on the *Neurobiology of Addiction* are a direct extension of our original book from 2006, *Neurobiology of Addiction* (Koob and Le Moal, 2006, Elsevier). As we embarked on updating the original book years ago, we quickly realized that a prodigious amount of new work had been done on the neurobiology of addiction during the ensuing 13 years. From 2006 until 2019, the number of PubMed citations for cocaine alone (41439 as of Feb 2019) had nearly doubled over the total number up until 2006 (~25000). This extraordinary progress in the field of the neurobiology of addiction required a different theoretical and practical approach to writing our second book.

From a theoretical perspective, we chose to use a heuristically identified domain model that originated in our seminal *Science* paper on addiction: "Drug abuse: hedonic homeostatic dysregulation" [1]. Here, based on the social psychology of self-regulation theory, experimental psychology, and psychiatry, we originally defined addiction as a cycle that consists of three stages: binge/intoxication, withdrawal/negative affect, and preoccupation/anticipation. Eventually, three corresponding domains and neurocircuits coalesced around these three stages: binge/intoxication (incentive salience/pathological habits domain, basal ganglia neurocircuits), withdrawal/negative affect (negative affect domain, extended amygdala), and preoccupation/anticipation (executive function, prefrontal cortex). A recent human clinical, behavioral, and self-report study confirmed these three neurofunctional domains, at least for alcohol use disorder [2]. Thus, the new revised book, *Neurobiology of Addiction*, is now organized along the *three stage/three domain* construct while retaining synthesis at the circuit, cellular, and molecular levels of analysis.

Our new book series incorporates components of an "experimental medicine" approach. Volume 1 of the series, *Introduction to Addiction*, includes three chapters. Chapter 1 (What is Addiction?) covers the disease construct and associated symptoms in humans at the three stages of the addiction cycle. Chapter 2 (Animal Models of Addiction) goes back to the laboratory and elaborates the various animal models that have been developed to explore the intricacies of each domain. Chapter 3 (Neurobiological Theories of Addiction) links neurobiology to pathophysiology. In this latter chapter, we sought to include as many of the different theoretical perspectives that we could find, including theories of different levels of intensity of addiction (based on the *Diagnostic and Statistical Manual of Mental Disorders*, 5th edition), different components of the addiction syndrome, and translational pathophysiology. The unique elements of addiction that are defined by different major drug classes—psychostimulants, alcohol, opioids, nicotine, and marijuana—are explored as separate volumes in this series.

From a practical perspective, the organization of the original book in different volumes was necessitated by the prodigious increase in publications from 2006 to present. We had prided ourselves in finding virtually all of the published work in 2006 and citing as much of it as possible. Most of the early cited literature has been retained in the present series, but such an approach of citing every study from 2006 to present was not humanly possible for the present series. As a result, for many of the topics, we rely on key seminal papers and review articles. For each seminal advance, where possible, we have included summary figures. We hope readers will see how the field has evolved at the level of refined techniques and consolidated theoretical approaches and apologize in advance to researchers who may have a key seminal paper that we missed.

We are very excited and encouraged about the tremendous advances that have been made in unveiling the neurobiology of addiction, both clinically and preclinically. We look forward to further insights that tomorrow's research will provide.

George F. Koob
Michael A. Arends
Mandy McCracken
Michel Le Moal

References

[1] Koob GF, Le Moal M. Drug abuse: hedonic homeostatic dysregulation. Science 1997;278:52–8.
[2] Kwako LE, Schwandt ML, Ramchandani VA, Diazgranados N, Koob GF, Volkow ND, Blanco C, Goldman D. Neurofunctional domains derived from deep behavioral phenotyping in alcohol use disorder. American Journal of Psychiatry 2019 [in press]. https://doi.org/10.1176/appi.ajp.2018.18030357444.

CHAPTER 1

What is addiction?

Contents

1. Definitions of addiction

1.1 Drug use, drug abuse, and drug addiction

Drug addiction is a chronically relapsing disorder that is characterized by (1) compulsion to seek and take the drug, (2) loss of control in limiting intake, and (3) emergence of a negative emotional state (e.g., dysphoria, anxiety, irritability) when access to the drug is prevented [2]. The occasional but limited use of an abusable drug is clinically distinct from escalated drug use, the loss of control over limiting drug intake, and the emergence of chronic compulsive drug seeking that characterize addiction. Historically, three types of drug use have been delineated: (1) occasional, controlled, or social use, (2) drug abuse or harmful use, and (3) drug addiction as characterized as either Substance Dependence (*Diagnostic and Statistical Manual of Mental Disorders*, fourth edition [DSM-IV]) or Dependence (International Statistical Classification of Diseases and Related Health Problems, 10th revision [ICD-10]). More current descriptions have elaborated a continuum of

Introduction to Addiction
ISBN 978-0-12-816863-9, https://doi.org/10.1016/B978-0-12-816863-9.00001-7

behavioral pathology, from drug use to addiction [2], in the context of the substance use disorder construct of DSM-5.

1.2 Diagnostic criteria for addiction

The diagnostic criteria for addiction as described in the DSM have evolved, from the first edition published in 1952 to the DSM-5 [3], with a shift from an emphasis on the criteria of tolerance and withdrawal to other criteria that are more directed at compulsive use. The criteria for Substance Use Disorders in the DSM-IV closely resembled those outlined in the *International Statistical Classification of Diseases and Related Health Problems*, 10th revision (ICD-10), for Drug Dependence ([4]; Table 1). The DSM-5 was published in 2013 [3], and the criteria for drug addiction were changed both conceptually and diagnostically. The new diagnostic criteria for addiction merged the abuse and dependence constructs (i.e., substance abuse and substance dependence) into one continuum that defines "substance use disorders" on a range of severity, from mild to moderate to severe, based on the number of criteria that are met out of 11. The severity of a substance use disorder (addiction) depends on how many of the established criteria are met by an individual. Mild Substance Use Disorder is the presence of 2–3 criteria, moderate is 4–5 criteria, and severe is ≥6 criteria. These criteria remain basically the same as in the DSM-IV and ICD-10, with the exception of the removal of "committing illegal acts" and the addition of a new "craving" criterion. For example, rather than differentiating subjects with alcohol dependence and subjects with alcohol abuse, the DSM-5 classification of Alcohol Use Disorder encompasses individuals who are afflicted by the disorder to different degrees, from "mild" (e.g., a typical college binge drinker who meets two criteria, such as alcohol is often taken in larger amounts or over a longer period than was intended, and there is a persistent desire or unsuccessful efforts to cut down or control alcohol use) to "severe" (e.g., a classic person with alcoholism who meets six or more criteria, such as a great deal of time spent in activities necessary to obtain alcohol, use alcohol, or recover from its effects, recurrent alcohol use resulting in a failure to fulfill major role obligations at work, school, or home, alcohol use despite knowledge of having a persistent or recurrent physical or psychological problem, continued alcohol use despite persistent social or interpersonal problems, tolerance, and withdrawal).

The terms Substance Use Disorder and Addiction will be used interchangeably throughout this book to refer to a usage process that moves from social drug use to compulsive use as defined above. Drug addiction is a disease and, more precisely, a *chronic* relapsing disease. The associated medical, social, and occupational difficulties that usually develop during the course of addiction do not disappear after detoxification. Addictive drugs produce changes in brain circuits that endure long after the person stops taking them. These prolonged neurochemical and neurocircuitry changes and the associated personal and social difficulties put former patients at risk for relapse. Abstinent rates were 12% in an untreated sample of alcohol-dependent subjects after 1 year [5].

Table 1 DSM-5, DSM-IV, and ICD-10 diagnostic criteria for abuse and dependence.

DSM-5	DSM-IV	ICD-10
Dependence		
A problematic pattern of substance use leading to clinically significant impairment or distress, as manifested by at least two of the following occurring within a 12 month period:	*A maladaptive pattern of substance use, leading to clinically significant impairment or distress as manifested by three or more of the following occurring at any time in the same 12-month period:*	*Three or more of the following have been experienced or exhibited at some time during the previous year:*
1. Tolerance is defined by either of the following: (a) a need for markedly increased amounts of substance to achieve intoxication or desired effect (b) a markedly diminished effect with continued use of the same amount of substance.	**1.** Need for markedly increased amounts of a substance to achieve intoxication or desired effect; or markedly diminished effect with continued use of the same amount of the substance.	**1.** Evidence of tolerance, such that increased doses are required in order to achieve effects originally produced by lower doses.
2. Withdrawal is manifested by either of the following: (a) the characteristic withdrawal syndrome for substance or (b) substance is taken to relieve or avoid withdrawal symptoms.	**2.** The characteristic withdrawal syndrome for a substance or use of a substance (or a closely related substance) to relieve or avoid withdrawal symptoms.	**2.** A physiological withdrawal state when substance use has ceased or been reduced as evidenced by: the characteristic substance withdrawal syndrome, or use of substance (or a closely related substance) to relieve or avoid withdrawal symptoms.
3. There is persistent desire or unsuccessful efforts to cut down or control substance use.	**3.** Persistent desire or one or more unsuccessful efforts to cut down or control substance use.	**3.** Difficulties in controlling substance use in terms of onset, termination, or levels of use.
4. Substance is often taken in larger amounts or over a longer period than was intended.	**4.** Substance used in larger amounts or over a longer period than the person intended.	None
5. Important social, occupational, or recreational activities are given up or reduced because of substance use.	**5.** Important social, occupational, or recreational activities given up or reduced because of substance use.	**4.** Progressive neglect of alternative pleasures or interests in favor of substance use; or

Continued

Table 1 DSM-5, DSM-IV, and ICD-10 diagnostic criteria for abuse and dependence.—cont'd

DSM-5	DSM-IV	ICD-10
6. A great deal of time is spent in activities necessary to obtain substance, use substance, or recover from its effects.	**6.** A great deal of time spent in activities necessary to obtain, to use, or to recover from the effects of substance used.	A great deal of time spent in activities necessary to obtain, to use, or to recover from the effects of substance use.
7. Continued substance use despite having persistent or recurrent social or interpersonal problems caused or exacerbated by the effects of substance.	**7.** Continued substance use despite knowledge of having a persistent or recurrent physical or psychological problem that is likely to be caused or exacerbated by use.	**5.** Continued substance use despite clear evidence of overtly harmful physical or psychological consequences.
None	None	**6.** A strong desire or sense of compulsion to use substance.
Abuse		
	A maladaptive pattern of substance use leading to clinically significant impairment or distress, as manifested by one (or more) of the following occurring with a 12 month period:	*A pattern of substance use that is causing damage to health.*
8. Substance use is continued despite knowledge of having a persistent or recurrent physical or psychological problem that is likely to have been caused or exacerbated by substance.	**1.** Recurrent substance use resulting in a failure to fulfill major role obligations at work, school, or home.	The damage may be physical or mental. The diagnosis requires that actual damage should have been caused to the mental or physical health of the user.
9. Recurrent use in situations in which it is physically hazardous.	**2.** Recurrent substance use in situations in which use is physically hazardous.	
None	**3.** Recurrent substance-related legal problems.	
10. Recurrent substance use resulting in a failure to fulfill major role obligations at work, school or home.	**4.** Continued substance use despite having persistent or recurrent social or interpersonal problems caused or exacerbated by the effects of the drug.	
11. Craving or a strong desire or urge to use alcohol (or other substance).	None	None

1.3 Frequency and cost of addiction

Data on the frequency of alcohol and substance use disorders over time can be challenging given the switch from the DSM-IV to DSM-5. One approach that is used by the National Survey on Drug Use and Health is to utilize a combination of the percentage of individuals who have drug abuse and drug dependence as defined by the DSM-IV because only recent data are available for the frequency of substance use disorders based on the DSM-5 criteria. Based on the criteria of the DSM-5, 20.8 million people aged 12 and older had a Substance Use Disorder (including alcohol and illicit drugs) in 2015, of which 7.7 million could be classified with illicit drug use disorder, and 15.7 million people could be classified with alcohol use disorder (U.S. [6]; Table 2). In 2015, 23.9% of the population (64.0 million people) used tobacco products in the past month, including cigarettes (52.0 million), cigars (12.5 million), smokeless tobacco (9.0 million), and pipe tobacco (2.3 million). In 2015, 4.0 million people aged 12 and older could be classified with marijuana use disorder, 896,000 could be classified with cocaine use disorder, and 591,000 could be classified with heroin use disorder. Fig. 1 illustrates the number of individuals with alcohol and substance use disorders from 2002 to 2017.

Table 2 Estimated numbers and percentages of persons of the US population aged 12 and older (268 million) who ever used particular substances or used them in the past year and, among those with past year use, the percentage who reached criteria for a DSM-IV substance use disorder.

	Ever used		Last-year use		Last-year use w/dependence		Last-year use w/abuse or dependence	
Substance	**Millions**	**%**	**Millions**	**%**	**Millions**	**%[a]**	**Millions**	**%[b]**
Cocaine	38.7	14.5	4.8	1.8	0.6	12.7	0.9	18.8
Stimulants	—	—	5.3	2.0	0.3	6.4	0.4	8.1
Methamphetamine	14.5	5.4	1.7	0.6	0.8	44.7	0.9	51.2
Heroin	5.1	1.9	0.8	0.3	0.6	71.3	0.6	73.8
Analgesics	—	—	12.5	4.7	1.6	12.4	2.0	16.0
Alcohol	216.8	81.0	175.9	65.7	7.8	4.4	15.7	8.9
Tobacco	171.1	63.9	78.3	29.2	—	—	—	—
Cigarettes	156.5	58.5	61.9	23.1	30.2	48.8	30.2	48.8
Cannabis	117.9	44.0	36.0	12.8	2.6	7.2	4.0	11.1
Ecstasy	18.3	6.8	2.6	1.0	—	—	—	—
Hallucinogens	40.9	15.3	4.7	1.8	0.1	2.0	0.3	5.7

Cigarette smokers with dependence are considered the number of people who smoke daily.
[a]% who had dependence among those who used in the previous year.
[b]% who had abuse or dependence among those who used in the last year.
Data from Substance Abuse and Mental Health Services Administration. Key substance use and mental health indicators in the United States: results from the 2015 National Survey on Drug Use and Health (HHS Publication No. SMA 16-4984, NSDUH Series H-51). Rockville: Substance Abuse and Mental Health Services Administration; 2016.

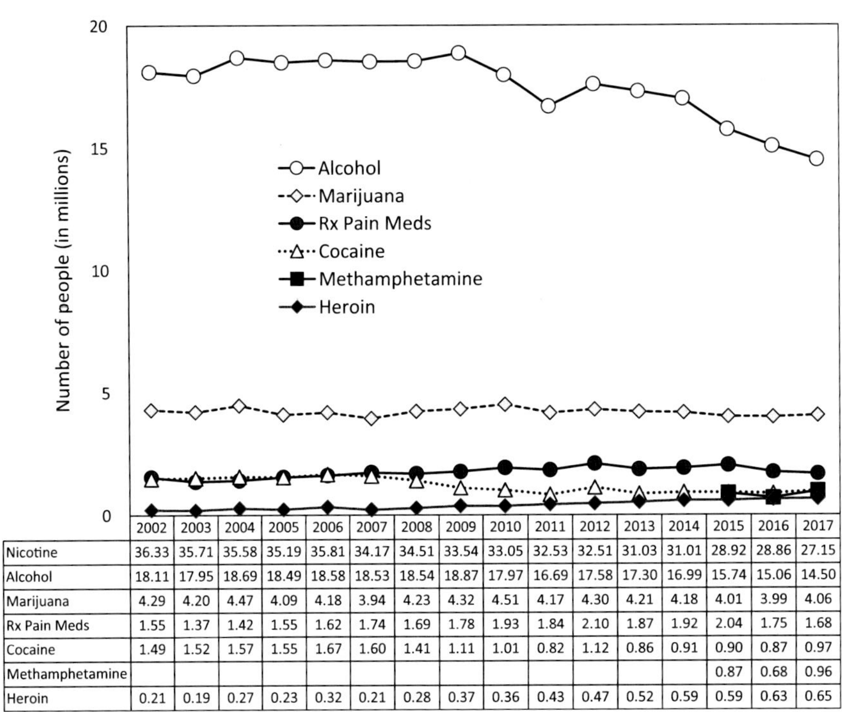

	2002	2003	2004	2005	2006	2007	2008	2009	2010	2011	2012	2013	2014	2015	2016	2017
Nicotine	36.33	35.71	35.58	35.19	35.81	34.17	34.51	33.54	33.05	32.53	32.51	31.03	31.01	28.92	28.86	27.15
Alcohol	18.11	17.95	18.69	18.49	18.58	18.53	18.54	18.87	17.97	16.69	17.58	17.30	16.99	15.74	15.06	14.50
Marijuana	4.29	4.20	4.47	4.09	4.18	3.94	4.23	4.32	4.51	4.17	4.30	4.21	4.18	4.01	3.99	4.06
Rx Pain Meds	1.55	1.37	1.42	1.55	1.62	1.74	1.69	1.78	1.93	1.84	2.10	1.87	1.92	2.04	1.75	1.68
Cocaine	1.49	1.52	1.57	1.55	1.67	1.60	1.41	1.11	1.01	0.82	1.12	0.86	0.91	0.90	0.87	0.97
Methamphetamine														0.87	0.68	0.96
Heroin	0.21	0.19	0.27	0.23	0.32	0.21	0.28	0.37	0.36	0.43	0.47	0.52	0.59	0.59	0.63	0.65

Figure 1 Trends in Alcohol and Substance Use disorders from 2002 to 2017 as the number of individuals afflicted (in millions). Note that tobacco is not illustrated graphically because it would skew the ordinate, but tobacco is listed in the table below the graph. Data from the Center for Behavioral Health Statistics and Quality and the National Survey on Drug Use and Health: Detailed Tables, 2002–17. Substance Abuse and Mental Health Services Administration, Rockville, MD. (Dr. Aaron White, National Institute on Alcohol Abuse and Alcoholism, personal communication). Note also that the population of the United States grew from 288 million in 2002 to 326 million in 2017.

The cost to society of drug abuse and drug addiction is prodigious in terms of both direct costs and indirect costs associated with secondary medical events, social problems, and the loss of productivity. In the United States alone, illicit drug use and addiction cost society $191 billion per year [7]. Alcohol misuse costs society $249 billion per year [8], and nicotine addiction costs society $300 billion [9]. In terms of health burden, tobacco is the leading cause of preventable death in the United States, and alcohol ranks fourth—both are key contributors to disability and reduced life expectancy [9a,10].

Much of the initial research of the neurobiology of drug addiction focused on the acute impact of drugs of abuse (analogous to comparing no drug use to drug use). The focus has shifted to chronic administration and the acute and long-term neuroadaptive changes that occur in the brain. Sound arguments have been made to support the hypothesis that addictions are similar to other chronic relapsing disorders, such as diabetes, asthma, and hypertension, in their chronic relapsing nature and treatment efficacy [11]. Current neuroscientific drug abuse research seeks to understand the cellular and molecular mechanisms that mediate the transition from occasional, controlled drug use to the loss of behavioral control over drug seeking and drug taking that defines chronic addiction [12].

1.4 Patterns of addiction

Different drugs produce different patterns of addiction, with an emphasis on different components of the addiction cycle (Fig. 2). Opioids are a classic drug of addiction, in which an evolving pattern of use includes intravenous or smoked drug taking, intense initial intoxication, the development of profound tolerance, the escalation of intake, and profound dysphoria, physical discomfort, and somatic withdrawal signs during abstinence.

Intense preoccupation with obtaining opioids (craving) develops and often precedes the somatic signs of withdrawal. This preoccupation is linked to stimuli that are associated

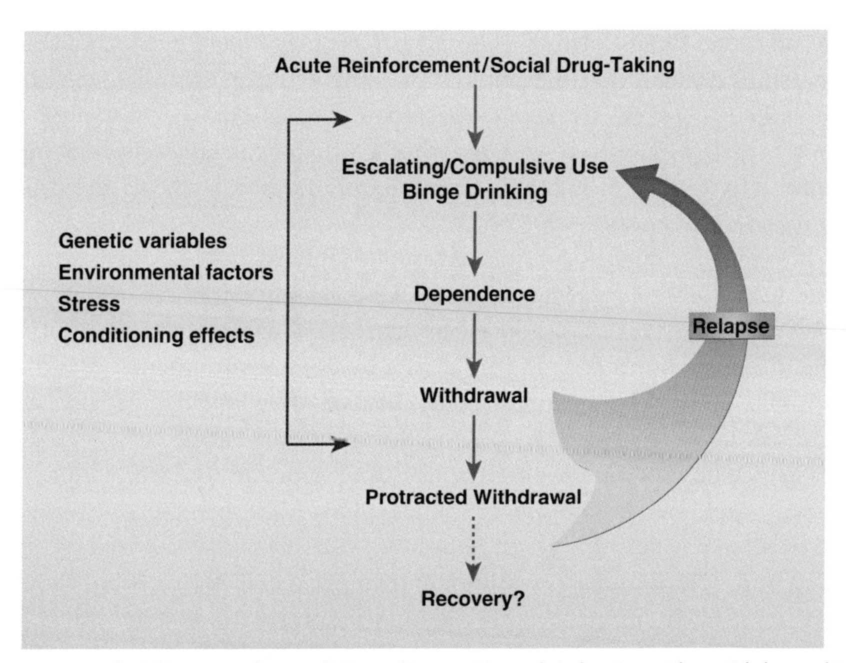

Figure 2 Stages of addiction to drugs of abuse. Drug-taking often begins with social drug taking and acute reinforcement and often, but not exclusively, moves in a pattern of use from escalating compulsive use to dependence, withdrawal, and protracted abstinence. During withdrawal and protracted abstinence, relapse to compulsive use is likely to occur with a repeat of the cycle. Genetic factors, environmental factors, stress, and conditioning all contribute to the vulnerability to enter the cycle of abuse/dependence and relapse within the cycle. *(Taken with permission from Koob GF, Le Moal M. Neurobiology of addiction. London: Academic Press; 2006 [181].)*

with obtaining the drug and stimuli that are associated with withdrawal and internal and external states of stress. A pattern develops in which the drug must be administered to avoid the severe dysphoria and discomfort of abstinence.

Alcohol use disorder or alcoholism follows a somewhat different pattern of drug taking that depends on the severity of the disorder. The initial intoxication is less intense than opioids, and the pattern of drug taking is often characterized by binges of alcohol intake that can be daily episodes or prolonged days of heavy drinking. A binge is currently defined by the U.S. National Institute on Alcohol Abuse and Alcoholism as consuming sufficient alcohol to achieve a blood alcohol level of 0.08 g% in a 2 h period, generally five standard drinks for males and four standard drinks for females. Alcoholism is characterized by a severe emotional and somatic withdrawal syndrome and intense craving for the drug that is often driven by negative emotional states but also by positive emotional states. Many individuals with alcoholism continue with such a binge/withdrawal pattern for extended periods; for others, the pattern evolves into opioid-like addiction, in which they must have alcohol available at all times to avoid the consequences of abstinence.

Nicotine addiction contrasts with the above patterns. Nicotine is associated with virtually no binge-like behavior in the *binge/intoxication* stage of the addiction cycle. Cigarette smokers who met the criteria for substance dependence under the DSM-IV and ICD-10 criteria are likely to smoke throughout the waking hours and experience negative emotional states (dysphoria, irritability, and intense craving) during abstinence. The pattern of intake is one of highly titrated intake of the drug during waking hours.

Psychostimulants, such as cocaine and amphetamines, show a pattern with a greater emphasis on the *binge/intoxication* stage. Such binges can last hours or days, often followed by a crash that is characterized by extreme dysphoria and inactivity. Intense craving and anxiety occur later and are driven by both environmental cues that signify the availability of the drug and internal states that are often linked to negative emotional states and stress.

Marijuana follows a pattern that is similar to opioids and tobacco, with a significant intoxication stage. As chronic use continues, individuals begin to show a pattern of chronic intoxication during waking hours. Withdrawal is characterized by dysphoria, irritability, and sleep disturbances. Although marijuana craving has been less studied to date, it is most likely linked to both cues and internal states that are often associated with negative emotional states and stress, similar to other drugs of abuse.

1.5 The "dependence" view of addiction

The term "dependence" within the conceptual framework of addiction has a confused history and as such was eliminated in the DSM-5. However, discussing evolution of the term is instructive. Historically, definitions of addiction began with definitions of dependence. Himmelsbach defined physical dependence as:

> *... an arbitrary term used to denote the presence of an acquired abnormal state wherein the regular administration of adequate amounts of a drug has, through previous prolonged use, become requisite to physiologic equilibrium. Since it is not yet possible to diagnose physical dependence objectively without withholding drugs, the* sine qua non *of physical dependence remains the demonstration of a characteristic abstinence syndrome*
>
> ***Himmelsbach [13]***

This definition eventually evolved into the definition for physical dependence: "intense physical disturbances when administration of a drug is suspended" [14]. However, this terminology clearly did not capture many of the aspects of an addictive process that does not show physical signs, necessitating creation of the term *psychic dependence* to capture the behavioral aspects of the symptoms of addiction:

> *A condition in which a drug produces 'a feeling of satisfaction and a psychic drive that require periodic or continuous administration of the drug to produce pleasure or to avoid discomfort' ...*
>
> ***Eddy et al. [44]***

Later definitions of addiction resembled a combination of physical and psychic dependence, with more of an emphasis on the psychic or motivational aspects of withdrawal, rather than on the physical symptoms of withdrawal:

> Addiction *from the Latin verb 'addicere,' to give or bind a person to one thing or another. Generally used in the drug field to refer to chronic, compulsive, or uncontrollable drug use, to the extent that a person (referred to as an 'addict') cannot or will not stop the use of some drugs. It usually implies a strong (Psychological) Dependence and (Physical) Dependence resulting in a Withdrawal Syndrome when use of the drug is stopped. Many definitions place primary stress on psychological factors, such as loss of self-control and over powering desires; i.e., addiction is any state in which one craves the use of a drug and uses it frequently. Others use the term as a synonym for physiological dependence; still others see it as a combination (of the two)*
>
> ***Nelson et al. [130]***

Unfortunately, the word *dependence* in this process has multiple meanings. Any drug can produce dependence if dependence is defined as the manifestation of a withdrawal syndrome upon the cessation of drug use. Meeting the ICD-10 criteria for Dependence or DSM-5 criteria for Substance Use Disorder requires much more than simply manifesting a withdrawal syndrome. For the purposes of this book, *dependence* refers to the manifestation of a withdrawal syndrome, and *addiction* refers to Dependence as defined by the ICD-10. The terms *dependence*, *addiction*, and *alcoholism* are held equivalent for this book. The term Substance Use Disorder is defined as a problematic pattern of drug use that leads to clinically significant impairment or distress, reflected by at least two of the 11 criteria within a 12 month period (see above). How this cluster of cognitive, behavioral, and physiological symptoms will be considered equivalent to "addiction" remains to be determined but likely reflects a moderate to severe substance use disorder [15]. Note that other views of addiction have focused on a multistep theory in which the transition to addiction is composed of at least three steps: recreational sporadic drug use, intensified, sustained, escalated use, and the loss of control over drug use [16]. A key part of the validation of the view of Piazza and Deroche-Gamonet focused on the argument that their animal model of the loss of control exhibits the same percentage of vulnerable animals (~15–20%). Their arguement is that this generally represents the percentage of vulnerable individuals in the human population and thus reflects face validity [16]. Finally, in rodents and humans, prior psychostimulant use can increase impulsive and risky and/or potentially harmful decision-making, but when rodents have the choice between a drug and a competing, nondrug option, such impulsivity does not necessarily translate into more drug use. One hypothesis to explain this phenomenon is the delayed drug reward hypothesis ([17]; see Neurobiological Theories of Addiction chapter 3, Volume 1).

1.6 Psychiatric view of addiction

From a psychiatric perspective, drug addiction has aspects of both impulse control disorders and compulsive disorders. Impulse control disorders are characterized by an increasing sense of tension or arousal before committing an impulsive act, pleasure, gratification, or relief at the time of committing the act, and regret, self-reproach, or guilt following the act (see early versions of the DSM). In contrast, compulsive disorders are characterized by anxiety and stress before committing a compulsive repetitive behavior and relief from the stress by performing the compulsive behavior. As an individual moves from an impulsive disorder to a compulsive disorder, a shift occurs from positive reinforcement to negative reinforcement that drives the motivated behavior ([18,19]; Fig. 3).

Drug addiction progresses from impulsivity to compulsivity in a collapsed cycle of addiction that consists of three stages: *preoccupation/anticipation*, *binge/intoxication*, and *withdrawal/negative affect*. Different theoretical perspectives from experimental psychology, social psychology, and neurobiology can be superimposed on these three stages, which are conceptualized as feeding into each other, becoming more intense, and ultimately leading to the pathological state known as addiction.

1.7 Neurobiological view of addiction

As described above, a three-stage framework can be used to explore the behavioral, neurobiological, and treatment perspectives of addiction: *binge/intoxication*, *withdrawal/negative affect*, and *preoccupation/anticipation*. One can also utilize this framework to understand the basic neuroanatomy and neurocircuitry of addiction (Fig. 4).

1.7.1 Binge/intoxication stage: basal ganglia

The *binge/intoxication* stage heavily involves the basal ganglia. The basal ganglia is considered a key part of the extrapyramidal motor system (which controls involuntary

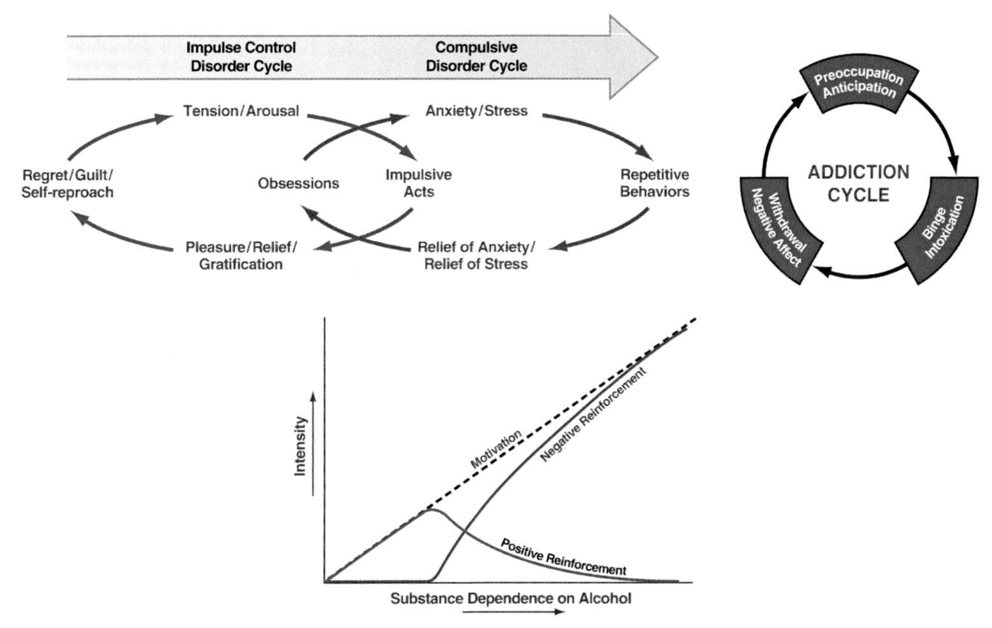

Figure 3 (Top left) Diagram showing the stages of impulse control disorder and compulsive disorder cycles related to the sources of reinforcement. In impulse control disorders, an increasing tension and arousal occurs before the impulsive act, with pleasure, gratification, or relief during the act. Following the act, there may or may not be regret or guilt. In compulsive disorders, there are recurrent and persistent thoughts (obsessions) that cause marked anxiety and stress followed by repetitive behaviors (compulsions) that are aimed at preventing or reducing distress [1]. Positive reinforcement (pleasure/gratification) is more closely associated with impulse control disorders. Negative reinforcement (relief of anxiety or relief of stress) is more closely associated with compulsive disorders. (Top right) Collapsing the cycles of impulsivity and compulsivity results in the addiction cycle, conceptualized as three major stages: *preoccupation/anticipation*, *binge/intoxication*, and *withdrawal/negative affect*. (Bottom) Change in the relative contribution of positive and negative reinforcement constructs during the development of drug dependence. *((Top right) Taken with permission from Koob GF. Neurobiology of addiction. In: Galanter M, Kleber HD, editors. Textbook of substance abuse treatment. 4th ed. Washington, DC: American Psychiatric Publishing; 2008. p. 3–16 [179], (Bottom) Taken with permission from Koob GF. Negative reinforcement in drug addiction: the darkness within. Current Opinion in Neurobiology 2013;23:559–63 [19].)*

movements or reflexes) and is historically associated with a number of key functions, including voluntary motor control, procedural learning related to routine behaviors or habits, and action selection. The basal ganglia includes the following structures: striatum, globus pallidus, substantia nigra, and subthalamic nucleus (Fig. 5).

The striatum can then be divided into the ventral striatum and dorsal striatum. The ventral striatum includes the nucleus accumbens, olfactory tubercle, ventral pallidum, and ventral tegmental area. This is a subarea of the basal ganglia that has gained recognition for its involvement in motivation and reward function. The ventral striatum is now

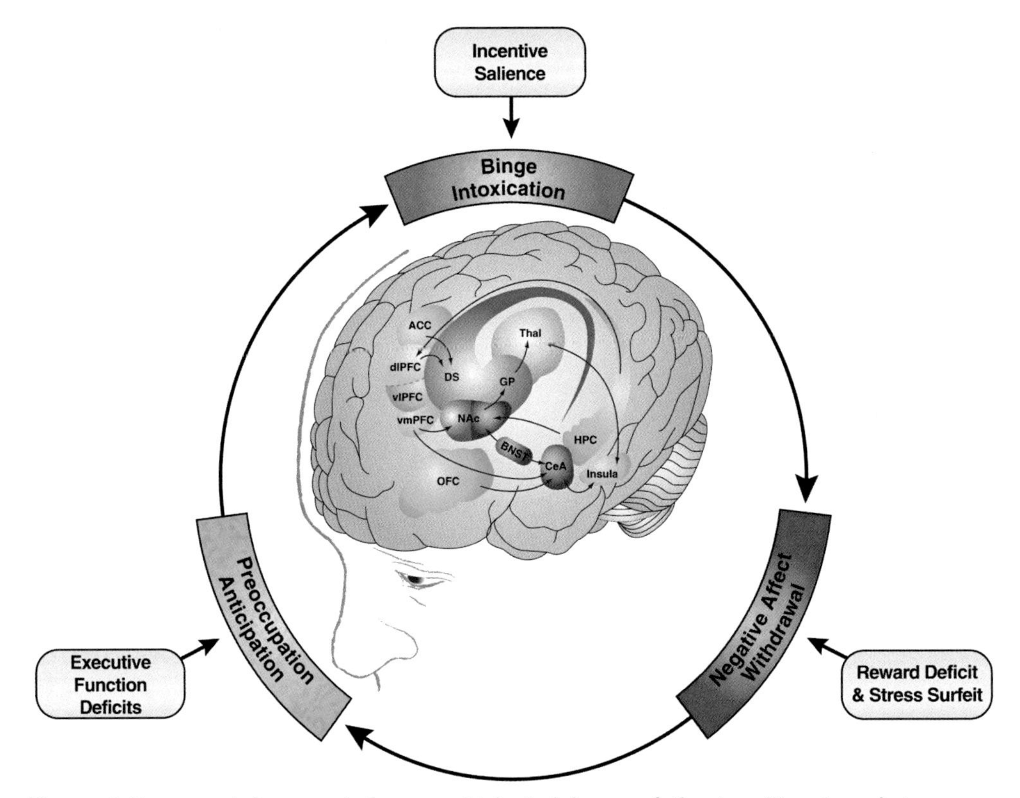

Figure 4 Conceptual framework for neurobiological bases of the transition to substance use disorders. Three stages of the addiction cycle correspond to three key superstructures of motivational circuitry. The *binge/intoxication* stage is mediated by the basal ganglia (*blue*). The *withdrawal/negative affect* stage is mediated by the extended amygdala (*red*). The *preoccupation/anticipation* stage is mediated by the prefrontal cortex (*green*). These stages and corresponding structures mediate the motivational changes that are associated with incentive salience, reward deficit/stress surfeit, and executive function disorder, respectively [174]. PFC, prefrontal cortex; Thal, thalamus; DS, dorsal striatum; GP, globus pallidus; NAC, nucleus accumbens; BNST, basal nucleus of the stria terminalis; Hippo, hippocampus; OFC, orbitofrontal cortex; AMG, amygdala; Insula, insular cortex.

considered a major integrative center for converting motivation to action. In the domain of addiction, it mediates the rewarding effects of drugs of abuse. The basal ganglia receives neurochemical inputs (or afferents) from the prefrontal cortex and midbrain dopamine system. The basal ganglia then sends neurochemical signals (or efferents) from the globus pallidus to the thalamus, which then relays motor and sensory signals to the cerebral cortex. The functions of the basal ganglia involve a series of cortical-striatal-pallidal-thalamic-cortical loops that encode habits related to compulsive behavior (Fig. 6).

Positive reinforcement with drugs of abuse occurs when the presentation of a drug increases the probability of a response to obtain the drug and usually refers to producing

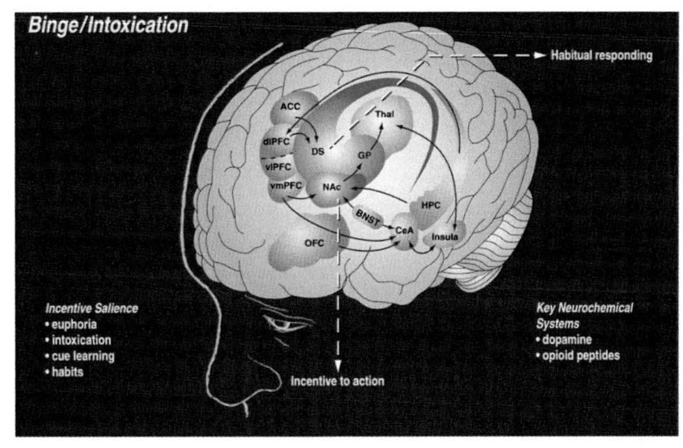

Figure 5 Neurocircuitry associated with the three stages of the addiction cycle: *binge/intoxication*. *(Modified with permission from George O, Koob GF. Control of craving by the prefrontal cortex. Proceedings of the National Academy of Sciences of the United States of America 2013;110:4165–6 [177], Koob GF, Volkow ND. Neurocircuitry of addiction. Neuropsychopharmacology Reviews 2010;35:217–38 [erratum: 35:1051] [174].)*

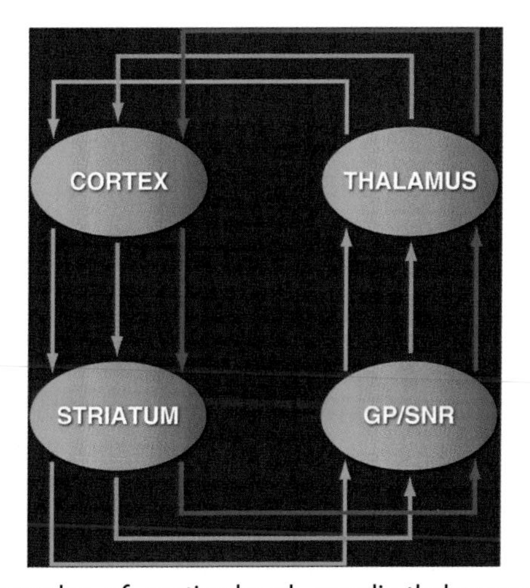

Figure 6 Functional topography of cortico-basal ganglia-thalamocortical circuits. Each color represents a different function. Each functional region of the cortex projects to a specific region of the striatum, represented by the same color. The striatum projects to a specific region of the globus pallidus/substantia nigra pars reticulata, also represented by the same color. The globus pallidus/substantia nigra pars reticulata projects to a specific thalamic region and back to the cortical region of origin. *(Taken with permission from Haber S, McFarland NR. The place of the thalamus in frontal cortical-basal ganglia circuits. The Neuroscientist 2001;7:315–24 [178].)*

a positive hedonic state. Animal models of the positive reinforcing or rewarding effects of drugs in the absence of withdrawal or deprivation are extensive and well-validated. These models include intravenous drug self-administration, conditioned place preference, and lower brain stimulation reward thresholds (see Animal Models chapter 2, Volume 1).

The acute reinforcing effects of drugs of abuse are mediated by brain structures that are connected by the medial forebrain bundle reward system, with a focus on the ventral tegmental area, nucleus accumbens, and amygdala. Much evidence supports the hypothesis that the mesolimbic dopamine system, projecting from the ventral tegmental area to the nucleus accumbens, is dramatically activated by psychostimulant drugs during limited-access self-administration. This system is critical for mediating the rewarding effects of cocaine, amphetamines, and nicotine. However, although acute administration of other drugs of abuse activates dopamine systems, opioids and alcohol have both dopamine-dependent and -independent rewarding effects (the Fig. 7).

μ Opioid receptors in both the nucleus accumbens and ventral tegmental area mediate the reinforcing effects of opioid drugs. Opioid peptides in the ventral striatum and amygdala mediate the acute reinforcing effects of alcohol, based largely on the effects of opioid receptor antagonists and knockout mice. γ-Aminobutyric acid (GABA) systems are activated pre- and postsynaptically in the amygdala by alcohol at intoxicating doses, and GABA receptor antagonists block alcohol self-administration. The administration of a specific nicotinic receptor, the $\alpha_4\beta_2$ subtype, either in the ventral tegmental area or nucleus accumbens mediates the reinforcing effects of nicotine via actions on the mesolimbic dopamine system. The cannabinoid CB_1 receptor, involving the activation of dopamine and opioid peptides in the ventral tegmental area and nucleus accumbens, mediates the reinforcing actions of Δ^9-tetrahydrocannabinol (marijuana).

Drugs of abuse have a profound effect on the response to previously neutral stimuli to which the drugs become paired, a phenomenon called the facilitation of "incentive salience." Early studies of the behavioral pharmacology of stimulants showed that these drugs facilitated "conditioned reinforcement." Psychostimulants caused rats to exhibit compulsive-like lever pressing in response to a cue that was previously paired with a water reward [20].

In a subsequent series of studies that recorded the electrical activity of ventral tegmental area dopamine neurons in primates during the repeated presentation of rewards and presentation of stimuli that were associated with reward, dopamine cells fired at the first exposure to the novel reward, but repeated exposure caused the neurons to stop firing during reward consumption and instead fire when they were exposed to stimuli that were *predictive* of the reward [21]. This suggested a role for phasic dopamine cell firing that leads to large and transient or short-term dopamine activity. Through the process of conditioning, previously neutral stimuli are linked to either a natural or drug reinforcer and acquire the ability to increase dopamine levels in the nucleus accumbens in anticipation of the reward, thus engendering strong motivation to seek the drug, termed incentive salience.

More recent conceptualizations of the role of dopamine in incentive salience focused on using the dopamine-reward prediction error (DA-RPE) theory, applied to addiction [22]. The independent, generally dramatic surge of dopamine that is induced by drugs of

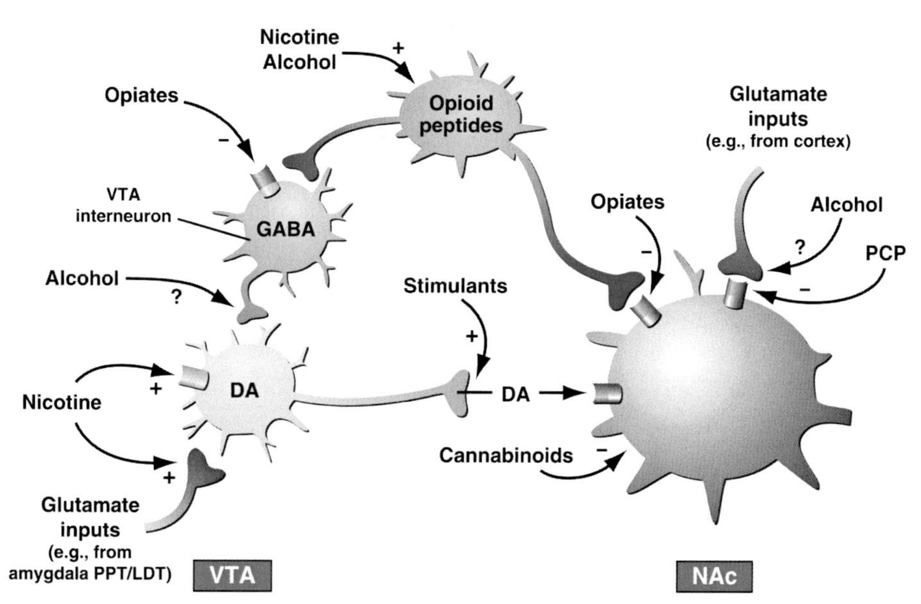

Figure 7 Simplified schematic of converging acute actions of drugs of abuse on the ventral tegmental area (VTA) and nucleus accumbens (NAc). Drugs of abuse, despite diverse initial actions, produce some common effects on the VTA and NAc. Stimulants directly increase dopaminergic transmission in the NAc. Opioids do the same indirectly: they inhibit γ-aminobutyric acid (GABA) interneurons in the VTA, which disinhibits VTA dopamine neurons. Opioids also directly act at opioid receptors on NAc neurons, and opioid receptors, like dopamine (DA) D_2 receptors, signal via G_i. Hence, the two mechanisms converge within some NAc neurons. The actions of the other drugs remain more conjectural. Nicotine appears to activate VTA dopamine neurons directly by stimulating nicotinic cholinergic receptors on those neurons and indirectly by stimulating its receptors on glutamatergic nerve terminals that innervate dopamine cells. Alcohol, by promoting $GABA_A$ receptor function, may inhibit GABAergic terminals in the VTA and hence disinhibit VTA dopamine neurons. It may similarly inhibit glutamatergic terminals that innervate NAc neurons. Many additional mechanisms (not shown) are proposed for alcohol. Cannabinoid mechanisms are complex and involve the activation of cannabinoid CB_1 receptors (which, like D_2 and opioid receptors, are G_i-linked) on glutamatergic and GABAergic nerve terminals in the NAc and on NAc neurons themselves. Phencyclidine (PCP) may act by inhibiting postsynaptic NMDA glutamate receptors in the NAc. Finally, there is some evidence that nicotine and alcohol may activate endogenous opioid pathways and that these and other drugs of abuse (such as opioids) may activate endogenous cannabinoid pathways (not shown). PPT/LDT, peduncular pontine tegmentum/lateral dorsal tegmentum. *(Modified with permission from Nestler EJ. Is there a common molecular pathway for addiction? Nature Neuroscience 2005;8:1445–9 [184].)*

abuse could contribute to drug taking and perhaps addiction. If drugs of abuse mimic a DA-RPE every time the drug is consumed, then over repeated drug use, the repetition of these dopamine signals would continue to reinforce drug-related cues and actions to pathological levels, biasing future decision-making toward drug choice ([22]; see Neurobiological Theories of Addiction chapter 3, Volume 1). The architecture of the DA-RPE circuit involves the regional propagation of DA-RPE from the more ventromedial to more dorsal and lateral striatal domains and may contribute to the shift in goal-directed to habitual responding that is outlined by other theoretical frameworks [22].

As noted previously, all drugs of abuse can initially elicit higher physiological dopamine release in the nucleus accumbens. This drug-induced dopamine signaling can eventually trigger neuroadaptations in other basal ganglia brain circuits that are related to habit formation. Key synaptic changes involve glutamate-modulated *N*-methyl-D-aspartate (NMDA) receptors and α-amino-3-hydroxy-5-methyl-4-isoxazolepropionic acid (AMPA) receptors in glutamatergic projections from the prefrontal cortex and amygdala to the ventral tegmental area and nucleus accumbens [23—25]. The power of initial dopamine release (and the activation of opioid peptide systems) upon initial drug taking begins the neuroadaptations that lead to tolerance and withdrawal and triggers the ability of drug-associated cues to increase dopamine levels in the dorsal striatum, a region that is involved in habit formation [26] and the strengthening of those habits as addiction progresses (see Neurobiological Theories of Addiction chapter 3, Volume 1). The recruitment of these circuits is significant for progression through the addiction cycle because such conditioned responses help explain the intense desire for the drug (craving) and compulsive use when individuals with addiction are exposed to drug-related cues. Conditioned responses within the incentive salience process can drive dopamine signaling to maintain the motivation to take the drug even when its pharmacological effects lessen.

1.7.2 Withdrawal/negative affect stage: extended amygdala

The *withdrawal/negative affect* stage involves key elements of the extended amygdala. The extended amygdala consists primarily of three structures: central nucleus of the amygdala, bed nucleus of the stria terminalis, and nucleus accumbens shell (for review of the anatomy of this structure, see Ref. [27]). The extended amygdala can also be divided into two major divisions: central division and medial division. These two divisions have important anatomical structural differences and dissociable afferent and efferent connections.

The central division of the extended amygdala includes the central nucleus of the amygdala, central sublenticular extended amygdala, lateral bed nucleus of the stria terminalis, and a transition area in the medial and caudal portions of the nucleus accumbens (Fig. 8). These structures in the central division have similar morphology (structure), immunohistochemistry (proteins associated with neurotransmission), and connectivity, and they receive afferent connections from limbic cortices, the hippocampus, the basolateral amygdala, the midbrain, and the lateral hypothalamus. The efferent connections from this complex include the posterior medial (sublenticular) ventral pallidum, ventral tegmental area, various brainstem projections, and a considerable projection to the lateral hypothalamus. The extended amygdala includes major components of brain stress systems that are associated with the negative reinforcement of dependence. The central division has also been found to receive cortical information and regulate the hypothalamic-pituitary-adrenal stress axis.

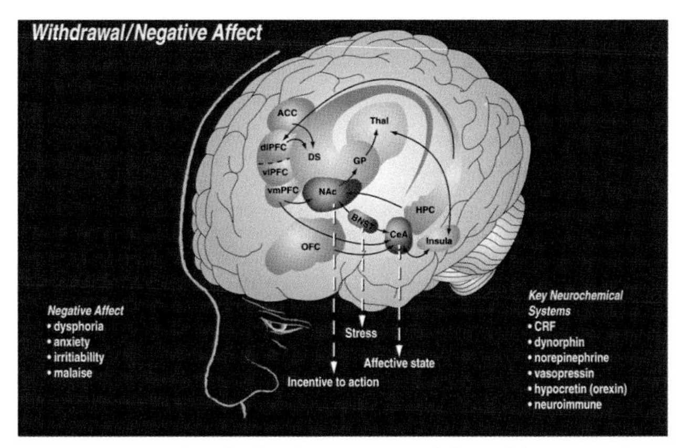

Figure 8 Neurocircuitry associated with the three stages of the addiction cycle: *withdrawal/negative affect. (Modified with permission from George O, Koob GF. Control of craving by the prefrontal cortex. Proceedings of the National Academy of Sciences of the United States of America 2013;110:4165–6 [177], Koob GF, Volkow ND. Neurocircuitry of addiction. Neuropsychopharmacology Reviews 2010;35:217–38 [erratum: 35:1051] [174].)*

The medial division of the extended amygdala consists of the medial bed nucleus of the stria terminalis, medial nucleus of the amygdala, and medial sublenticular extended amygdala. It appears to be more involved in sympathetic (fight-or-flight) and physiological responses and receives olfactory information. Most motivational experimental manipulations that modify the reinforcing effects of drugs of abuse through both positive and negative reinforcement appear to do so by impacting the central nucleus of the amygdala and lateral bed nucleus of the stria terminalis.

Negative reinforcement occurs when the removal of an aversive event increases the probability of a response. In the case of addiction, negative reinforcement involves the removal of a negative emotional state associated with withdrawal, such as dysphoria, anxiety, irritability, sleep disturbances, and hyperkatifeia. Such negative emotional states are thought to derive from two sources: within-system changes and between-system changes.

- *Within-system neuroadaptations in the reward system.* During the development of dependence, the brain systems in the ventral striatum that are important for the acute reinforcing effects of drugs of abuse, such as dopamine and opioid peptides, become compromised and begin to contribute to a negative reinforcement mechanism, in which the drug is administered to restore the decrease in function of the reward systems. Within-system changes within medium spiny neurons in the nucleus accumbens during acute withdrawal include a decrease in long-term potentiation, an increase in the trafficking of AMPA receptors to the surface of neurons, an increase in adenylate cyclase activity, and an increase in cyclic adenosine monophosphate (cAMP) response element binding protein (CREB) phosphorylation. Some of these changes may precede or drive between-system neuroadaptations. Neurochemical evidence of within-system neuroadaptations includes the observation that the chronic administration of all drugs of abuse decreases the function of the mesolimbic dopamine system. Decreases in neuronal firing rate in the mesolimbic dopamine system and decreases in serotonergic neurotransmission in the nucleus accumbens occur during drug withdrawal. Decreases in the firing of dopamine neurons in the ventral tegmental area have also been observed during withdrawal from opioids, nicotine, and alcohol. Imaging studies in drug-addicted humans have also consistently shown long-lasting decreases in the number of dopamine D_2 receptors in drug abusers compared with controls. Additionally, drug abusers have lower dopamine release in response to a pharmacological challenge with a stimulant drug. Decreases in the number of dopamine D_2 receptors, coupled with the decrease in dopaminergic activity in cocaine, nicotine, and alcohol abusers, results in the lower sensitivity of reward (incentive salience) circuits to stimulation by natural reinforcers ([28]; see Neurobiological Theories of Addiction chapter 3, Volume 1). These findings suggest an overall reduction of sensitivity of the dopamine component of reward circuitry to natural reinforcers and other drugs in drug-addicted individuals (Fig. 9).

- *Between-system neuroadaptations in the extended amygdala.* The neuroanatomical substrates for many of the motivational effects of drug dependence may also involve between-system neuroadaptations that occur in the ventral striatum and extended amygdala, which includes neurotransmitters that are associated with both the positive reinforcing effects of drugs of abuse and the brain stress systems that are involved in the negative reinforcement of dependence. Several neurotransmitters that are localized to the extended amygdala, such as corticotropin-releasing factor (CRF), norepinephrine, and dynorphin, are activated during states of stress and anxiety and during drug withdrawal. Antagonists of these neurochemical systems selectively block drug self-administration in dependent animals, suggesting a key role for these neurotransmitters in the ventral striatum and extended amygdala in the negative reinforcement that is associated with drug dependence.

Brain stress systems are not limited to CRF [29]. Multiple neurotransmitter systems converge on the extended amygdala to meet the needs of an organism to respond to an acute stressor but also to sustain a response to a chronic stressor (such as the cycle of repeated binge-withdrawal in addiction). Other modulatory brain neurotransmitter systems that have pro-stress actions also converge on the extended amygdala and include norepinephrine, vasopressin, substance P, hypocretin (orexin), and dynorphin, all of which may contribute to negative emotional states that are associated with drug withdrawal or protracted abstinence [30]. κ-Opioid receptor agonists (administered systemically) and dynorphins (administered intracerebrally) produce aversive-like effects in both animals and humans [31] and have been hypothesized to mediate negative emotional states that are associated with drug withdrawal [32]. High compulsive-like

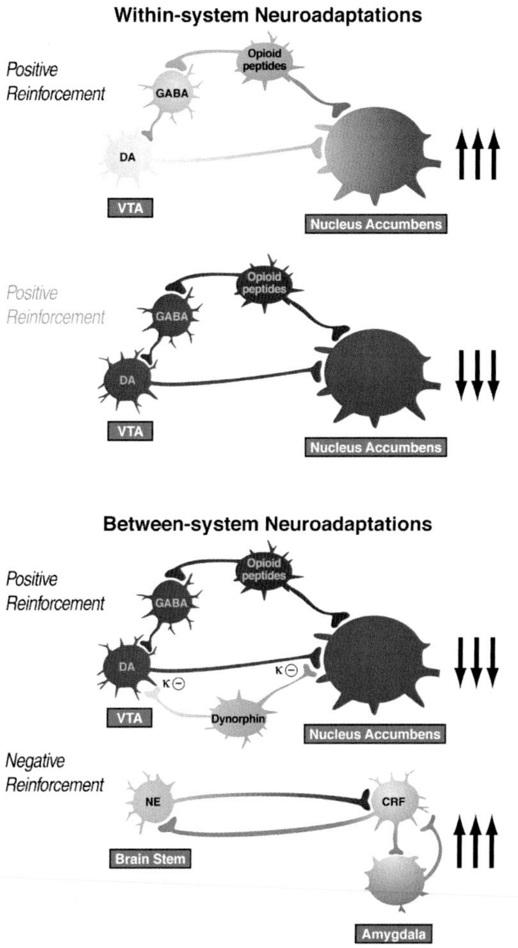

Within-system Neuroadaptations

Positive Reinforcement Opioid peptides GABA DA VTA Nucleus Accumbens

Positive Reinforcement Opioid peptides GABA DA VTA Nucleus Accumbens

Between-system Neuroadaptations

Positive Reinforcement Opioid peptides GABA DA κ⊖ κ⊖ VTA Dynorphin Nucleus Accumbens

Negative Reinforcement NE CRF Brain Stem Amygdala

Figure 9 Diagram of the hypothetical "within-system" and "between-system" changes that lead to the dark side of addiction. (Top) Circuitry for drug reward with major contributions from mesolimbic dopamine and opioid peptides that converge on the nucleus accumbens. During the *binge/intoxication* stage of the addiction cycle, the reward circuitry is excessively engaged. (Middle) Such excessive activation of the reward system triggers "within-system" neurobiological adaptations during the *withdrawal/negative affect* stage, including the activation of cyclic adenosine monophosphate (cAMP) and cAMP response element binding protein (CREB), downregulation of dopamine D_2 receptors, and decreased firing of ventral tegmental area (VTA) dopaminergic neurons. (Bottom) As dependence progresses and the *withdrawal/negative affect* stage is repeated, two major "between-system" neuroadaptations occur. One is the activation of dynorphin feedback that further decreases dopaminergic activity. The other is the recruitment of extrahypothalamic norepinephrine (NE)—corticotropin-releasing factor (CRF) systems in the extended amygdala. Facilitation of the brain stress system in the prefrontal cortex is hypothesized to exacerbate the between-system neuroadaptations while contributing to the persistence of the dark side into the *preoccupation/anticipation* stage of the addiction cycle. *(Taken with permission from Koob GF. Negative reinforcement in drug addiction: the darkness within. Current Opinion in Neurobiology 2013;23:559–63 [19].)*

drug intake that is associated with extended access to and dependence on methamphetamine, heroin, and alcohol is blocked by both systemic and intracerebral κ-opioid receptor antagonist administration. Two sites for these actions are the shell of the nucleus accumbens and amygdala, suggesting a κ-opioid receptor—dynorphin contribution within the extended amygdala to negative emotional states [32]. High compulsive-like alcohol drinking in dependent rats during withdrawal can also be blocked by a β-adrenergic receptor antagonist, α_1 adrenergic receptor antagonist, κ-opioid receptor antagonist, vasopressin 1b receptor antagonist, glucocorticoid receptor antagonist, and neuroimmune system antagonist [30,30a]. High compulsive-like heroin intake in the model of extended-access self-administration was blocked by a substance P antagonist and hypocretin-2 antagonist [33,34].

Similarly, one may hypothesize that the vulnerability to drive allostasis in addiction may not only derive from the activation of pro-stress neurotransmitter systems but also from anti-stress neurotransmitter systems (see derive Section 2.4). Anti-stress neurotransmitter systems may serve as neuroadaptive buffers to the pro-stress actions that are described above. Neurotransmitter/neuromodulatory systems that are implicated in anti-stress actions include neuropeptide Y (NPY), nociceptin, and endocannabinoids. Neuropeptide Y has powerful orexigenic and anxiolytic effects and have been hypothesized to act in opposition to the actions of CRF in addiction [35]. The activation of NPY in the central nucleus of the amygdala has opposite effects to CRF, in which NPY, injected into the brain, blocks the increase in GABA release in the central nucleus of the amygdala that is produced by alcohol, blocks high compulsive-like alcohol administration, and blocks the transition to excessive drinking with the development of dependence [36—41]. Nociceptin (also known as orphanin FQ) has anti-stress-like effects in animals [42,43]. Nociceptin and synthetic NOP receptor agonists have effects on GABA synaptic activity in the central nucleus of the amygdala that are similar to NPY and can block high alcohol consumption in a genetically selected line of rats that is known to be hypersensitive to stressors [44]. Evidence also implicates endocannabinoids in the regulation of affective states, in which reductions of cannabinoid CB_1 receptor signaling produce anxiogenic-like behavioral effects [45]. Blocking endocannabinoid clearance can also block some drug-seeking behaviors [46—48]. Thus, endocannabinoids may play a protective role in preventing drug dependence by buffering the stress activation that is associated with withdrawal.

Neuropharmacological studies that systemically administered neurotransmitter-modulating agents found that drugs that have either anti-stress or antidepressant-like activity in other animal models blocked the withdrawal-induced elevations of brain reward thresholds for most major drugs of abuse (Koob, 2017). Using nicotine as an example, a nicotinic receptor partial agonist, CRF_1 receptor antagonist [49—52], vasopressin 1b receptor antagonist [53], and $\alpha 1$ noradrenergic receptor antagonist [54] blocked the elevations of reward thresholds that were produced by withdrawal from chronic high-dose nicotine exposure. A CRF_1 receptor antagonist, injected systemically, also reversed the elevations of reward thresholds that were produced by alcohol withdrawal [54].

In summary, a multi-determined neurocircuitry promotes the activation of pro-stress neuromodulators and, combined with a weakening or inadequate anti-stress response, leads to negative emotional states that set up an allostatic hedonic load that drives negative reinforcement. Under this framework, a strong multi-determined buffer, if activated and sufficient to allow the pro-stress systems to recover, may help return the organism to homeostasis.

Heightened pain perception has long been observed in individuals with addiction to opioids [55] and more recently with addiction to alcohol [56]. In humans, withdrawal from opioids and alcohol can lower pain thresholds and exacerbate pain. Patients who are on methadone maintenance have low pain tolerance [57], and pain is one of the main triggers of relapse to addiction in methadone-maintained individuals. Former opioid-addicted individuals who were maintained on either methadone or the opioid receptor partial agonist buprenorphine presented an increase in sensitivity to cold pressor pain [58]. Others have found that a hyperalgesic state can persist for up to 5 months in abstinent individuals with opioid addiction, and addicted individuals with more pain sensitivity also exhibited greater cue-induced craving at this time point [59]. Subjects who were in acute withdrawal (24—72 h) from opioids or protracted abstinence (average of 30 months) exhibited decreases in pain thresholds and pain tolerance in the ischemic pain submaximal tourniquet procedure, and these effects were exacerbated by negative emotional states. Individuals in all groups (i.e., nonusers, ex-users, and withdrawn users) exhibited lower pain tolerance after viewing negative pictures compared with tolerance latencies that were observed after viewing positive and neutral pictures. Indeed, even acute opioid administration can produce hyperalgesia in humans [60]. Here, healthy non-opioid-dependent men who were tested in an acute opioid physical dependence paradigm exhibited the presence of hyperalgesia in response to experimental cold-pressor pain using three different pretreatment opioid administration protocols whereby acute physical dependence was precipitated by naloxone [60].

In acute alcohol withdrawal, patients also exhibited greater heat pain sensitivity to a noxious thermal stimulus [56]. Again, the perceived painful thermal sensation was more intense in patients who were experiencing negative affective states, in which pain tolerance correlated with their scores on the Beck Depression Inventory [56].

In animal models, withdrawal from the chronic self-administration of opioids and alcohol produced hyperalgesia (i.e., lowered pain thresholds; [61]). With opioids, hyperalgesia has been observed in numerous studies [62,63]. Such hyperalgesia has even been demonstrated with a single injection of heroin in rats [64]. Animals that were allowed extended access to intravenous heroin self-administration developed dependence and compulsive-like responding and exhibited hyperalgesia during withdrawal [65]. Hyperalgesia was partially blocked by the systemic administration of a CRF_1 receptor antagonist [65]. More compelling, every-other-day administration of a CRF receptor antagonist blocked the development of escalation of heroin intake and the development of hyperalgesia [66]. CRF_1 receptors mediate the pronociceptive effects of this peptide, and this relationship is mediated at least partially by the central nucleus of the amygdala [67,68]. CRF_1 receptors also mediate pain-related anxiety-like behavior [68a]. The antinociceptive effects of CRF_1 receptor antagonists have been demonstrated across several pain models, although this class of drugs does not alter various pain-related indices (e.g., audible or ultrasonic vocalizations or paw withdrawal thresholds) in non-injured animals (e.g., Ref. [67]).

In animal models of alcohol dependence, an alcohol (6.5%)-containing liquid diet was fed continuously to male rats. Hyperalgesia was produced during withdrawal, and such hyperalgesia took 4 weeks to develop [69]. In another, more binge-like paradigm, feeding an alcohol liquid diet (6.5%) in repeated cycles of 4 days of alcohol followed by 3 days without alcohol resulted in withdrawal-induced hyperalgesia that began at the end of one weekly cycle and reached a maximum during the fourth cycle. This withdrawal-induced hyperalgesia was reversibly inhibited by the intrathecal administration of an antisense oligodeoxynucleotide to protein kinase Cε [70]. Using an operant oral self-administration model, animals that were trained to self-administer alcohol and were made dependent escalated their intake and exhibited hyperalgesia during withdrawal [65]. This hyperalgesia was partially blocked by the systemic administration of a CRF_1 receptor antagonist, consistent with the results that were observed with opioid-dependent rats (see above). These results are consistent with the studies by Neugebauer and colleagues with regard to the role of the extrahypothalamic CRF stress system in pain modulation.

Integrating pain with the other symptoms of a negative emotional state led to the hypothesis that there can be hypersensitivity not only to physical pain but also to emotional pain. Such hypersensitivity was defined as hyperkatifeia ([71]; see Neurobiological Theories of Addiction chapter 3, Volume 1). Hyperkatifeia (derived from the Greek "katifeia" for dejection or negative emotional state) was defined as a greater intensity of the constellation of negative emotional/motivational symptoms and signs that are observed during withdrawal from drugs of abuse. Links have also been hypothesized to exist between the neural mechanisms that are responsible for hyperkatifeia and opioid-induced hyperalgesia [71].

For example, evidence suggests that the neural substrates of stress system neuroadaptations that are associated with addiction may overlap with substrates of emotional aspects of pain processing in such areas as the amygdala [72]. The spino (trigemino)-ponto-amygdaloid pathway projects from the dorsal horn of the spinal cord to the mesencephalic parabrachial area and then to the central nucleus of the amygdala. This pathway has been implicated in processing emotional components of pain perception [73,74].

Pain-responsive neurons are also abundant in the lateral part of the central nucleus of the amygdala [75], an area that may also be responsible for negative emotional responses to abused drugs [76]. As noted above, opioid withdrawal and alcohol withdrawal in animal models of compulsive-like self-administration produce greater anxiety-like responses and hyperalgesia, both of which are blocked by CRF receptor antagonists [65].

Thus, one hypothesis to explain the crosstalk between opioid addiction and alcohol addiction and possibly addictions to other drugs (e.g., marijuana, nicotine, etc.) and chronic pain syndromes is that some patients may be more prone to the development of hyperkatifeia during withdrawal. An allostatic view would suggest that opioid-induced hyperalgesia and hyperkatifeia would be much more likely to occur during chronic opioid administration if excessive opioids are administered. One could argue that because of overdosing, rapid escalation (overshooting), pharmacokinetic variables, or genetic sensitivity, the body will react to that perturbation with engagement of the opponent processes of hyperalgesia and hyperkatifeia that are mediated by significant crosstalk in such brain structures as the central nucleus of the amygdala. The repeated engagement of opponent processes without time for the system to reestablish homeostasis will engage the allostatic mechanisms that are described above. Such a framework suggests that the manifestation of opioid-induced hyperalgesia has important clinical implications: (*i*) the opioid has exceeded the amount that is effective for pain control, and (*ii*) susceptible individuals are at risk for developing hyperkatifeia, the unstable emotional and behavioral state that underlies addiction [71].

1.7.3 Preoccupation/anticipation stage: prefrontal cortex

The *preoccupation/anticipation* ("craving") stage involves key elements of the prefrontal cortex. The global function of the prefrontal cortex is to mediate executive function. *Executive function* can be conceptualized as the ability to organize thoughts and activities, prioritize tasks, manage time, and make decisions. To accomplish such complex tasks in the context of the neurobiology of addiction, the prefrontal cortex can be divided into two opposing systems: the Go system and the Stop system [12]. The Go system engages habit systems, possibly even subconsciously and automatically. The Stop system inhibits such systems. The result of the interactions between these two system produces the well-known impulsivity that is associated with the addiction process, during both the initiation of drug intake and relapse.

The Go system is hypothesized to involve the anterior cingulate cortex and dorsolateral prefrontal cortex. The anterior cingulate cortex facilitates the maintenance and selection of responses, particularly under high attentional demands, planning, self-initiation, and the self-monitoring of goal-directed behaviors. The functions of the dorsolateral prefrontal cortex involve working memory, planning, and strategy.

The Stop system is hypothesized to largely involve the ventrolateral prefrontal cortex and orbitofrontal cortex. The functions of the ventrolateral prefrontal cortex involve

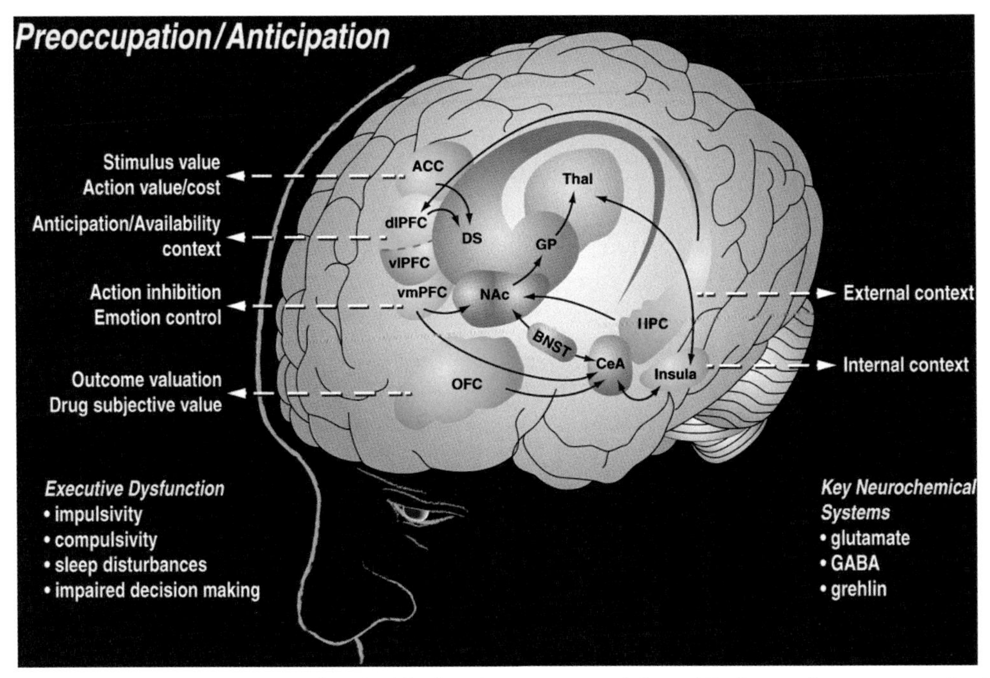

Figure 10 Neurocircuitry associated with the three stages of the addiction cycle: *preoccupation/anticipation. (Modified with permission from George O, Koob GF. Control of craving by the prefrontal cortex. Proceedings of the National Academy of Sciences of the United States of America 2013;110:4165–6 [177], Koob GF, Volkow ND. Neurocircuitry of addiction. Neuropsychopharmacology Reviews 2010;35: 217–38 [erratum: 35:1051] [174].)*

response inhibition, sustained attention, memory retrieval, rule generation, and shifting. The functions of the orbitofrontal cortex, including the ventromedial prefrontal cortex, include the assignment of value (valuation) and integration of reward and punishment (Fig. 10). The anterior cingulate cortex and dorsolateral prefrontal cortex in humans correspond to the anterior cingulate cortex and prelimbic cortex in rats, and the ventromedial prefrontal cortex and orbitofrontal cortex in humans correspond to the infralimbic cortex and orbitofrontal cortex in rats (Fig. 11).

The *preoccupation/anticipation* stage of the addiction cycle is a key element of relapse in humans, defining addiction as a chronic relapsing disorder. Although often linked to the construct of craving, the concept of craving *per se* has been difficult to measure in human clinical studies and often does not correlate with relapse. Nevertheless, the stage of the addiction cycle at which an individual reinstates drug-seeking behavior after abstinence remains a challenging focus of neurobiological studies and medications development.

Animal models of craving can be divided into two domains: (*i*) drug seeking that is induced by the drug or stimuli that are paired with drug taking (reward craving) and (*ii*) alcohol seeking that is induced by an acute stressor or state of stress (relief craving;

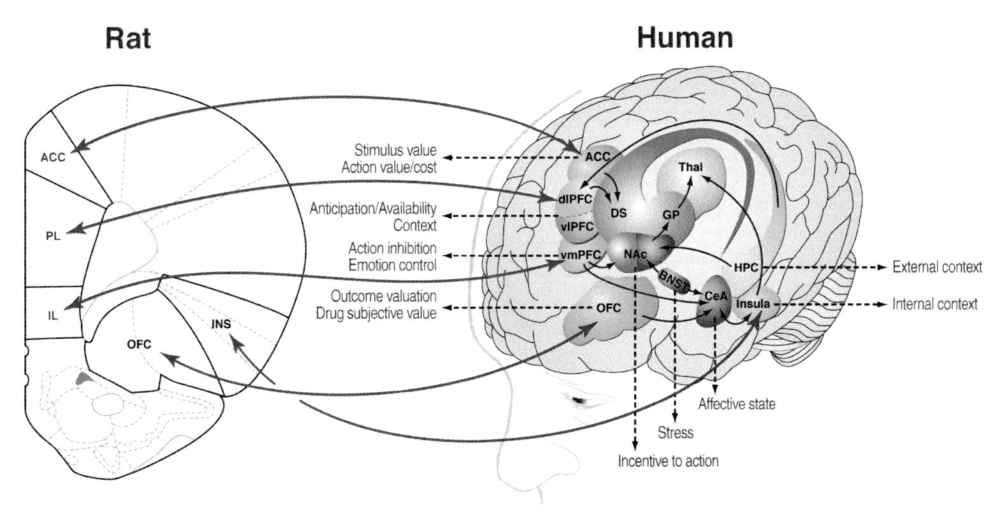

Figure 11 Correspondence between rat and human brain regions that are relevant to the addiction process. Rats are commonly studied to unveil the neurobiological mechanisms of addiction because they have a well-characterized central nervous system whose neurochemical and molecular pathways in subcortical areas correspond reasonably well to humans. ACC, anterior cingulate cortex; PL, prelimbic cortex; IL, infralimbic cortex; OFC, orbitofrontal cortex; INS, insula; dlPFC, dorsolateral prefrontal cortex; vlPFC, ventrolateral prefrontal cortex; DS, dorsal striatum; Thal, thalamus; GP, globus pallidus; NAC, nucleus accumbens; BNST, bed nucleus of the stria terminalis; CeA, central nucleus of the amygdala; HPC, hippocampus. *(Modified with permission from George O, Koob GF. Control of craving by the prefrontal cortex. Proceedings of the National Academy of Sciences of the United States of America 2013; 110:4165–6 [177].)*

Table 3). Drug-induced reinstatement appears to be localized to a medial prefrontal cortex/ventral striatum circuit that is mediated by the neurotransmitter glutamate. Cue-induced reinstatement appears to involve the basolateral amygdala, with a possible feed-forward mechanism that goes through the same prefrontal cortex system that is involved in drug-induced reinstatement. Neurotransmitter systems that are involved in drug-induced reinstatement include a glutamate projection from the frontal cortex to nucleus accumbens that is modulated by dopamine in the frontal cortex. Cue-induced reinstatement also involves a glutamate projection from the basolateral amygdala and ventral subiculum to the nucleus accumbens. Stress-induced reinstatement depends on the activation of both CRF and norepinephrine in the extended amygdala. Protracted abstinence, largely described in alcohol dependence models, involves both an overactive glutamatergic GO system and sensitized CRF systems. Brain CRF systems remain hyperactive during protracted abstinence, and this hyperactivity has motivational significance for excessive alcohol drinking.

Executive control over incentive salience is essential to maintain goal-directed behavior and the flexibility of stimulus-response associations. The prefrontal cortex sends glutamatergic projections directly to mesocortical dopamine neurons in the ventral

Table 3 Drug craving.

Drug craving	"Drug craving is the desire for the previously experienced effects of a psychoactive substance. This desire can become compelling and can increase in the presence of both internal and external cues, particularly with perceived substance availability. It is characterized by an increased likelihood of drug-seeking behavior and, in humans, or drug-related thoughts." [77]
Reward craving	• Induced by stimuli that have been paired with drug self-administration such as environmental cues. • Termed *conditioned positive reinforcement* in experimental psychology. • Animal model: Cue-induced reinstatement where a cue previously paired with access to a drug reinstates responding for a lever that has been extinguished.
Relief craving	• State of protracted abstinence in drug-dependent individuals weeks after acute withdrawal. • Conceptualized as a state change characterized by anxiety and dysphoria. • Animal model: Residual hypersensitivity to states of stress and environmental stressors that lead to relapse to drug-seeking behavior.

tegmental area, exerting excitatory control over dopamine in the prefrontal cortex. Thus, the ventral part of the prefrontal cortex (the STOP system) can inhibit incentive salience and suppress conditioned behavior when a salient cue is present. Appetitive stimuli activate the prefrontal cortex. Lesions of the prefrontal cortex can induce impulsivity. Withdrawal from alcohol is associated with an increase in glutamate release in the nucleus accumbens. The cue-induced reinstatement of psychostimulant-seeking behavior dramatically increases prefrontal cortex activity and glutamate release in the nucleus accumbens. The increase in activity of the prefrontal-glutamatergic system during relapse may elicit a dramatic glutamatergic response that may in turn mediate craving-like responses during the *preoccupation/anticipation* stage.

Behavioral procedures have been developed to reinstate alcohol self-administration using previously neutral stimuli that are paired with alcohol self-administration or that predict alcohol self-administration. Cue-induced reinstatement can be blocked by opioid receptor antagonists, dopamine D_1 and D_2 receptor antagonists, and glutamate receptor antagonists. Stress exposure can also reinstate responding for drugs in rats that are extinguished from drug-seeking behavior. Such stress-induced reinstatement can be blocked by CRF antagonists, dynorphin antagonists, and norepinephrine antagonists.

Human imaging studies reveal similar circuit dysregulation during the *preoccupation/anticipation* stage as demonstrated in animal models. A decrease in frontal cortex activity parallels deficits in executive function in neuropsychologically challenging tasks. Individuals with alcoholism exhibit impairments in the maintenance of spatial information, the disruption of decision making, and impairments in behavioral inhibition. Such frontal cortex-derived executive function disorders have been linked to the ineffectiveness of some behavioral treatments in individuals with alcoholism. Thus, individual differences in the prefrontal cortical control of incentive salience may explain individual differences in the vulnerability to addiction, see Section 2.8. The excessive attribution of incentive salience to drug-related cues and residual hypersensitivity of the brain stress systems may perpetuate excessive drug intake, compulsive behavior, and relapse.

Other "craving" theories also have a focus on "top-down" prefrontal cortex control over subcortical neurocircuits in the basal ganglia and extended amygdala. For example, the competing neurobehavioral decision systems (CNDS) theory accounts for self-control failure (defined as a failure to show delayed discounting) and integrates the neuroscience of addiction with developmental processes, socioeconomic status, and comorbidities [78]. These authors posited that neurobiological evidence clearly supports the hypothesis that the impulsive decision system is mediated by the "limbic" system (midbrain, amygdala, habenular commissure, and striatum) and paralimbic system (insula and nucleus accumbens; [79]). In contrast, the executive function system is argued to be mediated by parietal lobes and the prefrontal cortex [79]. The authors of CNDS theory argue that there is hyperactivation of the impulsive system and hypoactivation of the executive function system in stimulant addiction (see Neurobiological Theories of Addiction chapter 3, Volume 1). Similarly, the somatic marker theory of addiction has the premise of a "myopia for the future," such that there is lower emotional/somatic impact of natural reinforcers but an exaggeration of the emotional/somatic impact of the immediate prospects of obtaining drugs. This shift is hypothesized to involve dysfunction in the ventromedial prefrontal cortex system and hyperactivity of the amygdala [80] (see Neurobiological Theories of Addiction chapter 3, Volume 1).

2. Neuroadaptational views of addiction

2.1 Behavioral sensitization

Repeated exposure to many drugs of abuse, particularly psychostimulants, results in progressive and enduring enhancement of the motor-stimulant effect that is elicited by subsequent exposures to the drug. This phenomenon of *behavioral sensitization* was hypothesized to underlie some aspects of the neuroplasticity of drug addiction [81]. Behavioral or psychomotor sensitization, defined as an increase in locomotor activation that is produced by repeated drug administration, is more likely to occur with intermittent drug exposure, whereas tolerance is more likely to occur with continuous exposure. Sensitization may also grow with the passage of time. Stress and stimulant sensitization show cross-sensitization. The phenomenon of behavioral sensitization was observed and characterized in the 1970s and 1980s for various drugs [82–86] and helped identify some of the early neuroadaptations and neuroplasticity that are associated with repeated drug administration that can lead to substance use disorders. Psychomotor sensitization was invariably linked to sensitization of the activity of the mesolimbic dopamine system [87]. However, behavioral sensitization has rarely been observed in the human population and has largely been discounted as a valid animal model of addiction.

Nevertheless, the conceptualization of psychomotor sensitization in drug addiction evolved into one of the more prominent functional components of the *binge/intoxication* stage of the addiction cycle, termed incentive salience. Based on the behavioral sensitization framework, a shift in an incentive-salience state, described as *wanting* (as opposed to *liking*), was hypothesized to progressively increase with repeated drug exposure [87]. Pathologically strong *wanting* or craving was proposed to define compulsive use. This theory also stated that there is no causal relationship between the subjective pleasurable effects of the drug (drug *liking*) and the motivation to take the drug (drug *wanting*). The brain systems that are sensitized were argued to mediate a subcomponent of reward, termed *incentive salience* (i.e., the motivation to take the drug or drug *wanting*) rather than the pleasurable or euphoric effects of the drug. The psychological process of incentive salience was theorized to be responsible for instrumental drug-seeking and drug-taking behavior (*wanting*; [88]). This sensitized incentive salience process then produces compulsive patterns of drug use. Through associative learning, the enhanced incentive value becomes oriented specifically toward drug-related stimuli, leading to the escalation of drug seeking and taking. This theory further argued that the underlying sensitization of neural structures persists, making individuals with addiction vulnerable to long-term relapse.

As detailed as this theory is in terms of attempting to explain the transition from initial drug use to compulsive use, it has been undermined by multiple scientific observations. Individuals with addiction invariably show tolerance to the rewarding effects of drugs of abuse—*not* sensitization. Animal models of compulsive-like responding for drugs also show tolerance, not sensitization, and millions of individuals take psychostimulants as medications for attention-deficit/hyperactivity disorder—they do not show sensitization. Nonetheless, the one redeeming feature of the sensitization model is its focus on the incentive salience of drugs, which has both empirical and conceptual merit but may involve other mechanisms than those that are reflected in behavioral (locomotor) sensitization as outlined by the theory of Robinson and Berridge ([88]; see Neurobiological Theories of Addiction chapter 3, Volume 1).

2.2 Counteradaptation and opponent process

Counteradaptation hypotheses have long been proposed in an attempt to explain tolerance and withdrawal and the motivational changes that occur with the development of addiction. The initial acute effects of a drug are opposed or counteracted by homeostatic changes in systems that mediate the primary effects of the drug [89—91]. The origins of counteradaptive hypotheses can be traced back to much earlier work on physical dependence [13] and counteradaptive changes in physiological measures that are associated with acute and chronic opioid administration.

The opponent-process theory was developed in the 1970s [91–96]. Since then, it has been applied by many researchers to various situations, including drugs of abuse, fear conditioning, tonic immobility, ulcer formation, eating disorders, jogging, peer separation, glucose preference [91,93,95,96] and even parachuting.

The theory assumes that the brain contains many affective control mechanisms that serve as a kind of emotional immunization system that counteracts or opposes departures from emotional neutrality or equilibrium, regardless of whether they are aversive or pleasant [91]. The theory is basically a negative feed-forward control construct that is designed to keep mood in check although stimulation is strong. The system is conceptualized as being composed of three subparts that are organized in a temporal manner. Two opposing processes control a *summator*, which determines the controlling affect at a given moment. First, an unconditioned arousing stimulus triggers a primary affective process, termed the *a-process*, an unconditional reaction that translates the intensity, quality, and duration of the stimulus (e.g., the first, initial use of an opioid). Second, as a consequence of the *a-process*, the opposing *b-process* is evoked after a short delay, thus defining the opponent process. The *b-process* feeds a negative signal into the summator, subtracting the impact of the already existing *a-process* in the summator. These two responses are consequently and temporally linked (*a* triggers *b*) and depend on different neurobiological mechanisms. The *b-process* has a longer latency, but some data show that it may appear soon after the beginning of the stimulus in the course of the stimulus action [97] and have more inertia, slower recruitment, and a more sluggish decay. At a given moment, the pattern of affect will be simply the algebraic sum of these opposite influences, yielding the net product of the opponent process [95] with the passage of time (Fig. 12).

Importantly, with repetition of the stimulus, the dynamics or net product is a result of a progressive increase in the *b-process*. In other words, the *b-process* itself is sensitized with repeated drug use and appears more and more rapidly after the unconditional stimulus onset. It then persists longer after its initial, intended action (the unconditioned effect) and eventually masks the unconditioned effect (*a-process*), resulting in apparent tolerance [98,99]. Experimental data show that if the development of the *b-process* is blocked, then no tolerance can develop. The unconditioned effect of the drug does not change with repeated drug administration. The development of the *b-process* equals the development of a negative affective state and withdrawal symptoms, in opposition to the hedonic quality of the unconditioned stimulus. Importantly, the nature of the acquired motivation is specified by the nature of the *b-process* (i.e., an aversive affect in the case of drug abuse). The individual will work to reduce, terminate, or prevent the negative affect.

In this opponent-process theory from a drug addiction perspective, tolerance and dependence are inextricably linked [91]. Solomon argued that the first few self-

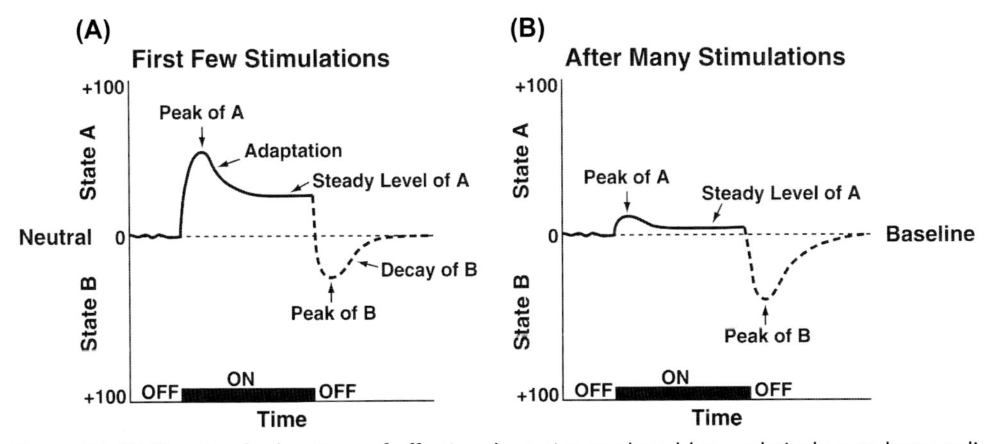

Figure 12 (A) The standard pattern of affective dynamics produced by a relatively novel unconditioned stimulus. (B) The standard pattern of affective dynamics produced by a familiar, frequently repeated unconditioned stimulus. *(Taken with permission from Solomon RL. The opponent-process theory of acquired motivation: the costs of pleasure and the benefits of pain. American Psychologist 1980;35:691–712 [95].)*

administrations of an opioid drug produce a pattern of motivational changes. The onset of the drug effect produces euphoria (*a-process*), followed by a subsequent decline in intensity. After the effects of the drug wear off, the *b-process* emerges as an aversive craving state. The *b-process* gets progressively larger over time, in effect contributing to or producing more complete tolerance to the initial euphoric effects of the drug.

2.3 Motivational view of addiction

Rather than focusing on the *physical* signs of dependence, our conceptual framework has focused on the *motivational* aspects of addiction. The emergence of a negative emotional state (dysphoria, anxiety, irritability) when access to the drug is prevented [100] is associated with the transition from drug use to addiction. The development of such a negative affective state can define dependence as it relates to addiction:

> *The notion of dependence on a drug, object, role, activity or any other stimulus-source requires the crucial feature of negative affect experienced in its absence. The degree of dependence can be equated with the amount of this negative affect, which may range from mild discomfort to extreme distress, or it may be equated with the amount of difficulty or effort required to do without the drug, object, etc.*
>
> ***Russell [101]***

A key common element of all drugs of abuse in animal models is the dysregulation of brain reward function that is associated with the cessation of chronic drug administration. Rapid acute tolerance and opponent-process-like actions against the hedonic effects of cocaine have been reported in humans who smoke coca paste ([102]; Fig. 13). After a

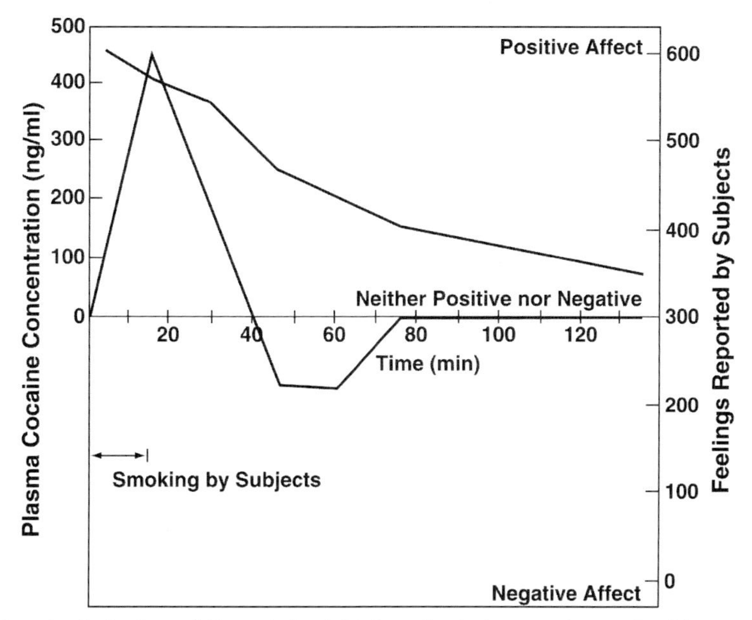

Figure 13 Dysphoric feelings followed the initial euphoria in experimental subjects who smoked cocaine paste, although the concentration of cocaine in the plasma of the blood remained relatively high. The dysphoria was characterized by anxiety, depression, fatigue, and a desire for more cocaine. The peak feelings were probably reached shortly before the peak plasma concentration, but the first psychological measurements were made later than the plasma assay. The temporal sequence of the peaks shown cannot be regarded as definitive. *(Taken with permission from Van Dyke C, Byck R. Cocaine. Scientific American 1982;246:128–41 [102].)*

single cocaine smoking bout, the onset and intensity of the "high" are very rapid via the smoked route of administration. Rapid tolerance is evident, in which the "high" decreases rapidly despite high blood levels of cocaine. Human subjects also report subsequent dysphoria, again despite high blood levels of cocaine. Intravenous cocaine produces similar patterns (a rapid "rush" followed by an increased "low"; [103]; Fig. 14).

With intravenous cocaine self-administration in animal models, elevations of brain reward threshold begin rapidly and can be observed within a single self-administration session ([104]; Fig. 15), bearing a striking resemblance to human subjective reports. These results demonstrate that the elevation of brain reward thresholds following prolonged access to cocaine fail to return to baseline levels, thus creating a progressive elevation of "baseline" reward thresholds and defining a new set point. These data provide compelling evidence of brain reward dysfunction in escalated cocaine self-administration and strong support for a hedonic allostasis model of drug addiction.

The hypothesis that the compulsive use of drugs is accompanied by the chronic perturbation of brain reward homeostasis has been tested in animal models of the escalation of drug intake with prolonged access. Animals that have access to intravenous

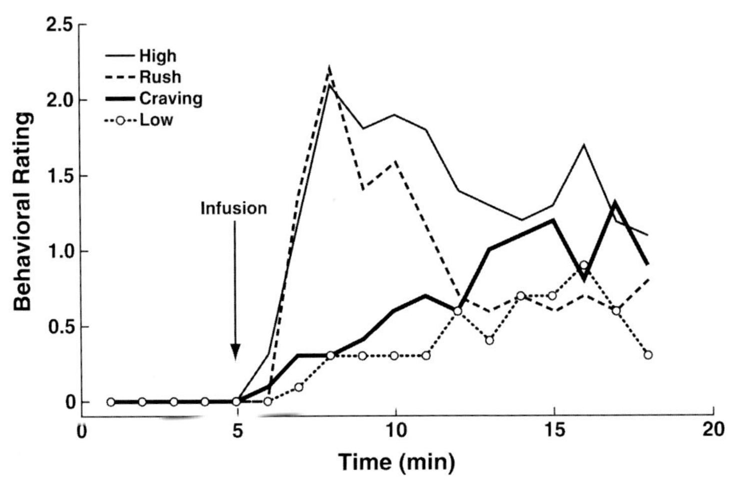

Figure 14 Average behavioral ratings after an infusion of cocaine (0.6 mg/kg over 30 s; $n = 9$). The rush, high, low, and craving ratings were averaged within each category for the subjects who had interpretable cocaine functional magnetic resonance imaging data after motion correction and behavioral ratings time-locked to the scanner. Both peak rush and peak high occurred 3 min post-infusion. Peak low (primary reports of dysphoria and paranoia) occurred 11 min post-infusion. Peak craving occurred 12 min post-infusion. No subject reported effects of the saline infusion on any of the four measures. Ratings obtained for rush, high, low, and craving measures were higher in subjects blinded to the 0.6 mg/kg cocaine dose compared with subjects unblinded to a 0.2 mg/kg cocaine dose. *(Taken with permission from Breiter HC, Gollub RL, Weisskoff RM, Kennedy DN, Makris N, Berke JD, Goodman JM, Kantor HL, Gastfriend DR, Riorden JP, Mathew RT, Rosen BR, Hyman SE. Acute effects of cocaine on human brain activity and emotion. Neuron 1997;19:591–611 [103].)*

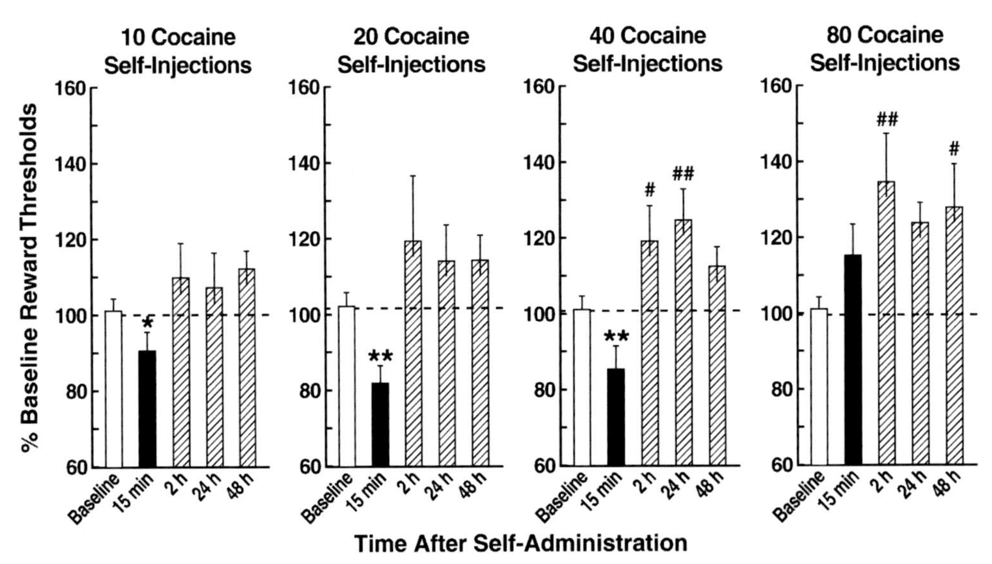

Figure 15 Rats ($n = 11$) were allowed to self-administer 10, 20, 40, and 80 injections of cocaine (0.25 mg per injection), and intracranial self-stimulation thresholds were measured 15 min, 2 h, 24 h, and 48 h after the end of each intravenous cocaine self-administration session. The *horizontal dotted line* in each plot represents 100% of baseline levels. The data are expressed as the mean percentage of baseline intracranial self-stimulation thresholds. $^{*}p < 0.05$, $^{**}p < 0.01$, compared with baseline; $^{\#}p < 0.05$, $^{\#\#}p < 0.01$, compared with baseline. *(Taken with permission from Kenny PJ, Polis I, Koob GF, Markou A. Low dose cocaine self-administration transiently increases but high dose cocaine persistently decreases brain reward function in rats. European Journal of Neuroscience 2003;17:191–5 [104].)*

cocaine self-administration exhibit increases in cocaine self-administration over days when given prolonged access (long-access [LgA] for 6 h [105−107]) compared with short access (ShA for 1 h). This differential exposure to cocaine also has dramatic effects on intracranial self-stimulation (ICSS) reward thresholds. ICSS thresholds progressively increase in LgA rats but not in ShA or control rats across successive self-administration sessions ([108]; see Volume 2, Psychostimulants). Elevations of baseline ICSS thresholds temporally precede and are highly correlated with the escalation of cocaine intake. Post-session elevations of ICSS reward thresholds then fail to return to baseline levels before the onset of subsequent self-administration sessions, thereby deviating progressively more from control levels. The progressive elevation of reward thresholds is associated with the dramatic escalation of cocaine consumption. After escalation occurs, an acute cocaine challenge facilitates brain reward responsiveness to the same degree as previously but results in higher absolute brain reward thresholds in LgA rats than in ShA rats. Similar results have been observed with extended access to methamphetamine [109] and heroin [110]. Rats that were allowed 6 h access to methamphetamine or 23 h access to heroin also presented a time-dependent elevation of reward thresholds that paralleled the increases in heroin intake (Fig. 16). Precipitated withdrawal elevated brain reward thresholds during extended-access nicotine intake [111].

2.4 Allostasis and neuroadaptation

More recently, opponent process theory has been expanded into the domains of the neurocircuitry and neurobiology of drug addiction from a physiological perspective. An allostatic model of the brain motivational systems has been proposed to explain persistent changes in motivation that are associated with vulnerability to relapse in addiction, and this model may generalize to other psychopathologies that are associated with dysregulated motivational systems. Allostasis from the addiction perspective has been defined as the process of maintaining apparent reward function stability through changes in brain reward mechanisms [100]. The allostatic state represents a chronic deviation of brain reward set point that is often *not* overtly observed while the individual is actively taking the drug. Thus, the allostatic view is that not only does the *b-process* get larger with repeated drug taking, but the reward set point from which the *a-process* and *b-process* are anchored gradually moves downward, creating an allostatic state ([100]; Fig. 17).

The allostatic state is fueled by the dysregulation of neurochemical elements of brain reward circuits and the activation of brain and hormonal stress responses. The hypothesized reward dysfunction appears to be common to all addictions and is not drug-specific. The established anatomical connections and manifestation of this allostatic state as compulsive drug taking and the loss of control over drug intake are critically based on the dysregulation of specific neurotransmitter function in neurocircuits of the ventral striatum and extended amygdala [112]. The chronic elevation of reward thresholds

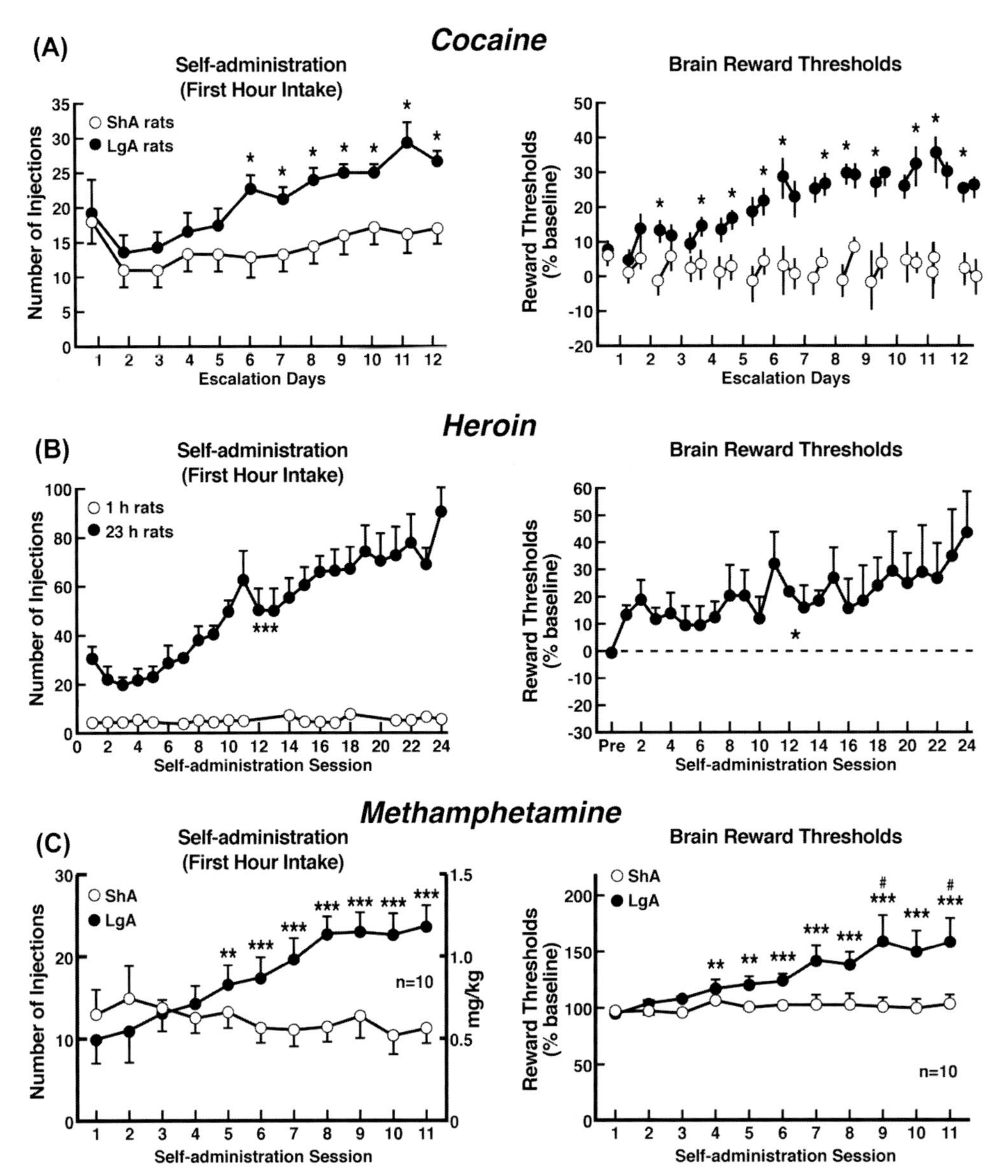

Figure 16 (A) Relationship between elevation of ICSS reward thresholds and escalation of cocaine intake. (*Left*) Percent change from baseline response latencies (3 h and 17–22 h after each self-administration session; first data point indicates 1 h before the first session). (*Right*) Percent change from baseline ICSS thresholds. $*p < 0.05$, compared with drug-naive and/or ShA rats (tests for simple main effects). (B) Unlimited daily access to heroin escalated heroin intake and decreased the excitability of brain reward systems. (*Left*) Heroin intake (±SEM; 20 μg per infusion) in rats during limited (1 h) or unlimited (23 h) self-administration sessions. $***p < 0.001$, main effect of access (1 or 23 h). (*Right*) Percent change from baseline ICSS thresholds (±SEM) in 23 h rats. Reward thresholds, assessed immediately after each daily 23 h self-administration session, became progressively more elevated as

[100] is viewed as a key element in the development of addiction that sets up other self-regulation failures and persistent vulnerability to relapse during protracted abstinence. The view that drug addiction and alcoholism are the pathology that results from an allostatic mechanism that usurps the circuits established for natural rewards provides an approach to identifying the neurobiological factors that produce the vulnerability to addiction and relapse.

2.5 Psychodynamic view of addiction

A psychodynamic view of addiction that can be integrated with the neurobiology of addiction was elaborated by Khantzian [113–115], with a focus on the factors that lead to vulnerability to addiction. This perspective is deeply rooted in the psychodynamic aspects of clinical practice that were developed from a contemporary perspective with regard to substance use disorders. The focus of this approach is on developmental difficulties, emotional disturbances, structural (ego) factors, personality organization, and building of the "self."

Two critical elements (disordered emotions and disordered self-care) and two contributory elements (disordered self-esteem and disordered relationships) were identified. These evolved into a self-medication hypothesis, in which individuals with substance use disorders take drugs as a means to cope with painful and threatening emotions. In this conceptualization, individuals with addiction experience states of subjective distress and suffering that may or may not be sufficient to meet DSM-5 criteria for a psychiatric diagnosis [3]. Individuals with addiction have feelings that are overwhelming and unbearable and may consist of an affective life that is absent and nameless. From this perspective, drug addiction is viewed as an attempt to medicate such a dysregulated affective state. Patient suffering is deeply rooted in disordered emotions, characterized at their extremes by unbearable painful affect or a painful sense of emptiness. Others cannot express personal feelings or cannot access emotions and may suffer from

exposure to self-administered heroin increased across sessions. $*p < 0.05$, main effect of heroin on reward thresholds. (C) Escalation of methamphetamine self-administration and ICSS in rats. Rats were daily allowed to receive ICSS in the lateral hypothalamus 1 h before and 3 h after intravenous methamphetamine self-administration with either 1- or 6-h access. (*Left*) Methamphetamine self-administration during the first hour of each session. (*Right*) ICSS measured 1 h before and 3 h after methamphetamine self-administration. $*p < 0.05$, $**p < 0.01$, $***p < 0.001$, compared with session 1; $^{\#}p < 0.05$, compared with LgA 3 h after. *((A) Taken with permission from Ahmed SH, Kenny PJ, Koob GF, Markou A. Neurobiological evidence for hedonic allostasis associated with escalating cocaine use. Nature Neuroscience 2002;5:625–6 [108], (B) Taken with permission from Kenny PJ, Chen SA, Kitamura O, Markou A, Koob GF. Conditioned withdrawal drives heroin consumption and decreases reward sensitivity. Journal of Neuroscience 2006;26:5894–900 [110], (C) Taken with permission from Jang CG, Whitfield T, Schulteis G, Koob GF, Wee S. A dysphoric-like state during early withdrawal from extended access to methamphetamine self-administration in rats. Psychopharmacology 2013;225:753–63 [109].)*

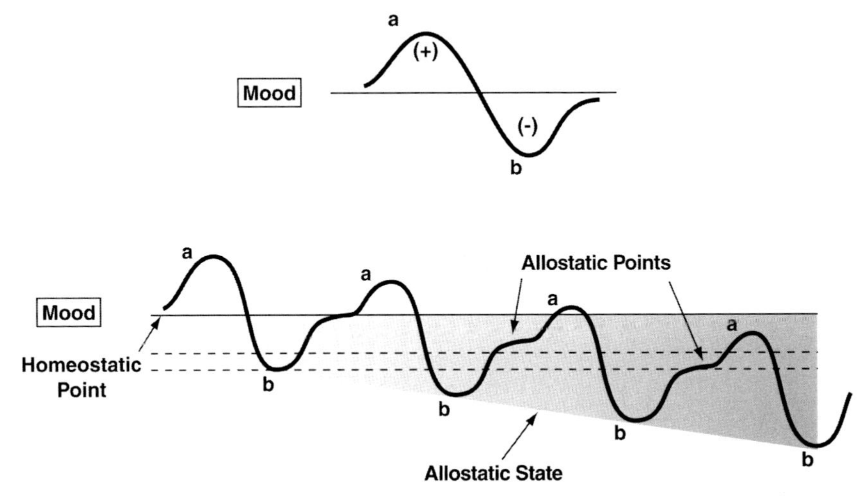

Figure 17 Diagram illustrating an extension of the [91] opponent-process model of motivation to outline the conceptual framework of the allostatic hypothesis. Both panels represent the affective response to the presentation of a drug. (Top) This diagram represents the initial experience of a drug with no prior drug history. The *a-process* represents a positive hedonic or positive mood state, and the *b-process* represents the negative hedonic or negative mood state. The affective stimulus (state) has been argued to be a sum of both an *a-process* and a *b-process*. An individual who experiences a positive hedonic mood state from a drug of abuse with sufficient time between readministering the drug is hypothesized to retain the *a-process*. In other words, an appropriate counteradaptive opponent-process (*b-process*) that balances the activational process (*a-process*) does not lead to an allostatic state. (Bottom) The changes in the affective stimulus (state) in an individual with repeated frequent drug use that may represent a transition to an allostatic state in the brain reward systems and, by extrapolation, a transition to addiction. The apparent *b-process* never returns to the original homeostatic level before drug taking is reinitiated, thus creating a progressively greater allostatic state in the brain reward system. The counteradaptive opponent-process (*b-process*) does not balance the activational process (*a-process*) but in fact shows residual hysteresis. Although these changes are exaggerated and condensed over time in the present conceptualization, the hypothesis is that even after detoxification during a period of prolonged abstinence, the reward system is still bearing allostatic changes. In the nondependent state, reward experiences are normal, and the brain stress systems are not greatly engaged. During the transition to the state known as addiction, the brain reward system is in a major underactivated state while the brain stress system is highly activated. The following definitions apply: *allostasis*, the process of achieving stability through change; *allostatic state*, a state of chronic deviation of the regulatory system from its normal (homeostatic) operating level; *allostatic load*, the cost to the brain and body of the deviation, accumulating over time, and reflecting in many cases pathological states and accumulation of damage. *(Taken with permission from Koob GF, Le Moal M. Drug addiction, dysregulation of reward, and allostasis. Neuropsychopharmacology 2001;24:97—129 [100].)*

alexithymia, defined as "a marked difficulty to use appropriate language to express and describe feelings and to differentiate them from bodily sensation" [116].

Such self-medication may be drug-specific. Patients may preferentially use drugs that fit the nature of their painful affective states. Opioids might effectively reduce psychopathological states of violent anger or rage. Others who suffer from anhedonia, anergia, or the lack of feelings might prefer the activating properties of psychostimulants. Still others who sense themselves as being flooded by their feelings or cut off from their feelings entirely may opt for repeated moderate doses of alcohol or depressants in a medicinal effort to express feelings that they are otherwise unable to communicate [115,117,118]. The common element of the self-medication hypothesis is that each drug class serves as an antidote or "replacement for a defect in the psychological structure" [119]. The paradox is that using drugs to self-medicate emotional pain will eventually perpetuate it by perpetuating a life that revolves around drugs.

Disordered self-care combines with a disordered emotional life to become a principal determinant of substance use disorders. Self-care deficits reflect an inability to ensure one's self-preservation and are characterized by an inability to anticipate or avoid harmful or dangerous situations and an inability to use appropriate judgment and feelings as guides in the face of adversity. Thus, self-care deficits reflect an inability to appropriately experience emotions and fully recognize the consequences of dangerous behaviors. The core element of this psychodynamic perspective is a dysregulated emotional system in individuals who are vulnerable to addiction.

This psychodynamic approach integrates well with the critical role of dysregulated brain reward and stress systems that have been revealed by studies of the neurobiology of addiction. From a neurobiological perspective, additional harm to the personality can be produced by the direct effects of the drugs themselves, thus perpetuating or actually *creating* such character flaws [120].

2.6 Social psychological and self-regulation views of addiction

At the social psychology level, failures in self-regulation have been argued to be the root of major social pathologies [121]. Important self-regulation elements are involved in different stages of addiction, including other pathological behaviors, such as compulsive gambling and binge eating [121]. Failures in self-regulation can lead to addiction in the case of drug use or an addiction-like pattern with nondrug behaviors. Underregulation, reflected by strength deficits, a failure to establish standards, conflicting standards, and attentional failures, and misregulation (misdirected attempts to self-regulate) can contribute to the development of addiction-like behavioral patterns (Fig. 18).

The transition to addiction can be facilitated by lapse-activated causal patterns (patterns of behavior that contribute to the transition from an initial lapse in self-regulation to a large-scale breakdown), thus leading to spiraling distress [121]. In some cases, the first self-regulation failure can lead to emotional distress, setting the stage for a cycle of repeated failures to self-regulate and where each violation brings additional negative affect, resulting in spiraling distress [121]. For example, a failure of strength may lead to initial drug use or relapse, and other self-regulation failures can be recruited to provide entry into or prevent exit from the addiction cycle.

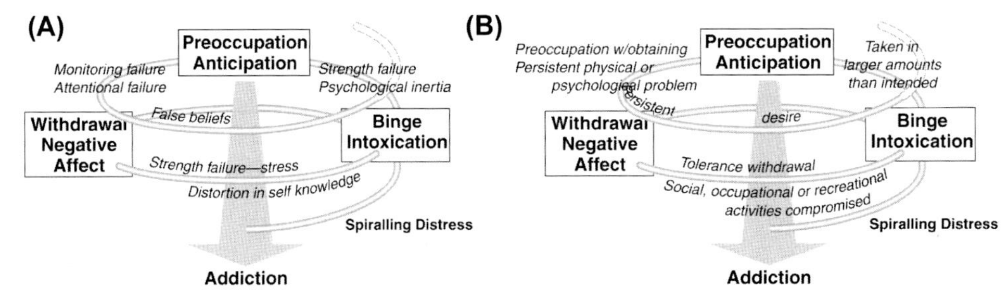

Figure 18 Diagram describing the spiraling distress/addiction cycle from two conceptual perspectives: social psychological, psychiatric, dysadaptational, and neurobiological. Notice that the addiction cycle is conceptualized as a spiral that increases in amplitude with repeated experience, ultimately resulting in the pathological state known as addiction. (A) The three major components of the addiction cycle—*preoccupation/anticipation*, *binge/intoxication*, and *withdrawal/negative affect*—and some of the sources of potential self-regulation failure in the form of underregulation and misregulation. (B) The same three major components of the addiction cycle with the different criteria for substance dependence incorporated from the DSM-IV. *(Taken with permission from Koob GF, Le Moal M. Drug abuse: hedonic homeostatic dysregulation. Science 1997;278:52–8 [2].)*

At the neurobehavioral level, such dysregulation may be reflected by deficits in information-processing, attention, planning, reasoning, self-monitoring, inhibition, and self-regulation, many of which involve functioning of the frontal lobe [122,123]. Executive function deficits, self-regulation problems, and frontal lobe dysfunction or pathology constitute risk factors for biobehavioral disorders, including drug abuse [124]. Deficits in frontal cortex regulation in children or young adolescents predict later drug and alcohol consumption, especially in children who are raised in families with histories of drug and biobehavioral problems [124,125]. Smaller volumes of the frontal cortex predict relapse in individuals with alcohol use disorder during recovery (90 days after discharge from 6 weeks of inpatient treatment [126]; Fig. 19).

2.7 Vulnerability to addiction

A wide range of factors may predispose individuals to addiction, many of which provide insights into the etiology of the disorder. These include comorbidities with other psychiatric disorders, such as anxiety, affective, personality, and psychotic disorders, and neurobehavioral traits, such as impulsivity, development (adolescence), psychosocial stress, and gender. The neurobiological bases for predisposing factors that are common to the neurobiological changes that are associated with addiction provide insights into the neuropathology that is associated with the development of addiction. Individual differences in temperament, social development, comorbidity, protective factors, and genetics are areas of intense research. A detailed discussion of these contributions to addiction are beyond the scope of this book, but each of these factors presumably interacts with the neurobiological processes that are discussed herein. A reasonable assertion is

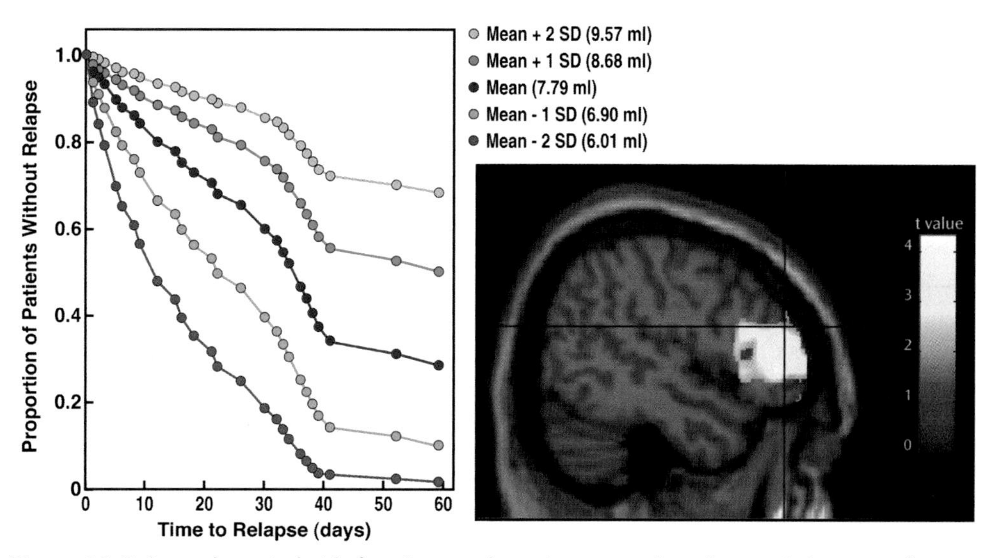

Figure 19 Estimated survival risk functions and receiver operating characteristic curves for gray matter volumes in specific significant brain regions in alcohol-dependent patients ($n = 44$). The figure shows the estimated survival risk functions (with mean age, IQ, and baseline total amount of alcohol consumed held constant) for mean gray matter volumes as well as for volumes one and two standard deviations above and below the mean for the medial frontal cluster (panel A; cluster $\chi^2 = 6.7$, $p < 0.009$; hazard ratio = 0.52, 95% confidence interval = 0.31–0.85). Although the survival function was a 90-day analysis, the graph is cut off at day 60 because all alcohol-dependent patients with gray matter volumes two standard deviations below the mean relapsed by day 60. For patients with volumes two standard deviations above the mean in the medial frontal cluster, the estimated survival function at day 60 spans a 0.68 (68%) proportion of surviving relapse, whereas for patients with volumes two standard deviations below the mean, the estimated survival function at day 60 for both regions spans only a 0.02% chance of surviving relapse. The high-resolution (T_1-weighted) structural magnetic resonance imaging scan shows significant clusters of gray matter volume deficit in an alcohol-dependent patient relative to healthy comparison subjects (family-wise error $p < 0.025$) in the right lateral prefrontal cortex with crosshairs at Montreal Neurological Institute coordinates x = 51, y = 40, z = 19 (Brodmann's area 46; dorsolateral prefrontal cortex). *(Taken with permission from Rando K, Hong KI, Bhagwagar Z, Li CS, Bergquist K, Guarnaccia J, Sinha R. Association of frontal and posterior cortical gray matter volume with time to alcohol relapse: a prospective study. American Journal of Psychiatry 2011;168:183–92 [126].)*

that the initiation of drug abuse is more associated with social and environmental factors, whereas the progression to a substance use disorder (addiction) is more associated with neurobiological factors [127]. Personality traits and some temperament clusters that have been identified with vulnerability include such factors as impulsivity, novelty- and sensation-seeking, conduct disorder, and negative affect [128,129,131,132]. From the perspective of comorbid psychiatric disorders, the strongest associations are found with mood disorders, anxiety disorders, antisocial personality disorders, and conduct disorders [133]. Subjects with attention-deficit/hyperactivity disorder have a higher likelihood of having a substance use disorder. The association between attention-deficit/hyperactivity disorder and drug abuse has been hypothesized to be explained largely by the higher comorbidity with conduct disorder in these individuals [134]. Independent of this association, little firm data indicate a higher risk that is caused by the pharmacological treatment of attention-deficit/hyperactivity disorder with stimulants [135]. In fact, in children with attention-deficit/hyperactivity disorder who are treated with psychostimulants and closely monitored, psychostimulant and other adjunct medications can be safely used to treat attention-deficit/hyperactivity disorder, and such treatment may also improve outcome of substance use disorders [136].

Developmental factors are important components of vulnerability, with strong evidence that adolescent exposure to alcohol, tobacco, and drugs of abuse leads to a significantly higher likelihood of drug dependence and drug-related problems in adulthood. Individuals who experience their first intoxication at 16 years of age or younger are more likely to drink and drive, ride with an intoxicated driver, be seriously injured when drinking, become heavy drinkers, and develop substance dependence on alcohol ([137]; Fig. 20). Similarly, people who smoke their first cigarette at 14–16 years of age are 1.6-times more likely to become dependent than people who begin to smoke when they are older [138,139]. Others have argued that regular smoking during adolescence raises the risk for adult smoking by a factor of 16 compared with nonsmoking during adolescence [140]. The age at which smoking begins influences the total years of smoking [141], the number of cigarettes smoked [142], and the likelihood of quitting ([140,143]; Fig. 21). When the prevalence of lifetime illicit or nonmedical drug use and dependence was estimated for each year of onset of drug use from ages $\leq$13 years and $\geq$21 years, the early onset of drug use was a significant predictor of the subsequent development of drug abuse ([144]; Fig. 22). Overall, the lifetime prevalence of substance dependence (measured by DSM-IV criteria) among people who began using drugs under the age of 14 was 34%. This percentage dropped to 14% for those who began using at age 21 or older [144].

The adolescent period is associated with specific stages and pathways of drug involvement [145]. Initiation usually begins with legal drugs (alcohol and tobacco). Involvement with illicit drugs occurs later in the developmental sequence, and marijuana is often the bridge between licit and illicit drugs. However, although this sequence is common, it does not represent an inevitable progression. Only a very small percentage of young people progress from one stage to the next and on to late-stage illicit drug use or dependence [146]. However, adolescence is a period during which a confluence of factors converge to produce changes in allostatic load that ultimately convey a strong vulnerability to addiction ([147]; Fig. 23).

The neurobiological consequences of such drug-induced allostatic load may involve compromised frontal cortex function. The adolescent brain undergoes widespread changes in structure and function, both within individual regions and in the connections

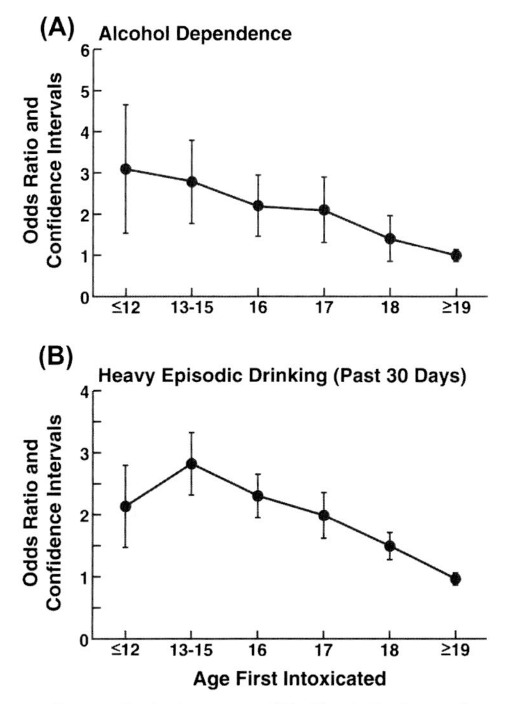

Figure 20 Representative college alcohol survey. (A) Alcohol dependence according to age first intoxicated. (B) Past 30 days of heavy episodic drinking according to age first intoxicated. After controlling for personal and demographic characteristics and respondent age, the odds of meeting alcohol dependence criteria were 3.1-times greater for those who were first drunk at or prior to age 12 compared with drinkers who were first drunk at age 19 or older. The relationship between the early onset of being drunk and heavy episodic drinking in college persisted even after further controlling for alcohol dependence. Respondents who were first drunk at or prior to age 12 were 2.1-times more likely to report recent heavy episodic drinking than college drinkers who were first drunk at age 19 or older. *(Taken with permission from Hingson R, Heeren T, Zakocs R, Winter M, Wechsler H. Age of first intoxication, heavy drinking, driving after drinking and risk of unintentional injury among U.S. college students. Journal of Studies on Alcohol 2003;64:23–31 [137].)*

between them. Studies have shown that a reduction of cortical gray matter begins in preadolescence and continues until early adulthood into the mid-20s, possibly reflecting a normal pruning process [148,149]. Equally compelling are data that show that white matter volume increases over the course of adolescence, presumably reflecting connectivity changes, including axonal extension and myelination ([148,150]; Fig. 24).

Accumulating evidence suggests that the complex changes that underlie neurodevelopment render the adolescent brain particularly vulnerable to the deleterious effects of alcohol and possibly other drugs [151]. Heavy alcohol use during adolescence is associated with a range of neurobehavioral sequelae, including impairments in visuospatial processing, attention, and memory [152,153] and a higher risk for future alcohol use

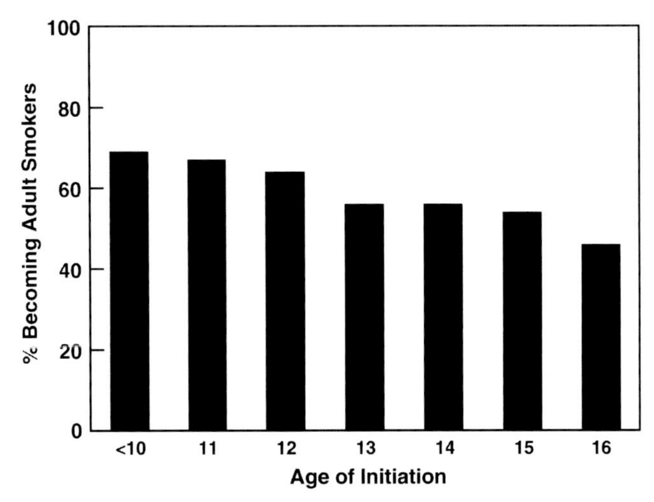

Figure 21 Percentage of adolescent regular smokers who became adult regular smokers as a function of grade of smoking initiation. The subjects consisted of all consenting 6th to 12th graders within a Midwestern county school system in the United States who were present in school on the day of testing. All 6th to 12th grade classrooms (excluding special education) were surveyed annually between 1980 and 1983. A potential pool of 5799 individuals had been assessed at least once during adolescence between 1980 and 1983. At the time of follow-up, 25 of these subjects were deceased, 175 refused participation, and 4156 provided data (72%). The subjects were predominantly Caucasian (96%), equally divided by sex (49% male, 51% female), and an average of 21.8 years old. Of the respondents, 71% had never been married, and 26% were currently married; 58% had completed at least some college by the time of follow-up; 32% were still students; and 43% had a high school education. For nonstudents, occupational status ranged from 29% in factory, crafts, and labor occupations, to 39% in professional, technical, and managerial occupations. At follow-up, the overall rate of smoking at least weekly was 26.7%. *(Taken with permission from Chassin L, Presson CC, Sherman SJ, Edwards DA. The natural history of cigarette smoking: predicting young-adult smoking outcomes from adolescent smoking patterns. Health Psychology 1990;9:701–16 [140].)*

disorders. In a powerful longitudinal study, adolescents who had engaged in episodes of heavy drinking presented faster declines in volumes in specific neocortical gray matter regions and smaller increases in regional white matter volumes relative to non-drinking adolescents (Fig. 25). In this longitudinal study, 59 adolescent controls and 75 adolescent heavy drinkers were followed from 13 to 24 years of age with at least two scans for each subject. The authors suggested that the pattern of lower gray matter may reflect "accelerated declining volume trajectories or premature cortical gray matter decline similar to adult alcoholics or even 'normal' aging" [154].

The frontal cortex controls not only impulsivity via projections to the basal ganglia but also compulsivity via projections to the extended amygdala [12,155]. Under this framework, ventromedial prefrontal cortex projections to the basal ganglia and extended amygdala have been conceptualized as a STOP system and as such control assessments of the incentive value of choices and the suppression of affective responses to negative

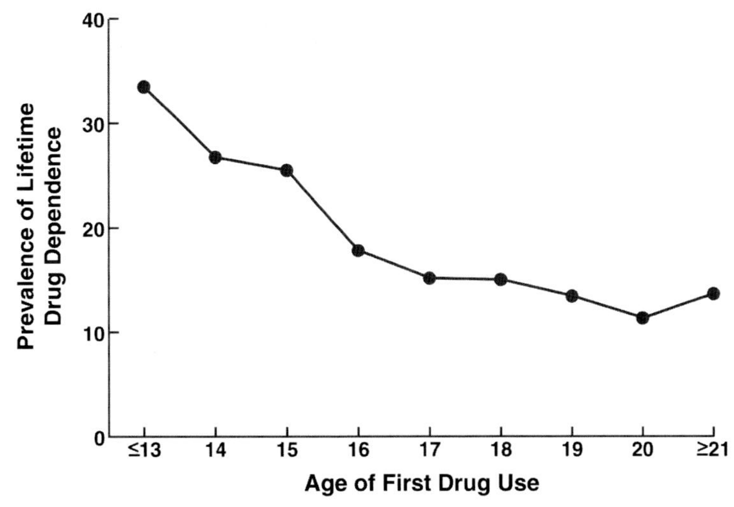

Figure 22 Prevalence of lifetime drug dependence by age at first drug use. The prevalence of lifetime dependence decreased steeply with increasing age of onset of drug use. Overall, the prevalence of lifetime dependence among those who started using drugs under the age of 14 years was approximately 34%, dropping sharply to 15.1% for those who initiated use at age 17, to approximately 14% among those who initiated use at age 21 or older. *(Taken with permission from Grant BF, Dawson DA. Age of onset of drug use and its association with DSM-IV drug abuse and dependence: results from the National Longitudinal Alcohol Epidemiologic Survey. Journal of Substance Abuse 1998;10: 163–73 [144].)*

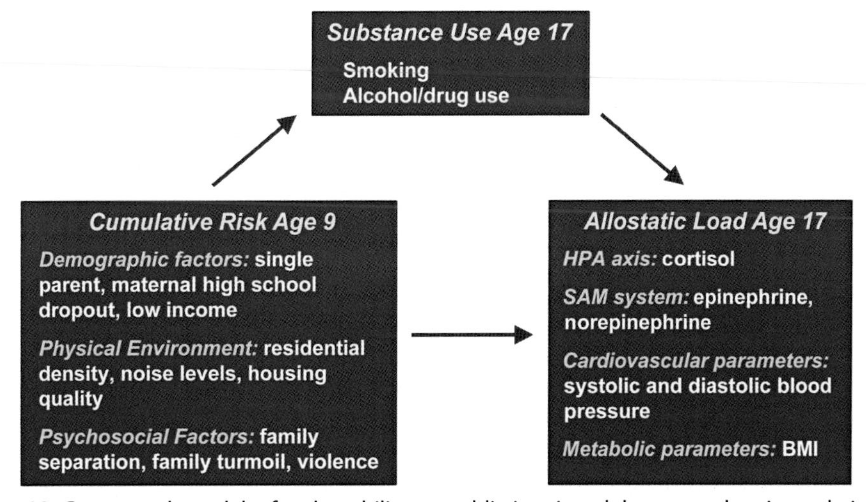

Figure 23 Conceptual model of vulnerability to addiction in adolescence showing relationship between childhood stress, adolescent substance use, and allostatic load. *(Taken with permission from Doan SN, Dich N, Evans GW. Childhood cumulative risk and later allostatic load: mediating role of substance use. Health Psychology November 2014;33(11):1402–9 [147].)*

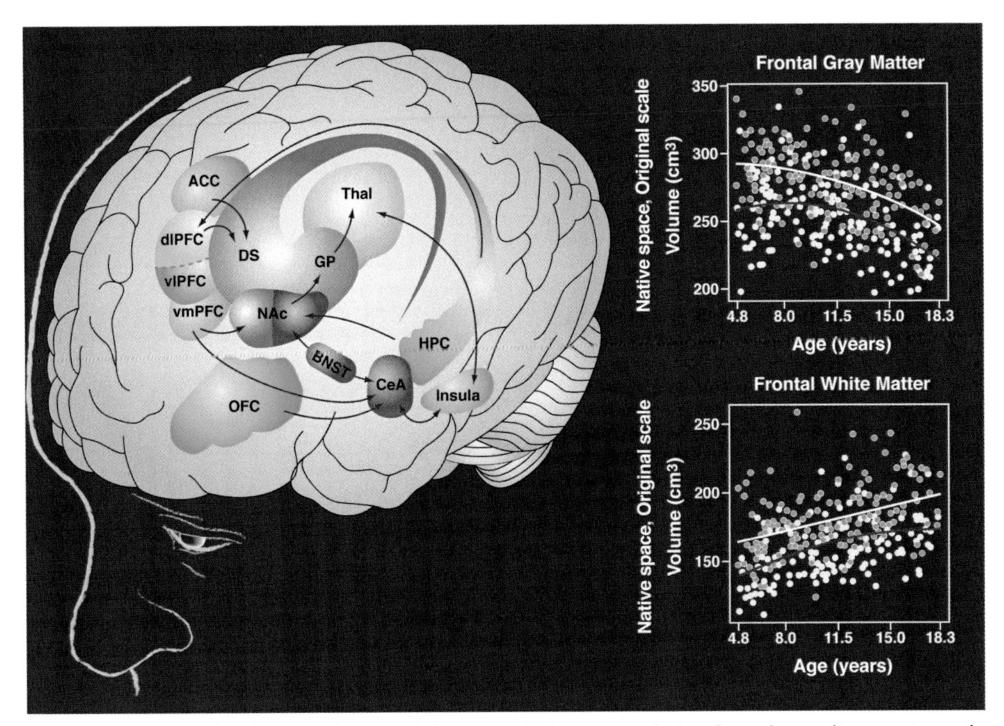

Figure 24 Frontal lobe changes during adolescence. Using a population-based sampling strategy, the National Institutes of Health Magnetic Resonance Imaging Study of Normal Brain Development compiled a longitudinal normative reference database of neuroimaging and correlated clinical/behavioral data from a demographically representative sample of healthy children and adolescents aged newborn through early adulthood. The present paper reports brain volume data for 325 children, ages 4.5–18 years, from the first cross-sectional time point. Measures included volumes of whole-brain gray matter (GM) and white matter (WM), left and right lateral ventricles, frontal, temporal, parietal and occipital lobe GM and WM, subcortical GM (thalamus, caudate, putamen, and globus pallidus), cerebellum, and brainstem [150]. Individual data points and best-fitting cross-sectional age curves for males and females are shown separately for the frontal cortex volumes without consideration of any other covariates. *(Taken with permission from Brain Development Cooperative Group. Total and regional brain volumes in a population-based normative sample from 4 to 18 years: the NIH MRI Study of Normal Brain Development. Cerebral Cortex January 2012;22(1):1–12 [150].)*

emotional signals. For example, in posttraumatic stress disorder, there is hypoactivity in the ventromedial prefrontal cortex and hyperactivity in the central nucleus of the amygdala that reflect an inhibitory connection between the ventromedial prefrontal cortex and central nucleus of the amygdala [156], and this system is hypoactive in subjects with alcohol use disorder (see Volume 3, Alcohol).

Genetic contributions to addiction can result from complex genetic differences that range from alleles that control drug metabolism to hypothesized genetic control over drug sensitivity and environmental influences [157]. The classic approach to studying

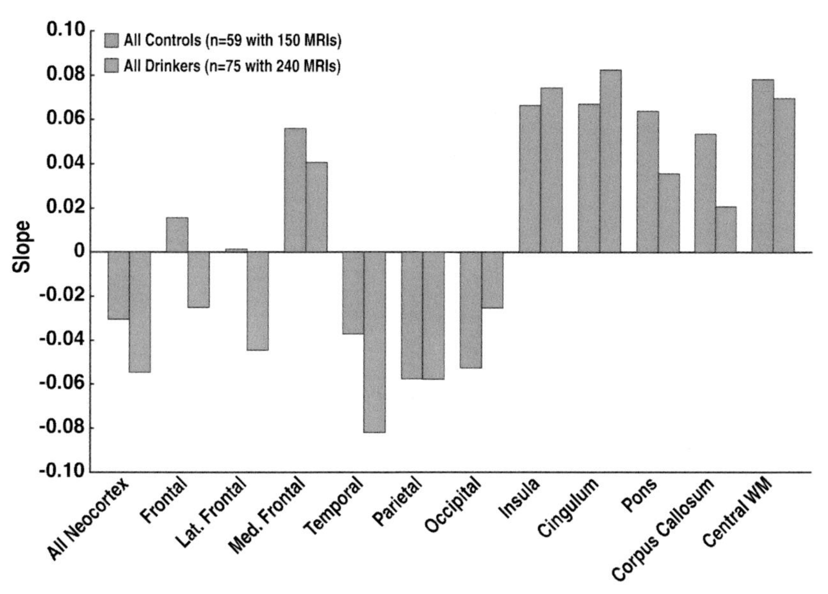

Figure 25 This study used magnetic resonance imaging to examine brain gray and white matter volume trajectories in adolescents who transitioned to heavy drinking compared with those who did not. Gray and white matter volume trajectories were evaluated in 134 adolescents, of whom 75 transitioned to heavy drinking and 59 remained light drinkers or nondrinkers over roughly 3.5 years. Each underwent magnetic resonance imaging scanning two to six times between ages 12 and 24 and was followed for up to 8 years. Heavy-drinking adolescents showed accelerated gray matter reduction in cortical lateral frontal and temporal volumes and attenuated white matter growth of the corpus callosum and pons relative to nondrinkers. Longitudinal analysis enabled the detection of accelerated typical volume decline in frontal and temporal cortical volumes and attenuated growth in principle white matter structures in adolescents who started to drink heavily. Mean slopes of each region of interest are shown for all controls and all drinkers. Negative slopes are in the direction of declining volume over age. The findings provide further evidence that heavy drinking during adolescence alters the trajectory of brain development. *(Taken with permission from Squeglia LM, Tapert SF, Sullivan EV, Jacobus J, Meloy MJ, Rohlfing T, Pfefferbaum A. Brain development in heavy-drinking adolescents. American Journal of Psychiatry June 2015;172(6):531–42 [154].)*

complex genetic traits is to examine co-occurrence or comorbidity in monozygotic versus dizygotic twins who are reared together or apart or in family studies with biological relatives. Twin and adoption studies provide researchers with estimates of the extent to which genetics influence a given phenotype, termed *heritability*. Genetic factors may account for approximately 40% of the total variability of the phenotype (Table 4). In no case does heritability account for 100% of variability, which argues strongly for gene-environment interactions, including the specific stages of the addiction cycle, developmental factors, and social factors.

Genetic factors can also convey protection against drug abuse. For example, certain Asian populations who lack one or more alleles of the acetaldehyde dehydrogenase

Table 4 Heritability estimates for drug dependence.

	Males	Females
Cocaine	44%	65%
Heroin (opiates)	43%	–
Marijuana	33%	79%
Tobacco	53%	62%
Alcohol	49% (40–60%)	64%
Addiction overall	*40%*	

Male cocaine, heroin, marijuana: [158].
Male nicotine: [159].
Female cocaine: [160].
Female marijuana: [161].
Female nicotine: [162].
Male alcohol: [163–165].
Female alcohol: [164].
Addiction overall: [157].

gene present significantly less vulnerability to alcoholism [166–169]. A similar genetic defect in metabolizing nicotine has been discovered in which faster metabolizers of nicotine have higher smoking rates and may also convey a vulnerability to dependence in [170–172].

2.8 Individual differences, neuroclinical assessment, and precision medicine

Clinical heterogeneity is a major challenge to the diagnosis, prevention, and treatment of addictive disorders. As noted above in both the DSM and ICD, addictions are categorical diagnoses that are based on symptom counts. The DSM-5 estimates the level of severity, also based on symptom counts. Because a diagnosis of addictive disorder under ICD and DSM criteria requires that the patient meet a limited number of criteria from a larger list of partially inter-correlated criteria, there is considerable within-diagnosis heterogeneity. Any patient can reach the endpoints that are represented by these behaviorally focused criteria via different routes and beginning from very different and even polar opposite starting points of vulnerability. For example, both anxiety (internalizing behavior: enhanced affective response, prior sensitization to stress/trauma) and risk-taking (externalizing behavior: impulsivity, enhanced responses to novelty, novelty seeking) can predispose to addiction liability that arises from genetic risk factors and exposures. The diagnostic criteria for alcohol use disorder and other addictive disorders focus on overt behavioral symptoms and consequences of use rather than underlying neurobiological differences that lead to vulnerability and can define progression.

Given the significant advances in our understanding of the neurocircuitry of addiction and the neurobehavioral differences that are involved in liability and the capacity to

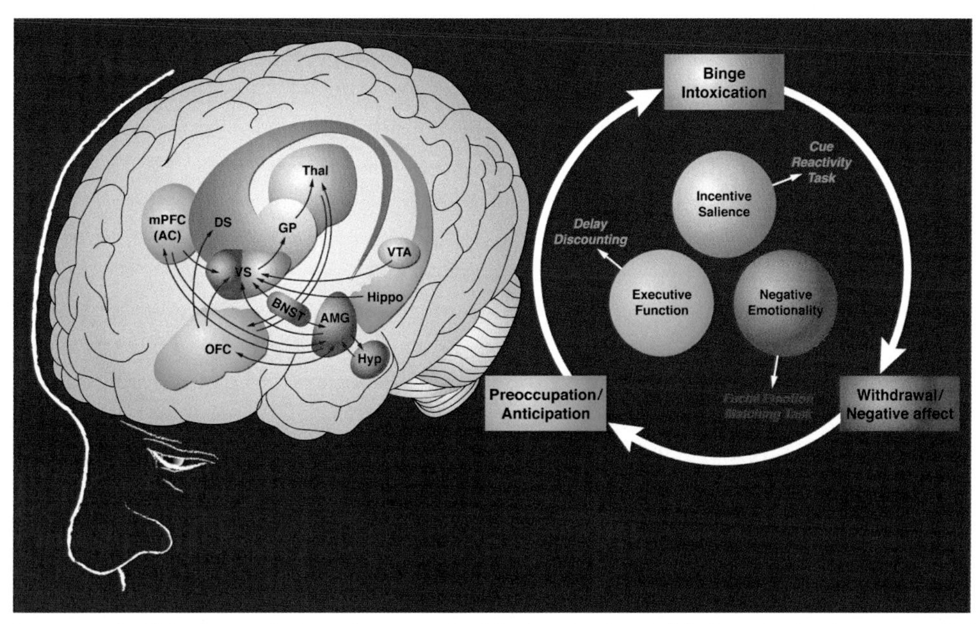

Figure 26 (Left) Pathways for key elements of addiction circuitry. Addiction circuitry is composed of structures involved in the three stages of the addiction cycle: *binge/intoxication* (ventral striatum, dorsal striatum, thalamus), *withdrawal/negative affect* (ventral striatum, bed nucleus of the stria terminalis, central nucleus of the amygdala), *preoccupation/anticipation* (prefrontal cortex, orbitofrontal cortex, hippocampus). AMG, amygdala; BNST, bed nucleus of the stria terminalis; DS, dorsal striatum; GP, globus pallidus; Hippo, hippocampus; Hyp, hypothalamus; Insula, insular cortex; OFC, orbitofrontal cortex; PFC, prefrontal cortex; Thal, thalamus; VS, ventral striatum; VTA, ventral tegmental area. (Right) Addiction neuroclinical assessment domains that correspond to the three stages of the addiction cycle and the neurocircuitry on the left. Representative human laboratory tasks that may apply to each domain, respectively, are indicated by the *arrows*. *(Left panel modified with permission from Zald DH, Kim SW. The orbitofrontal cortex. In: Salloway SP, Malloy PF, Duffy JD, editors. The frontal lobes and neuropsychiatric illness. Washington, DC: American Psychiatric Press; 2001. p. 33–70 [188], (Right) From Kwako LE, Momenan R, Grodin EN, Litten RZ, Koob GF, Goldman G. Addictions neuroclinical assessment: a reverse translational approach. Neuropharmacology 2017;122:254–264 [182], see also Koob GF. Theoretical frameworks and mechanistic aspects of alcohol addiction: alcohol addiction: alcohol addiction as a reward deficit disorder. In: Sommer WH, Spanagel R, editors. Behavioral neurobiology of alcohol addiction. Current topics in behavioral neuroscience, vol. 13. Berlin: Springer-Verlag; 2013. p. 3–30 [180].)*

measure the activity and output of the brain for relevant domains, an argument can be made that it is an appropriate time to leverage knowledge toward a more refined approach to diagnosis and treatment [173]. As a result, a heuristic framework for the Addictions Neuroclinical Assessment (ANA) was proposed that incorporates key functional domains that were derived from the neurocircuitry of addiction [173]. Three domains—executive function, incentive salience, and negative emotionality—that are tied to different phases of the addiction cycle form the core functional elements of alcohol use disorder, which were used to conceptualize the ANA ([173]; Fig. 26).

The measurement of these domains in epidemiologic, genetic, clinical, and treatment studies will provide the underpinnings for an understanding of cross-population and temporal variation in addictions, shared mechanisms in addictive disorders, the impact of changing environmental influences, and gene identification. The ANA approach may be key to reconceptualizing the nosology of substance use disorders on the basis of process and etiology, an advance that can lead to better prevention and treatment. It is practical to implement such a deep neuroclinical assessment using a combination of neuroimaging and performance measures.

Such a hypothetical approach aligns with similar initiatives in mental health (e.g., the Research Domain Criteria [RDoC] program at the National Institute of Mental Health) and in medicine more generally (e.g., the National Institutes of Health Precision Medicine Initiative Cohort Program). One major distinction between the ANA and these initiatives is that the former is focused on a particular disease category (i.e., addiction) rather than across various diseases.

3. Summary

This chapter defines addiction as a chronic relapsing disorder that is characterized by compulsive drug seeking, the loss of control in limiting intake, and the emergence of a negative emotional state when access to the drug is prevented. The definition of addiction is derived from evolution of the concept of dependence and the nosology of addiction diagnosis. A distinction is made between drug use and substance use disorders (formerly abuse and dependence), with individuals termed as suffering from "addiction." Addiction affects a large percentage of society and has enormous monetary costs. Addiction evolves over time, moving from impulsivity to compulsivity and ultimately being composed of three stages—*preoccupation/anticipation*, *binge/intoxication*, and *withdrawal/negative affect*—that worsen over time and involve allostatic changes in the brain reward and stress systems.

Two primary sources of reinforcement, positive and negative reinforcement, have been hypothesized to play a role in this allostatic process. The construct of negative reinforcement is defined as drug taking that alleviates a negative emotional state. The negative emotional state that drives such negative reinforcement is hypothesized to derive from the dysregulation of key neurochemical elements that are involved in the brain reward and stress systems within the ventral striatum, extended amygdala, and frontal cortex. Specific neurochemical elements in these structures include decreases in reward system function (within-system opponent processes), recruitment of the classic stress axis that is mediated by CRF in the extended amygdala, and the recruitment of aversive dynorphin-κ opioid systems in the ventral striatum, and extended amygdala (both between-system opponent processes). Acute withdrawal from all major drugs of abuse elevates reward thresholds, decreases mesolimbic dopamine activity, increases anxiety-like responses, increases extracellular levels of CRF in the central nucleus of the amygdala, and increases dynorphin in the ventral striatum. Corticotropin-releasing factor receptor antagonists block anxiety-like responses that are associated with withdrawal. They also block the elevations of reward thresholds that are produced by withdrawal from drugs of abuse and block compulsive-like drug taking during extended access. Excessive drug intake also activates CRF in the medial prefrontal cortex, paralleled by deficits in executive function that may facilitate the transition to compulsive-like responding. Neuropeptide Y, a powerful anti-stress neurotransmitter, affects compulsive-like responding for alcohol similarly to a CRF_1 receptor antagonist. The excessive activation of dopamine receptors in the nucleus accumbens via the release of mesocorticolimbic dopamine or opioid peptide activation of opioid receptors also activates the dynorphin-κ opioid system, which in turn can decrease dopaminergic activity in the mesocorticolimbic dopamine system. Blockade of the κ-opioid system can also block dysphoric-like effects that are associated with withdrawal from drugs of abuse and block the development of compulsive-like responding during extended access to drugs of abuse, suggesting another powerful brain stress system that contributes to compulsive drug seeking. Thus, the brain reward systems become compromised, and the brain stress systems become activated by acute excessive drug intake. These changes become sensitized during repeated withdrawal, continue into protracted abstinence, and contribute to the development and persistence of addiction. The loss of reward function and recruitment of brain stress systems provide a powerful neurochemical basis for the negative emotional states that are responsible for the negative reinforcement that drives the compulsivity of addiction.

Motivational, psychodynamic, social psychological, and vulnerability factors all contribute to the etiology of addiction, but this book focuses on the neuroadaptational changes that occur during the addiction cycle. A theoretical framework is described that derives from early homeostatic theories and subsequent opponent process theories to provide a framework for understanding the neurobiology of addiction. This framework is followed in each volume of this book series for each major drug class (Psychostimulants, Opioids, Alcohol, Nicotine, and Cannabinoids).

References

[1] American Psychiatric Association. Diagnostic and statistical manual of mental disorders. 4th ed. Washington, DC: American Psychiatric Press; 1994.

[2] Koob GF, Le Moal M. Drug abuse: hedonic homeostatic dysregulation. Science 1997;278:52–8.

[3] American Psychiatric Association. Diagnostic and statistical manual of mental disorders. 5th ed. Washington, DC: American Psychiatric Publishing; 2013.

[4] World Health Organization. International statistical classification of diseases and related health problems, 10th revision. Geneva: World Health Organization; 1992.

[5] Weisner C, Matzger H, Kaskutas LA. How important is treatment? One-year outcomes of treated and untreated alcohol-dependent individuals. Addiction 2003;98:901–11.

[6] Substance Abuse and Mental Health Services Administration. Key substance use and mental health indicators in the United States: results from the 2015 National Survey on Drug Use and Health (HHS Publication No. SMA 16-4984, NSDUH Series H-51). Rockville: Substance Abuse and Mental Health Services Administration; 2016.
[7] National Institute on Drug Abuse. Trends and statistics. 2017. https://www.drugabuse.gov/related-topics/trends-statistics.
[8] Centers for Disease Control and Prevention. Excessive drinking is draining the U.S. Economy. 2016. https://www.cdc.gov/features/costsofdrinking.
[9] U.S. Department of Health and Human Services. The health consequences of smoking—50 years of progress. A report of the surgeon general. Atlanta: U.S. Department of Health and Human Services, Centers for Disease Control and Prevention, National Center for Chronic Disease Prevention and Health Promotion, Office on Smoking and Health; 2014.
[9a] Centers for Disease Control and Prevention. Tobacco-related mortality. Atlanta: Centers for Disease Control and Prevention; 2017. https://www.cdc.gov/tobacco/data_statistics/fact_sheets/health_effects/tobacco-related_mortality/.
[10] Stahre M, Roeber J, Kanny D, Brewer RD, Zhang X. Contribution of excessive alcohol consumption to deaths and years of potential life lost in the United States. Preventing Chronic Disease 2014; 11:130293.
[11] McLellan AT, Lewis DC, O'Brien CP, Kleber HD. Drug dependence, a chronic medical illness: implications for treatment, insurance, and outcomes evaluation. Journal of the American Medical Association 2000;284:1689−95.
[12] Koob GF, Volkow ND. Neurobiology of addiction: a neurocircuitry analysis. Lancet Psychiatry 2016;3:760−73.
[13] Himmelsbach CK. Can the euphoric, analgetic, and physical dependence effects of drugs be separated? IV With reference to physical dependence. Federation Proceedings 1943;2:201−3.
[14] Eddy NB, Halbach H, Isbell H, Seevers MH. Drug dependence: its significance and characteristics. Bulletin of the World Health Organization 1965;32:721−33.
[15] Volkow ND, Koob GF. Drug addiction: the neurobiology of behavior gone awry. In: Miller SC, Fiellin DA, Rosenthal RN, Saitz R, editors. Principles of addiction medicine. 6th ed. 2017 [in press].
[16] Piazza PV, Deroche-Gamonet V. A multistep general theory of transition to addiction. Psychopharmacology 2013;229:387−413.
[17] Ahmed SH. Individual decision-making in the causal pathway to addiction: contributions and limitations of rodent models. Pharmacology Biochemistry and Behavior 2018;164:22−31.
[18] Koob GF. Allostatic view of motivation: implications for psychopathology. In: Bevins R, Bardo MT, editors. Motivational factors in the etiology of drug abuse. Nebraska symposium on motivation, vol. 50. Lincoln, NE: University of Nebraska Press; 2004. p. 1−18.
[19] Koob GF. Negative reinforcement in drug addiction: the darkness within. Current Opinion in Neurobiology 2013;23:559−63.
[20] Robbins TW. Relationship between reward-enhancing and stereotypical effects of psychomotor stimulant drugs. Nature 1976;264:57−9.
[21] Schultz W, Dayan P, Montague PR. A neural substrate of prediction and reward. Science 1997;275: 1593−9.
[22] Keiflin R, Janak PH. Dopamine prediction errors in reward learning and addiction: from theory to neural circuitry. Neuron 2015;88:247−63.
[23] Kalivas PW. The glutamate homeostasis hypothesis of addiction. Nature Reviews Neuroscience 2009;10:561−72.
[24] Luscher C, Malenka RC. Drug-evoked synaptic plasticity in addiction: from molecular changes to circuit remodeling. Neuron 2011;69:650−63.
[25] Wolf ME, Ferrario CR. AMPA receptor plasticity in the nucleus accumbens after repeated exposure to cocaine. Neuroscience & Biobehavioral Reviews 2010;35:185−211.
[26] Belin D, Balado E, Piazza PV, Deroche-Gamonet V. Pattern of intake and drug craving predict the development of cocaine addiction-like behavior in rats. Biological Psychiatry 2009;65:863−8.

[27] Alheid GF, De Olmos JS, Beltramino CA. Amygdala and extended amygdala. In: Paxinos G, editor. The rat nervous system. 2nd ed. San Diego: Academic Press; 1995. p. 495–578.
[28] Ashok AH, Mizuno Y, Volkow ND, Howes OD. Association of stimulant use with dopaminergic alterations in users of cocaine, amphetamine, or methamphetamine: a systematic review and meta-analysis. JAMA Psychiatry 2017;74:511–9.
[29] Koob GF. The dark side of emotion: the addiction perspective. Eur J Pharmacol 2015;753:73–87.
[30] Koob GF. A role for brain stress systems in addiction. Neuron 2008;59:11–34.
[30a] Koob GF. Antireward, compulsivity, and addiction: seminal contributions of Dr. Athina Markou to motivational dysregulation in addiction. Psychopharmacology 2017;234:1315–32.
[31] Shippenberg TS, Zapata A, Chefer VI. Dynorphin and the pathophysiology of drug addiction. Pharmacology & Therapeutics 2007;116:306–21.
[32] Chavkin C, Koob GF. Dynorphin, dysphoria and dependence: the stress of addiction. Neuropsychopharmacology 2016;41:373–4.
[33] Barbier E, Vendruscolo LF, Schlosburg JE, Edwards S, Juergens N, Park PE, Misra KK, Cheng K, Rice KC, Schank J, Schulteis G, Koob GF, Heilig M. The NK1 receptor antagonist L822429 reduces heroin reinforcement. Neuropsychopharmacology 2013;38:976–84.
[34] Schmeichel BE, Barbier E, Misra KK, Contet C, Schlosburg JE, Grigoriadis D, Williams JP, Karlsson C, Pitcairn C, Heilig M, Koob GF, Vendruscolo LF. Hypocretin receptor 2 antagonism dose-dependently reduces escalated heroin self-administration in rats. Neuropsychopharmacology 2015;40:1123–9.
[35] Heilig M, Koob GF. A key role for corticotropin-releasing factor in alcohol dependence. Trends in Neurosciences 2007;30:399–406.
[36] Gilpin NW, Misra K, Herman MA, Cruz MT, Koob GF, Roberto M. Neuropeptide Y opposes alcohol effects on gamma-aminobutyric acid release in amygdala and blocks the transition to alcohol dependence. Biological Psychiatry 2011;69:1091–9.
[37] Gilpin NW, Misra K, Koob GF. Neuropeptide Y in the central nucleus of the amygdala suppresses dependence-induced increases in alcohol drinking. Pharmacology Biochemistry and Behavior 2008;90:475–80.
[38] Gilpin NW, Stewart RB, Murphy JM, Li TK, Badia-Elder NE. Neuropeptide Y reduces oral ethanol intake in alcohol-preferring (P) rats following a period of imposed ethanol abstinence. Alcoholism Clinical and Experimental Research 2003;27:787–94.
[39] Thorsell A, Rapunte-Canonigo V, O'Dell L, Chen SA, King A, Lekic D, Koob GF, Sanna PP. Viral vector-induced amygdala NPY overexpression reverses increased alcohol intake caused by repeated deprivations in Wistar rats. Brain 2007;130:1330–7.
[40] Thorsell A, Slawecki CJ, Ehlers CL. Effects of neuropeptide Y and corticotropin-releasing factor on ethanol intake in Wistar rats: interaction with chronic ethanol exposure. Behavioural Brain Research 2005;161:133–40.
[41] Thorsell A, Slawecki CJ, Ehlers CL. Effects of neuropeptide Y on appetitive and consummatory behaviors associated with alcohol drinking in Wistar rats with a history of ethanol exposure. Alcoholism Clinical and Experimental Research 2005;29:584–90.
[42] Ciccocioppo R, Economidou D, Fedeli A, Massi M. The nociceptin/orphanin FQ/NOP receptor system as a target for treatment of alcohol abuse: a review of recent work in alcohol-preferring rats. Physiology and Behavior 2003;79:121–8.
[43] Martin-Fardon R, Zorrilla EP, Ciccocioppo R, Weiss F. Role of innate and drug-induced dysregulation of brain stress and arousal systems in addiction: focus on corticotropin-releasing factor, nociceptin/orphanin FQ, and orexin/hypocretin. Brain Research 2010;1314:145–61.
[44] Economidou D, Hansson AC, Weiss F, Terasmaa A, Sommer WH, Cippitelli A, Fedeli A, Martin-Fardon R, Massi M, Ciccocioppo R, Heilig M. Dysregulation of nociceptin/orphanin FQ activity in the amygdala is linked to excessive alcohol drinking in the rat. Biological Psychiatry 2008;64:211–8.
[45] Serrano A, Parsons LH. Endocannabinoid influence in drug reinforcement, dependence and addiction-related behaviors. Pharmacology & Therapeutics 2011;132:215–41.

[46] Adamczyk P, McCreary AC, Przegalinski E, Mierzejewski P, Bienkowski P, Filip M. The effects of fatty acid amide hydrolase inhibitors on maintenance of cocaine and food self-administration and on reinstatement of cocaine-seeking and food-taking behavior in rats. Journal of Physiology and Pharmacology 2009;60:119–25.
[47] Forget B, Coen KM, Le Foll B. Inhibition of fatty acid amide hydrolase reduces reinstatement of nicotine seeking but not break point for nicotine self-administration: comparison with CB_1 receptor blockade. Psychopharmacology 2009;205:613–24.
[48] Scherma M, Panlilio LV, Fadda P, Fattore L, Gamaleddin I, Le Foll B, Justinová Z, Mikics E, Haller J, Medalie J, Stroik J, Barnes C, Yasar S, Tanda G, Piomelli D, Fratta W, Goldberg SR. Inhibition of anandamide hydrolysis by cyclohexyl carbamic acid 3′-carbamoyl-3-yl ester (URB597) reverses abuse-related behavioral and neurochemical effects of nicotine in rats. Journal of Pharmacology and Experimental Therapeutics 2008;327:482–90.
[49] Bruijnzeel AW, Ford J, Rogers JA, Scheick S, Ji Y, Bishnoi M, Alexander JC. Blockade of CRF1 receptors in the central nucleus of the amygdala attenuates the dysphoria associated with nicotine withdrawal in rats. Pharmacology Biochemistry and Behavior 2012;101:62–8.
[50] Bruijnzeel AW, Prado M, Isaac S. Corticotropin-releasing factor-1 receptor activation mediates nicotine withdrawal-induced deficit in brain reward function and stress-induced relapse. Biological Psychiatry 2009;66:110–7.
[51] Bruijnzeel AW, Zislis G, Wilson C, Gold MS. Antagonism of CRF receptors prevents the deficit in brain reward function associated with precipitated nicotine withdrawal in rats. Neuropsychopharmacology 2007;32:955–63.
[52] Marcinkiewcz CA, Prado MM, Isaac SK, Marshall A, Rylkova D, Bruijnzeel AW. Corticotropin-releasing factor within the central nucleus of the amygdala and the nucleus accumbens shell mediates the negative affective state of nicotine withdrawal in rats. Neuropsychopharmacology 2009;34:1743–52.
[53] Qi X, Guzhva L, Ji Y, Bruijnzeel AW. Chronic treatment with the vasopressin 1b receptor antagonist SSR149415 prevents the dysphoria associated with nicotine withdrawal in rats. Behavioural Brain Research 2015;292:259–65.
[54] Bruijnzeel AW, Bishnoi M, van Tuijl IA, Keijzers KF, Yavarovich KR, Pasek TM, Ford J, Alexander JC, Yamada H. Effects of prazosin, clonidine, and propranolol on the elevations in brain reward thresholds and somatic signs associated with nicotine withdrawal in rats. Psychopharmacology 2010;212:485–99.
[55] Ho A, Dole VP. Pain perception in drug-free and in methadone-maintained human ex-addicts. Proceedings of the Society for Experimental Biology and Medicine 1979;162:392–5.
[56] Jochum T, Boettger MK, Burkhardt C, Juckel G, Bär KJ. Increased pain sensitivity in alcohol withdrawal syndrome. European Journal of Pain 2010;14:713–8.
[57] Doverty M, White JM, Somogyi AA, Bochner F, Ali R, Ling W. Hyperalgesic responses in methadone maintenance patients. Pain 2001;90:91–6.
[58] Compton P, Charuvastra VC, Ling W. Pain intolerance in opioid-maintained former opiate addicts: effect of long-acting maintenance agent. Drug and Alcohol Dependence 2001;63:139–46.
[59] Ren ZY, Shi J, Epstein DH, Wang J, Lu L. Abnormal pain response in pain-sensitive opiate addicts after prolonged abstinence predicts increased drug craving. Psychopharmacology 2009;204:423–9.
[60] Compton P, Athanasos P, Elashoff D. Withdrawal hyperalgesia after acute opioid physical dependence in nonaddicted humans: a preliminary study. The Journal of Pain 2003;4:511–9.
[61] Egli M, Koob GF, Edwards S. Alcohol dependence as a chronic pain disorder. Neuroscience & Biobehavioral Reviews 2012;36:2179–92.
[62] Martin WR, Gilbert PE, Jasinski DR, Martin CD. An analysis of naltrexone precipitated abstinence in morphine-dependent chronic spinal dogs. Journal of Pharmacology and Experimental Therapeutics 1987;240:565–70.
[63] Tilson HA, Rech RH, Stolman S. Hyperalgesia during withdrawal as a means of measuring the degree of dependence in morphine dependent rats. Psychopharmacologia 1973;28:287–300.

[64] Laulin JP, Larcher A, Celerier E, Le Moal M, Simonnet G. Long-lasting increased pain sensitivity in rat following exposure to heroin for the first time. European Journal of Neuroscience 1998;10:782–5.
[65] Edwards S, Vendruscolo LF, Schlosburg JE, Misra KK, Wee S, Park PE, Schulteis G, Koob GF. Development of mechanical hypersensitivity in rats during heroin and ethanol dependence: alleviation by CRF1 receptor antagonism. Neuropharmacology 2012;62:1142–51.
[66] Park PE, Schlosburg JE, Vendruscolo LF, Schulteis G, Edwards S, Koob GF. Chronic CRF1 receptor blockade reduces heroin intake escalation and dependence-induced hyperalgesia. Addiction Biology 2015;20:275–84.
[67] Fu Y, Neugebauer V. Differential mechanisms of CRF1 and CRF2 receptor functions in the amygdala in pain-related synaptic facilitation and behavior. Journal of Neuroscience 2008;28:3861–76.
[68] Ji G, Neugebauer V. Differential effects of CRF1 and CRF2 receptor antagonists on pain-related sensitization of neurons in the central nucleus of the amygdala. Journal of Neurophysiology 2007;97:3893–904.
[68a] Ji D, Gilpin NW, Richardson HN, Rivier CL, Koob GF. Effects of naltrexone, duloxetine, and a corticotropin-releasing factor type 1 receptor antagonist on binge-like alcohol drinking in rats. Behavioural Pharmacology 2008;19:1–12.
[69] Dina OA, Barletta J, Chen X, Mutero A, Martin A, Messing RO, Levine JD. Key role for the epsilon isoform of protein kinase C in painful alcoholic neuropathy in the rat. Journal of Neuroscience 2000;20:8614–9.
[70] Dina OA, Messing RO, Levine JD. Ethanol withdrawal induces hyperalgesia mediated by PKCε. European Journal of Neuroscience 2006;24:197–204.
[71] Shurman J, Koob GF, Gutstein HB. Opioids, pain, the brain, and hyperkatifeia: a framework for the rational use of opioids for pain. Pain Medicine 2010;11:1092–8.
[72] Neugebauer V. The amygdala: different pains, different mechanisms. Pain 2007;127:1–2.
[73] Bester H, Menendez L, Besson JM, Bernard JF. Spino (trigemino) parabrachiohypothalamic pathway: electrophysiological evidence for an involvement in pain processes. Journal of Neurophysiology 1995;73:568–85.
[74] Price DD. Psychological and neural mechanisms of the affective dimension of pain. Science 2000;288:1769–72.
[75] Neugebauer V, Li W. Processing of nociceptive mechanical and thermal information in central amygdala neurons with knee-joint input. Journal of Neurophysiology 2002;87:103–12.
[76] Funk CK, O'Dell LE, Crawford EF, Koob GF. Corticotropin-releasing factor within the central nucleus of the amygdala mediates enhanced ethanol self-administration in withdrawn, ethanol-dependent rats. Journal of Neuroscience 2006;26:11324–32.
[77] United Nations International Drug Control Programme. Informal expert group meeting on the craving mechanism (report no. V92-54439T). Geneva: United Nations International Drug Control Programme and World Health Organization; 1992.
[78] Bickel WK, Jarmolowicz DP, Mueller ET, Gatchalian KM, McClure SM. Are executive function and impulsivity antipodes? A conceptual reconstruction with special reference to addiction. Psychopharmacology 2012;221(3):361–87.
[79] Bickel WK, Miller ML, Yi R, Kowal BP, Lindquist DM, Pitcock JA. Behavioral and neuroeconomics of drug addiction: competing neural systems and temporal discounting processes. Drug and Alcohol Dependence 2007;90S:S85–91.
[80] Verdejo-García A, Bechara A. A somatic marker theory of addiction. Neuropharmacology 2009;56(Suppl. 1):48–62.
[81] Vanderschuren LJ, Kalivas PW. Alterations in dopaminergic and glutamatergic transmission in the induction and expression of behavioral sensitization: a critical review of preclinical studies. Psychopharmacology 2000;151:99–120.
[82] Babbini M, Gaiardi M, Bartoletti M. Persistence of chronic morphine effects upon activity in rats 8 months after ceasing the treatment. Neuropharmacology 1975;14:611–4.

[83] Bartoletti M, Gaiardi M, Gubellini C, Babbini M. Further evidence for a motility substitution test as a tool to detect the narcotic character of new drugs in rats. Neuropharmacology 1983;22:177–81.
[84] Bartoletti M, Gaiardi M, Gubellini G, Bacchi A, Babbini M. Long-term sensitization to the excitatory effects of morphine: a motility study in post-dependent rats. Neuropharmacology 1983;22:1193–6.
[85] Eichler AJ, Antelman SM. Sensitization to amphetamine and stress may involve nucleus accumbens and medial frontal cortex. Brain Research 1979;176:412–6.
[86] Kolta MG, Shreve P, De Souza V, Uretsky NJ. Time course of the development of the enhanced behavioral and biochemical responses to amphetamine after pretreatment with amphetamine. Neuropharmacology 1985;24:823–9.
[87] Robinson TE, Berridge KC. The neural basis of drug craving: an incentive-sensitization theory of addiction. Brain Research Reviews 1993;18:247–91.
[88] Robinson TE, Berridge KC. Addiction. Annual Review of Psychology 2003;54:25–53.
[89] Poulos CX, Cappell H. Homeostatic theory of drug tolerance: a general model of physiological adaptation. Psychological Review 1991;98:390–408.
[90] Siegel S. Evidence from rats that morphine tolerance is a learned response. Journal of Comparative and Physiological Psychology 1975;89:498–506.
[91] Solomon RL, Corbit JD. An opponent-process theory of motivation: 1. Temporal dynamics of affect. Psychological Review 1974;81:119–45.
[92] D'Amato MR. Derived motives. Annual Review of Psychology 1974;25:83–106.
[93] Hoffman HS, Solomon RL. An opponent-process theory of motivation: III. Some affective dynamics in imprinting. Learning and Motivation 1974;5:149–64.
[94] Koob GF, Bloom FE. Cellular and molecular mechanisms of drug dependence. Science 1988;242: 715–23.
[95] Solomon RL. The opponent-process theory of acquired motivation: the costs of pleasure and the benefits of pain. American Psychologist 1980;35:691–712.
[96] Solomon RL, Corbit JD. An opponent-process theory of motivation. II. Cigarette addiction. Journal of Abnormal Psychology 1973;81:158–71.
[97] Larcher A, Laulin JP, Celerier E, Le Moal M, Simonnet G. Acute tolerance associated with a single opiate administration: involvement of N-methyl-D-aspartate-dependent pain facilitatory systems. Neuroscience 1998;84:583–9.
[98] Colpaert FC. System theory of pain and of opiate analgesia: no tolerance to opiates. Pharmacological Reviews 1996;48:355–402.
[99] Laulin JP, Celerier E, Larcher A, Le Moal M, Simonnet G. Opiate tolerance to daily heroin administration: an apparent phenomenon associated with enhanced pain sensitivity. Neuroscience 1999;89: 631–6.
[100] Koob GF, Le Moal M. Drug addiction, dysregulation of reward, and allostasis. Neuropsychopharmacology 2001;24:97–129.
[101] Russell MAH. What is dependence? In: Edwards G, editor. Drugs and drug dependence. Lexington, MA: Lexington Books; 1976. p. 182–7.
[102] Van Dyke C, Byck R. Cocaine. Scientific American 1982;246:128–41.
[103] Breiter HC, Gollub RL, Weisskoff RM, Kennedy DN, Makris N, Berke JD, Goodman JM, Kantor HL, Gastfriend DR, Riorden JP, Mathew RT, Rosen BR, Hyman SE. Acute effects of cocaine on human brain activity and emotion. Neuron 1997;19:591–611.
[104] Kenny PJ, Polis I, Koob GF, Markou A. Low dose cocaine self-administration transiently increases but high dose cocaine persistently decreases brain reward function in rats. European Journal of Neuroscience 2003;17:191–5.
[105] Ahmed SH, Koob GF. Transition from moderate to excessive drug intake: change in hedonic set point. Science 1998;282:298–300.
[106] Deroche-Gamonet V, Belin D, Piazza PV. Evidence for addiction-like behavior in the rat. Science 2004;305:1014–7.
[107] Mantsch JR, Yuferov V, Mathieu-Kia AM, Ho A, Kreek MJ. Effects of extended access to high versus low cocaine doses on self-administration, cocaine-induced reinstatement and brain mRNA levels in rats. Psychopharmacology 2004;175:26–36.

[108] Ahmed SH, Kenny PJ, Koob GF, Markou A. Neurobiological evidence for hedonic allostasis associated with escalating cocaine use. Nature Neuroscience 2002;5:625–6.
[109] Jang CG, Whitfield T, Schulteis G, Koob GF, Wee S. A dysphoric-like state during early withdrawal from extended access to methamphetamine self-administration in rats. Psychopharmacology 2013; 225:753–63.
[110] Kenny PJ, Chen SA, Kitamura O, Markou A, Koob GF. Conditioned withdrawal drives heroin consumption and decreases reward sensitivity. Journal of Neuroscience 2006;26:5894–900.
[111] Harris AC, Pentel PR, Burroughs D, Staley MD, Lesage MG. A lack of association between severity of nicotine withdrawal and individual differences in compensatory nicotine self-administration in rats. Psychopharmacology 2011;217:153–66.
[112] Koob GF, Sanna PP, Bloom FE. Neuroscience of addiction. Neuron 1998;21:467–76.
[113] Khantzian EJ. The self-medication hypothesis of affective disorders: focus on heroin and cocaine dependence. American Journal of Psychiatry 1985;142:1259–64.
[114] Khantzian EJ. Self-regulation and self-medication factors in alcoholism and the addictions: similarities and differences. In: Galanter M, editor. Combined alcohol and other drug dependence. Recent developments in alcoholism, vol. 8. New York: Plenum Press; 1990. p. 255–71.
[115] Khantzian EJ. The self-medication hypothesis of substance use disorders: a reconsideration and recent applications. Harvard Review of Psychiatry 1997;4:231–44.
[116] Sifneos PE. Alexithymia, clinical issues, politics and crime. Psychotherapy and Psychosomatics 2000; 69:113–6.
[117] Khantzian EJ. The 1994 distinguished lecturer in substance abuse. Journal of Substance Abuse Treatment 1995;12:157–65.
[118] Khantzian EJ, Wilson A. Substance abuse, repetition, and the nature of addictive suffering. In: Wilson A, Gedo JE, editors. Hierarchical concepts in psychoanalysis: theory, research, and clinical practice. New York: Guilford Press; 1993. p. 263–83.
[119] Kohut H. In: The analysis of the self. The psychoanalytic study of the child, vol. 4. New York: International Universities Press; 1971.
[120] Koob GF. The neurobiology of self-regulation failure in addiction: an allostatic view [commentary on Khantzian, "Understanding addictive vulnerability: an evolving psychodynamic perspective"]. Neuro-psychoanalysis 2003;5:35–9.
[121] Baumeister RF, Heatherton TF, Tice DM, editors. Losing control: how and why people fail at self-regulation. San Diego: Academic Press; 1994.
[122] Giancola PR, Moss HB, Martin CS, Kirisci L, Tarter RE. Executive cognitive functioning predicts reactive aggression in boys at high risk for substance abuse: a prospective study. Alcoholism Clinical and Experimental Research 1996;20:740–4.
[123] Giancola PR, Zeichner A, Yarnell JE, Dickson KE. Relation between executive cognitive functioning and the adverse consequences of alcohol use in social drinkers. Alcoholism Clinical and Experimental Research 1996;20:1094–8.
[124] Dawes MA, Tarter RE, Kirisci L. Behavioral self-regulation: correlates and 2 year follow-ups for boys at risk for substance abuse. Drug and Alcohol Dependence 1997;45:165–76.
[125] Aytaclar S, Tarter RE, Kirisci L, Lu S. Association between hyperactivity and executive cognitive functioning in childhood and substance use in early adolescence. Journal of the American Academy of Child & Adolescent Psychiatry 1999;38:172–8.
[126] Rando K, Hong KI, Bhagwagar Z, Li CS, Bergquist K, Guarnaccia J, Sinha R. Association of frontal and posterior cortical gray matter volume with time to alcohol relapse: a prospective study. American Journal of Psychiatry 2011;168:183–92.
[127] Glantz MD, Pickens RW, editors. Vulnerability to drug abuse. Washington, DC: American Psychological Association; 1992.
[128] Glantz MD, Weinberg NZ, Miner LL, Colliver JD. The etiology of drug abuse: mapping the paths. In: Glantz MD, Hartel CR, editors. Drug abuse: origins and interventions. Washington, DC: American Psychological Association; 1999. p. 3–45.

[129] Tarter RE, Blackson T, Brigham J, Moss H, Caprara GV. The association between childhood irritability and liability to substance use in early adolescence: a 2-year follow-up study of boys at risk for substance abuse. Drug and Alcohol Dependence 1995;39:253–61.
[130] Nelson JE, Pearson HW, Sayers M, Glynn TJ, editors. Guide to drug abuse research terminology. Rockville, MD: National Institute on Drug Abuse; 1982.
[131] Wills TA, Vaccaro D, McNamara G. Novelty seeking, risk taking, and related constructs as predictors of adolescent substance use: an application of Cloninger's theory. Journal of Substance Abuse 1994;6: 1–20.
[132] Windle M, Windle RC. The continuity of behavioral expression among disinhibited and inhibited childhood subtypes. Clinical Psychology Review 1993;13:741–61.
[133] Glantz MD, Hartel CR, editors. Drug abuse: origins and interventions. Washington, DC: American Psychological Association; 1999.
[134] Biederman J, Wilens T, Mick E, Faraone SV, Weber W, Curtis S, Thornell A, Pfister K, Jetton JG, Soriano J. Is ADHD a risk factor for psychoactive substance use disorders? Findings from a four-year prospective follow-up study. Journal of the American Academy of Child & Adolescent Psychiatry 1997;36:21–9.
[135] Biederman J, Wilens T, Mick E, Spencer T, Faraone SV. Pharmacotherapy of attention-deficit/hyperactivity disorder reduces risk for substance use disorder. Pediatrics 1999;104:e20.
[136] Luo SX, Levin FR. Towards precision addiction treatment: new findings in co-morbid substance use and attention-deficit hyperactivity disorders. Current Psychiatry Reports March 2017;19(3):14.
[137] Hingson R, Heeren T, Zakocs R, Winter M, Wechsler H. Age of first intoxication, heavy drinking, driving after drinking and risk of unintentional injury among U.S. college students. Journal of Studies on Alcohol 2003;64:23–31.
[138] Breslau N, Fenn N, Peterson EL. Early smoking initiation and nicotine dependence in a cohort of young adults. Drug and Alcohol Dependence 1993;33:129–37.
[139] Everett SA, Warren CW, Sharp D, Kann L, Husten CG, Crossett LS. Initiation of cigarette smoking and subsequent smoking behavior among U.S. high school students. Preventive Medicine 1999;29: 327–33.
[140] Chassin L, Presson CC, Sherman SJ, Edwards DA. The natural history of cigarette smoking: predicting young-adult smoking outcomes from adolescent smoking patterns. Health Psychology 1990;9: 701–16.
[141] Escobedo LG, Marcus SE, Holtzman D, Giovino GA. Sports participation, age at smoking initiation, and the risk of smoking among US high school students. Journal of the American Medical Association 1993;269:1391–5.
[142] Taioli E, Wynder EL. Effect of the age at which smoking begins on frequency of smoking in adulthood. New England Journal of Medicine 1991;325:968–9.
[143] Ershler J, Leventhal H, Fleming R, Glynn K. The quitting experience for smokers in sixth through twelfth grades. Addictive Behaviors 1989;14:365–78.
[144] Grant BF, Dawson DA. Age of onset of drug use and its association with DSM-IV drug abuse and dependence: results from the National Longitudinal Alcohol Epidemiologic Survey. Journal of Substance Abuse 1998;10:163–73.
[145] Kandel DB, Jessor R. The gateway hypothesis revisited. In: Kandel DB, editor. Stages and pathways of drug involvement: examining the gateway hypothesis. New York: Cambridge University Press; 2002. p. 365–72.
[146] Kandel DB. Stages in adolescent involvement in drug use. Science 1975;190:912–4.
[147] Doan SN, Dich N, Evans GW. Childhood cumulative risk and later allostatic load: mediating role of substance use. Health Psychology November 2014;33(11):1402–9.
[148] Giedd JN, Blumenthal J, Jeffries NO, Castellanos FX, Liu H, Zijdenbos A, Paus T, Evans AC, Rapoport JL. Brain development during childhood and adolescence: a longitudinal MRI study. Nature Neuroscience 1999;2:861–3.
[149] Gogtay N, Giedd JN, Lusk L, Hayashi KM, Greenstein D, Vaituzis AC, Nugent 3rd TF, Herman DH, Clasen LS, Toga AW, Rapoport JL, Thompson PM. Dynamic mapping of human

cortical development during childhood through early adulthood. Proceedings of the National Academy of Sciences of the United States of America 2004;101:8174–9.
[150] Brain Development Cooperative Group. Total and regional brain volumes in a population-based normative sample from 4 to 18 years: the NIH MRI Study of Normal Brain Development. Cerebral Cortex January 2012;22(1):1–12.
[151] Squeglia LM, Jacobus J, Tapert SF. The influence of substance use on adolescent brain development. Clinical EEG and Neuroscience January 2009;40(1):31–8.
[152] Hanson KL, Medina KL, Padula CB, Tapert SF, Brown SA. Impact of adolescent alcohol and drug use on neuropsychological functioning in young adulthood: 10-year outcomes. Journal of Child & Adolescent Substance Abuse 2011;20:135–54.
[153] Squeglia LM, Spadoni AD, Infante MA, Myers MG, Tapert SF. Initiating moderate to heavy alcohol use predicts changes in neuropsychological functioning for adolescent girls and boys. Psychology of Addictive Behaviors 2009;23:715–22.
[154] Squeglia LM, Tapert SF, Sullivan EV, Jacobus J, Meloy MJ, Rohlfing T, Pfefferbaum A. Brain development in heavy-drinking adolescents. American Journal of Psychiatry June 2015;172(6): 531–42.
[155] Jentsch JD, Taylor JR. Impulsivity resulting from frontostriatal dysfunction in drug abuse: implications for the control of behavior by reward-related stimuli. Psychopharmacology 1999;146: 373–90.
[156] Pitman RK, Rasmusson AM, Koenen KC, Shin LM, Orr SP, Gilbertson MW, Milad MR, Liberzon I. Biological studies of post-traumatic stress disorder. Nature Reviews Neuroscience November 2012;13(11):769–87.
[157] Uhl GR, Grow RW. The burden of complex genetics in brain disorders. Archives of General Psychiatry 2004;61:223–9.
[158] Tsuang MT, Lyons MJ, Eisen SA, Goldberg J, True W, Lin N, Meyer JM, Toomey R, Faraone SV, Eaves L. Genetic influences on DSM-III-R drug abuse and dependence: a study of 3,372 twin pairs. American Journal of Medical Genetics 1996;67:473–7.
[159] Carmelli D, Swan GE, Robinette D, Fabsitz RR. Heritability of substance use in the NAS-NRC twin registry. Acta Geneticae Medicae et Gemellologiae 1990;39:91–8.
[160] Kendler KS, Prescott CA. Cocaine use, abuse and dependence in a population-based sample of female twins. British Journal of Psychiatry 1998;173:345–50.
[161] Kendler KS, Prescott CA. Cannabis use, abuse, and dependence in a population-based sample of female twins. American Journal of Psychiatry 1998;155:1016–22.
[162] Kendler KS, Neale MC, Sullivan P, Corey LA, Gardner CO, Prescott CA. A population-based twin study in women of smoking initiation and nicotine dependence. Psychological Medicine 1999;29: 299–308.
[163] Liu IC, Blacker DL, Xu R, Fitzmaurice G, Lyons MJ, Tsuang MT. Genetic and environmental contributions to the development of alcohol dependence in male twins. Archives of General Psychiatry 2004;61:897–903.
[164] McGue M, Pickens RW, Svikis DS. Sex and age effects on the inheritance of alcohol problems: a twin study. Journal of Abnormal Psychology 1992;101:3–17.
[165] Prescott CA, Kendler KS. Genetic and environmental contributions to alcohol abuse and dependence in a population-based sample of male twins. American Journal of Psychiatry 1999; 156:34–40.
[166] Goedde HW, Agarwal DP, Harada S. The role of alcohol dehydrogenase and aldehyde dehydrogenase isozymes in alcohol metabolism, alcohol sensitivity and alcoholism. In: Rattazzi MC, Scandalios JG, Whitt GS, editors. Cellular localization, metabolism, and physiology. Isozymes: current topcs in biological and medical research, vol. 8. New York: Alan R. Liss; 1983. p. 175–93.
[167] Goedde HW, Agarwal DP, Harada S, Meier-Tackmann D, Ruofu D, Bienzle U, Kroeger A, Hussein L. Population genetic studies on aldehyde dehydrogenase isozyme deficiency and alcohol sensitivity. The American Journal of Human Genetics 1983;35:769–72.
[168] Higuchi S, Matsushita S, Murayama M, Takagi S, Hayashida M. Alcohol and aldehyde dehydrogenase polymorphisms and the risk for alcoholism. American Journal of Psychiatry 1995;152:1219–21.

[169] Mizoi Y, Tatsuno Y, Adachi J, Kogame M, Fukunaga T, Fujiwara S, Hishida S, Ijiri I. Alcohol sensitivity related to polymorphism of alcohol-metabolizing enzymes in Japanese. Pharmacology Biochemistry and Behavior 1983;18(Suppl. 1):127–33.
[170] Sellers EM, Tyndale RF, Fernandes LC. Decreasing smoking behaviour and risk through CYP2A6 inhibition. Drug Discovery Today 2003;8:487–93.
[171] Tyndale RF, Sellers EM. Variable CYP2A6-mediated nicotine metabolism alters smoking behavior and risk. Drug Metabolism and Disposition 2001;29:548–52.
[172] Tyndale RF, Sellers EM. Genetic variation in CYP2A6-mediated nicotine metabolism alters smoking behavior. Therapeutic Drug Monitoring 2002;24:163–71.
[173] Kwako LE, Spagnolo PA, Schwandt ML, Thorsell A, George DT, Momenan R, Rio DE, Huestis M, Anizan S, Concheiro M, Sinha R, Heilig M. The corticotropin releasing hormone-1 (CRH1) receptor antagonist pexacerfont in alcohol dependence: a randomized controlled experimental medicine study. Neuropsychopharmacology 2015;40:1053–63.
[174] Koob GF, Volkow ND. Neurocircuitry of addiction. Neuropsychopharmacology Reviews 2010;35: 217–38 [erratum: 35: 1051].
[175] Epping-Jordan MP, Watkins SS, Koob GF, Markou A. Dramatic decreases in brain reward function during nicotine withdrawal. Nature 1998;393:76–9.
[176] Gardner EL, Vorel SR. Cannabinoid transmission and reward-related events. Neurobiology of Disease 1998;5:502–33.
[177] George O, Koob GF. Control of craving by the prefrontal cortex. Proceedings of the National Academy of Sciences of the United States of America 2013;110:4165–6.
[178] Haber S, McFarland NR. The place of the thalamus in frontal cortical-basal ganglia circuits. The Neuroscientist 2001;7:315–24.
[179] Koob GF. Neurobiology of addiction. In: Galanter M, Kleber HD, editors. Textbook of substance abuse treatment. 4th ed. Washington, DC: American Psychiatric Publishing; 2008. p. 3–16.
[180] Koob GF. Theoretical frameworks and mechanistic aspects of alcohol addiction: alcohol addiction: alcohol addiction as a reward deficit disorder. In: Sommer WH, Spanagel R, editors. Behavioral neurobiology of alcohol addiction. Current topics in behavioral neuroscience, vol. 13. Berlin: Springer-Verlag; 2013. p. 3–30.
[181] Koob GF, Le Moal M. Neurobiology of addiction. London: Academic Press; 2006.
[182] Kwako LE, Momenan R, Grodin EN, Litten RZ, Koob GF, Goldman D. Addictions neuroclinical assessment: a reverse translational approach. Neuropharmacology 2017;122:254–64.
[183] Markou A, Koob GF. Post-cocaine anhedonia: an animal model of cocaine withdrawal. Neuropsychopharmacology 1991;4:17–26.
[184] Nestler EJ. Is there a common molecular pathway for addiction? Nature Neuroscience 2005;8: 1445–9.
[185] Paterson NE, Myers C, Markou A. Effects of repeated withdrawal from continuous amphetamine administration on brain reward function in rats. Psychopharmacology 2000;152:440–6.
[186] Schulteis G, Markou A, Cole M, Koob G. Decreased brain reward produced by ethanol withdrawal. Proceedings of the National Academy of Sciences of the United States of America 1995;92:5880–4.
[187] Schulteis G, Markou A, Gold LH, Stinus L, Koob GF. Relative sensitivity to naloxone of multiple indices of opiate withdrawal: a quantitative dose-response analysis. Journal of Pharmacology and Experimental Therapeutics 1994;271:1391–8.
[188] Zald DH, Kim SW. The orbitofrontal cortex. In: Salloway SP, Malloy PF, Duffy JD, editors. The frontal lobes and neuropsychiatric illness. Washington, DC: American Psychiatric Press; 2001. p. 33–70.

CHAPTER 2

Animal models of addiction

Contents

Introduction to Addiction
ISBN 978-0-12-816863-9 https://doi.org/10.1016/B978-0-12-816863-9.00002-9

1. Definitions and validation of animal models

1.1 Definitions of drug addiction relevant to animal models

Drug addiction has been conceptualized as a disorder that progresses from impulsivity to compulsivity in a collapsed cycle that consists of three stages: (*i*) *binge/intoxication*, (*ii*) *withdrawal/negative affect*, and (*iii*) *preoccupation/anticipation*. Although much focus in animal studies has been on synaptic sites and transduction mechanisms in the central nervous system at which drugs of abuse act initially to produce their positive reinforcing effects, new animal models of the negative reinforcing effects of dependence have been developed to explore the ways in which the central nervous system adapts to drug use.

The definition of drug addiction to be used in the present book draws on several different meanings (see Chapter 1): (*i*) compulsion to seek and take the drug, (*ii*) loss of control in limiting intake, and (*iii*) emergence of a negative emotional state (e.g., dysphoria, anxiety, irritability) when access to the drug is prevented. Much of the recent progress in understanding the mechanisms of addiction has derived from studies of animal models of addiction on specific drugs, such as opioids, stimulants, and alcohol. Most importantly, although no animal model of addiction fully emulates the human condition, animal models permit investigations of specific elements of the addiction process. Such elements can be defined by models of different systems, models of psychological constructs (e.g., positive and negative reinforcement), models of different stages of the addiction cycle, and models of actual symptoms of addiction as outlined by psychiatric nosology ([1,2]; see Chapter 1). For the purposes of this chapter, animal models are categorized by specific stages of the addiction cycle and the actual symptoms of addiction that are outlined in psychiatry.

The constructs of reinforcement and motivation are crucial parts of all animal models of addiction, almost by definition, because the drug is self-administered, its delivery describes reinforcement, and the drive to take the drug again describes motivation. A reinforcer can be operationally defined as "any event whose presentation increases the probability of a response." This definition can also be used to signify a definition of reward, and the two words are often used interchangeably. However, reward often connotes some additional emotional value, such as pleasure. Multiple powerful sources of reinforcement have long been identified during the course of drug addiction that provide the motivation for compulsive use and the loss of control over intake [3].

Motivation as a concept has many definitions and generally involves the constructs of both "motor" and "motive." Donald Hebb argued that motivation is "stimulation that arouses activity of a particular kind" [4], and C.P. Richter argued that "spontaneous activity arises from certain underlying physiological origins and such 'internal' drives are reflected in the amount of general activity" [5]. Dalbir Bindra defined motivation as a "rough label for the relatively persisting states that make an animal initiate and maintain actions leading to particular outcomes or goals" [6]. A more behavioristic view is that motivation is "the property of energizing of behavior that is proportional to the amount and quality of the reinforcer" [7]. Finally, a more neurobehavioral view is that motivation is a "set of neural processes that promote actions in relation to a particular class of environmental objects" [6].

The primary pharmacological effect of a drug can produce a direct effect through positive reinforcement or negative reinforcement (e.g., self-medication or relief from aversive symptoms of abstinence). The secondary pharmacological effects of the drug can also have motivating properties. Conditioned positive reinforcement involves the pairing of previously neutral stimuli with acute positive reinforcing effects of drugs, and conditioned negative reinforcement involves the pairing of previously neutral stimuli with the aversive stimulus effects of withdrawal or abstinence.

An approach to the development of animal models that has gained wide acceptance is that animal models are most likely to have construct or predictive validity when they reflect only the specific signs or symptoms that are associated with the psychopathological condition [8]. Animal models of a complete syndrome of a psychiatric disorder are unlikely to be possible either conceptually or practically. Certain areas of the human condition are obviously difficult to model in animals (e.g., kleptomania, psychopathology, child abuse, etc.). From a practical standpoint, psychiatric disorders are based on a classification of syndromes and diseases that are complex and constantly evolving in diagnosis (e.g., compare the *Diagnostic and Statistical Manual of Mental Disorders*, 4th edition [DSM-IV; [9]] and 5th edition [DSM-5; [10]]) and most certainly involve multiple subtypes and diverse etiologies (see Chapter 1). Models that attempt to reproduce entire syndromes require multiple endpoints, thus making it very difficult in practice to study underlying mechanisms.

Under such a framework of mimicking only the specific signs or symptoms of psychopathological conditions, specific "observables" (i.e., symptoms one can observe) that have been identified in addiction provide a focus for studies in animals. The reliance of animal models on a given observable also eliminates a fundamental problem that is associated with animal models of psychopathology, namely the frustration of attempting to provide complete validation of the whole syndrome. More definitive information that is related to a specific domain of addiction can be generated, and thus one can increase the confidence of cross-species validity. This framework also leads to a more pragmatic approach to the study of the neurobiological mechanisms of the behavior in question. Animal models allow one to study and understand the neurobiological bases of a given symptom, with cross-translation of such information with human laboratory models and actual clinical studies [11].

In the present chapter, these observables are organized within the *binge/intoxication*, *withdrawal/negative affect*, and *preoccupation/anticipation* (craving) stages of the addiction cycle. However, later in the chapter these observables are linked to the actual DSM-5 [10] criteria for addiction and further linked to human laboratory models of addiction. The particular behavior that is being assessed in an animal model may or may not even be symptomatic of the disorder but must be defined objectively and observed reliably. Indeed, the specific behavior may be found in both pathological and nonpathological states but still have predictive validity. A good example of such a situation is the widespread use of drug reinforcement or drug rewarded behavior as an animal model of addiction. Drug reinforcement does not necessarily lead to addiction (e.g., social drinking of alcohol), but the self-administration of alcohol has major predictive validity for the *binge/intoxication* stage of addiction, and it is difficult to imagine addiction without alcohol serving as a reinforcer. Challenges remain for the study of individual differences in the vulnerability to addiction and the converse resilience to addiction.

1.2 Validation of animal models of drug addiction

Animal models are necessary and critical for understanding the neuropharmacological mechanisms that are involved in the development of addiction. Although there are no complete animal models of addiction that fully recapitulate the human condition, animal models do exist for many elements of the syndrome. An animal model can be viewed as an experimental preparation that is developed for the purpose of studying a given phenomenon that is found in humans. The most relevant conceptualization of validity for animal models of addiction in the context of the neurobiology of addiction is the concept of construct validity [12]. Construct validity refers to the interpretability, "meaningfulness," or explanatory power of each animal model and incorporates most other measures of validity where multiple measures or dimensions are associated with conditions that are known to affect the construct [13]. An alternative conceptualization of construct validity is the requirement that the models meet the construct of functional equivalence, defined as "assessing how controlling variables influence outcome in the model and the target disorders" [14]. The most efficient process for evaluating functional equivalence has been argued to be through common experimental manipulations that should have similar effects in the animal model and the target disorder [14]. This process is very similar to the broad use of the construct of predictive validity (see below). Face validity is often the starting point in animal models, in which animal syndromes are produced that resemble those that are found in humans to study selected parts of the human syndrome but is limited by necessity [15]. Reliability refers to the stability and consistency with which the variable of interest can be measured and is achieved when, following objective repeated measurement of the variable, small within- and between-subject variability is noted, and the phenomenon is readily reproduced under similar circumstances (for review, see Ref. [16]). The construct of predictive validity refers to the model's ability to lead to accurate predictions about the human phenomenon based on the response of the model system. Predictive validity is most used often in the narrow sense in animal models of psychiatric disorders to refer to the ability of the model to identify pharmacological agents with potential therapeutic value in humans [15,17]. However, when predictive validity is more broadly extended to understanding the physiological mechanism of action of psychiatric disorders, it incorporates other types of validity (i.e., etiological, convergent or concurrent, discriminant) that are considered important for animal models and approaches the concept of construct validity [18]. The present chapter will describe animal models that have been shown to be reliable and in many cases to have construct validity for various stages of the addictive process.

2. Animal models of the binge/intoxication stage of the addiction cycle

2.1 Intravenous and oral drug self-administration

Animals and humans will readily self-administer drugs in the nondependent state. Drugs of abuse have powerful positive reinforcing properties. Animals will perform many different tasks and procedures to obtain drugs, even when not dependent. The drugs that have positive reinforcing effects, measured by direct self-administration, the lowering of brain stimulation reward thresholds, and conditioned place preference, in rodents and primates correspond very well with the drugs that have high abuse potential in humans.

Intravenous drug self-administration has proven to be a powerful tool for exploring the neurobiology of positive reinforcement. The self-administration of cocaine and heroin intravenously in rodents produces a characteristic pattern of behavior that lends itself to pharmacological and neuropharmacological studies. Rats on a simple schedule of continuous reinforcement, such as a fixed ratio-1 (FR1) schedule, where one press of a lever or nosepoke delivers one drug delivery, will develop a highly stable pattern of drug self-administration in a limited-access situation ([19]; Fig. 1).

However, as the unit dose is decreased, animals increase their self-administration rate, apparently compensating for decreases in the unit dose. Conversely, as the unit dose is increased, animals reduce their self-administration rate. Thus, manipulations that increase the self-administration rate on this FR schedule resemble decreases in the unit dose and may be interpreted as decreases in the reinforcing potency of the drug under study.

As would be predicted by the unit dose-response model, low to moderate doses of dopamine receptor antagonists increase cocaine self-administration that is maintained on this schedule in a manner that is similar to decreasing the unit dose of cocaine, suggesting that the partial blockade of dopamine receptors by competitive receptor antagonists reduces the reinforcing potency of cocaine. Conversely, dopamine receptor agonists

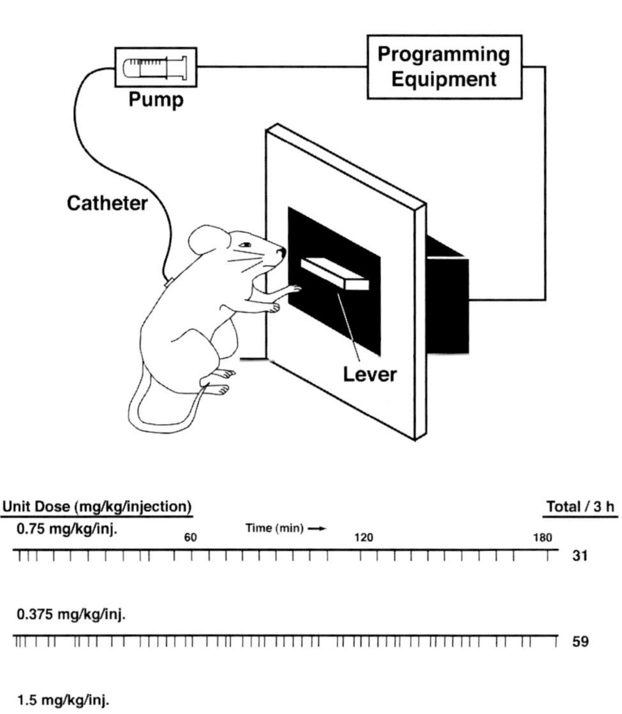

Figure 1 (Top) Procedure for intravenous self-administration in the rat. (Bottom) Event record and dose-response relationship that relates the dose of cocaine to the number of infusions. Rats that are implanted with intravenous catheters and trained to self-administer cocaine with limited access (3 h/day) will exhibit stable and regular drug intake over each daily session. No obvious tolerance or dependence develops. Rats are generally maintained on a low-requirement, fixed-ratio (FR) schedule for intravenous infusion of the drug, such as an FR1 or FR5. In an FR1 schedule, one lever press is required to deliver one intravenous infusion of cocaine. In an FR5 schedule, five lever presses are required to deliver one infusion of cocaine. A special aspect of using an FR schedule is that the rats appear to regulate the amount of drug self-administered. Lowering the dose from the training level of 0.75 mg/kg/injection increases the number of self-administered infusions and vice versa. *(Taken with permission from Caine SB, Lintz R, Koob GF. Intravenous drug self-administration techniques in animals. In: Sahgal A, editor. Behavioural neuroscience: a practical approach, vol. 2. Oxford: IRL Press; 1993. p. 117–43. [19])*

decrease cocaine self-administration in a manner that is similar to increasing the unit dose of cocaine, suggesting that the effects of dopamine receptor agonists together with cocaine self-administration can be additive, perhaps because of their mutual activation of the same neural substrates [20].

The use of different schedules of reinforcement in intravenous self-administration can provide important control procedures for nonspecific motor actions (e.g., increases in exploratory activity and locomotion) and motivational effects (e.g., the loss of reinforcement efficacy) and include a progressive-ratio schedule of reinforcement, second-order schedules, and multiple schedules.

To directly evaluate the reinforcing efficacy of a self-administered drug, one can use a progressive-ratio schedule of reinforcement. Here, the response requirements for each successive reinforcer delivery in the case of drug self-administration is increased, and a breakpoint is determined at which an animal no longer responds or suspends responding at a certain response requirement. Alternatively, one can simply measure the highest ratio that is achieved in a given session. This schedule is effective in determining the relative reinforcing strength of different reinforcers, including drugs. The unit dose-response model demonstrates that increasing the unit dose of self-administered drugs increases the breakpoint on a progressive-ratio schedule, and dopamine receptor antagonists, such as SCH 23390, have been shown to decrease the breakpoint for cocaine self-administration (Fig. 2). Performance on a progressive-ratio schedule can be linked to the following DSM-IV criterion for addiction: "a great deal of time spent in activities necessary to obtain the substance."

Oral self-administration almost exclusively involves alcohol because of the obvious face validity of oral alcohol self-administration and includes two-bottle choice and operant self-administration. Historically, home cage drinking and preference have been used to characterize genetic differences in drug preference, most often alcohol preference [21], and to explore the effects of pharmacological treatments on drug intake and preference. Here, a choice is offered between a drug solution and alternative solutions, one of which is often water, and the proportion of drug intake relative to total intake is calculated as a preference ratio. For two-bottle choice testing in mice or rats, the animals are singly housed, and a bottle that contains alcohol and a bottle that contains water are simultaneously placed in the cage. Most commonly, animals are allowed free choice of these drinking solutions for successive 24 h periods with simultaneous free access to food.

Two variants of the two-bottle choice paradigm can result in binge-like drinking in rodents. In the intermittent access to alcohol model, originally characterized by Wise [22] and revisited by Simms et al. [23,24], rats are given intermittent 24 h access to alcohol using a two-bottle choice procedure (alcohol versus water). Rats in the intermittent access to alcohol model progressively escalate their alcohol intake over the course of 2–3 weeks to reach blood alcohol levels (BALs) that are relevant to binge drinking in humans (50–100 mg%; [22,23]). However, BALs in 30-min two-bottle choice sessions in the intermittent 20% alcohol animals were significantly lower (~60 mg% in Wistar rats) than in dependent animals (see *Withdrawal/Negative Affect* stage below).

A variant of binge-like drinking in mice is termed drinking-in-the-dark, in which mice drink intoxicating amounts of alcohol with limited access [25]. Drinking-in-the-dark involves mice that are allowed to drink 20% alcohol 3 h after lights out for 2 h for 3 consecutive days and on the fourth day are allowed 4 h access, also 3 h after lights out [25].

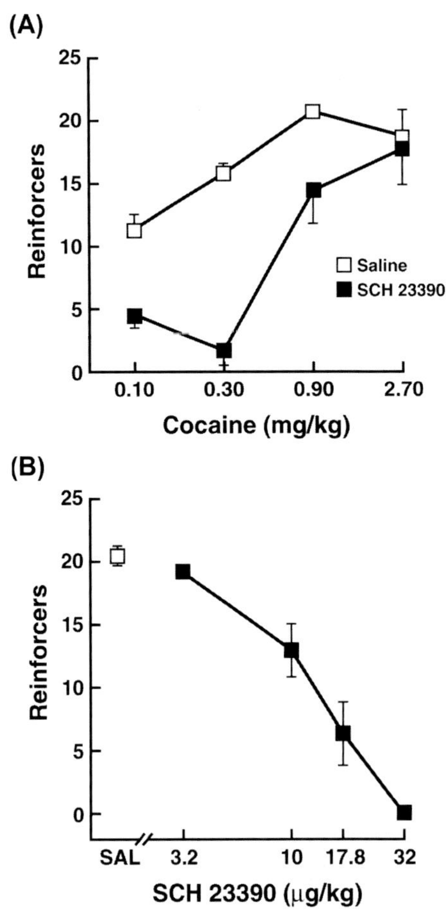

Figure 2 Effect of pretreatment with the dopamine receptor antagonist SCH 23390 in rats on the number of reinforcers obtained under a progressive-ratio schedule of reinforcement. (A) Each point represents the average number of reinforcers obtained in a session with saline or SCH 23390 (10 μg/kg, s. c.) pretreatment. (B) Each point represents the average number of reinforcers obtained in a session with subcutaneous pretreatment with saline or SCH 23390. Saline or SCH 23390 was tested against the training dose of cocaine (0.90 mg/kg). *(Taken with permission from Depoortere RY, Li DH, Lane JD, Emmett-Oglesby MW. Parameters of self-administration of cocaine in rats under a progressive-ratio schedule. Pharmacology Biochemistry and Behavior 1993;45:539–48 [231].)*

For operant alcohol self-administration, rats can be trained to lever press for alcohol using various techniques, but usually some attempt is made initially to overcome the aversive effects of initial exposure to alcohol, either by slowly increasing the concentration of alcohol or by adding a sweet solution. Using a sweetened solution fading procedure or 24 h access overnight, animals can be trained to lever press for concentrations of alcohol up to 40% (v/v). They will perform FR and progressive-ratio schedules and obtain significant BALs in a 30 min session (Fig. 3).

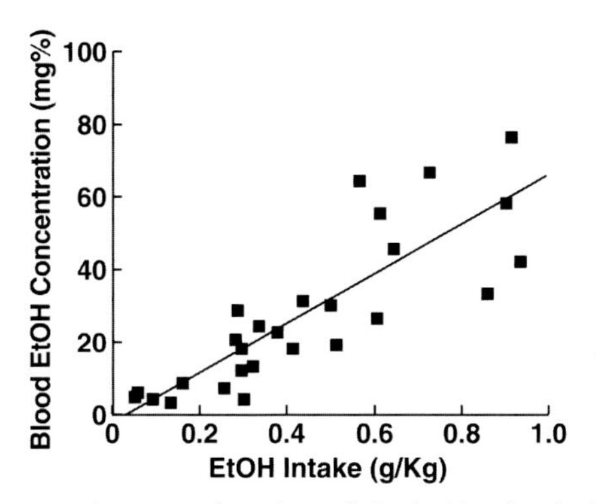

Figure 3 Blood alcohol concentrations as a function of alcohol intake during a baseline session in the two-leverl free-choice operant task. The variation in these measures in this distribution reflects the nature of responding that is typically observed in a group of nonselected heterogeneous Wistar rats. *(The data were derived from Rassnick S, Pulvirenti L, Koob GF. SDZ 205,152, a novel dopamine receptor agonist, reduces oral ethanol self-administration in rats. Alcohol 1993;10:127—32 [238].)*

The operant self-administration of oral alcohol has also been validated as a measure of the reinforcing effects of alcohol in primates. The advantages of the operant approach are that the effort to obtain the substance can be separated from the consummatory response (e.g., drinking), and intake can be charted easily over time. Additionally, different schedules of reinforcement can be used to change baseline parameters.

2.2 Brain stimulation reward (intracranial self-stimulation)

Animals will perform a variety of tasks to self-administer short (~250 ms) electrical trains of stimulation to many different brain areas [26]. The highest rates and preference for electrical stimulation follow the course of the medial forebrain bundle that courses bidirectionally from the midbrain to basal forebrain (for review, see Ref. [27]). The study of the neuroanatomical and neurochemical substrates of intracranial self-stimulation (ICSS) has led to the hypothesis that ICSS directly activates neuronal circuits that are activated by conventional reinforcers (e.g., food, water, and sex) and that ICSS may reflect the direct electrical stimulation of the brain systems that are involved in motivated behavior. Drugs of abuse lower ICSS thresholds, in which less electrical stimulation is needed for the animal to perceive the stimulation as rewarding. Conversely, withdrawal from drugs of abuse elevates reward thresholds, meaning that more stimulation is needed to perceive the stimulation as rewarding. There is a good correspondence between the ability of drugs to lower ICSS thresholds and their abuse potential [28,29]. Two ICSS procedures that have been used extensively to measure changes in reward threshold are the rate-frequency curve-shift procedure [30] and the discrete-trial, current-intensity procedure [29,31]. Neither of these procedures is confounded by influences on motor/performance capability (Figs. 4—6).

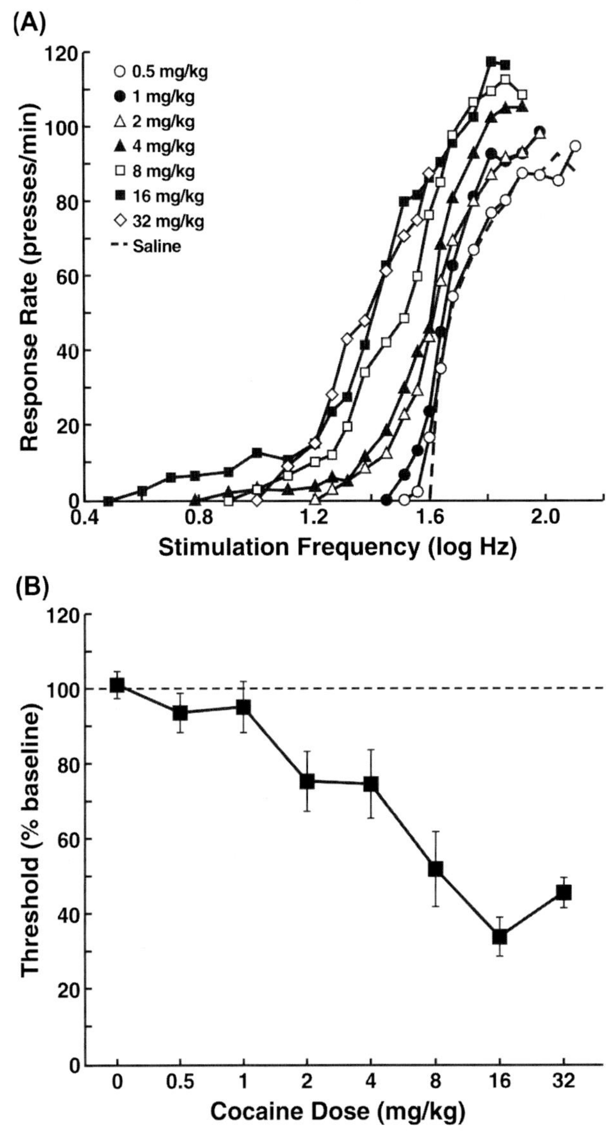

Figure 4 (A) Lever-pressing in rats during the first 15 min after a cocaine injection as a function of the frequency of electrical stimulation of the lateral hypothalamus and cocaine dose. (B) Self-stimulation frequency threshold (expressed as a percentage of baseline) as a function of cocaine dose. The data are expressed as the means from the first threshold determination in the first hour after the injection. Each reference (baseline) value is the mean from the two threshold determinations taken just before the respective drug test. *(Taken with permission from Bauco P, Wise RA. Synergistic effects of cocaine with lateral hypothalamic brain stimulation reward: lack of tolerance or sensitization. Journal of Pharmacology and Experimental Therapeutics 1997;283:1160–7 [228].)*

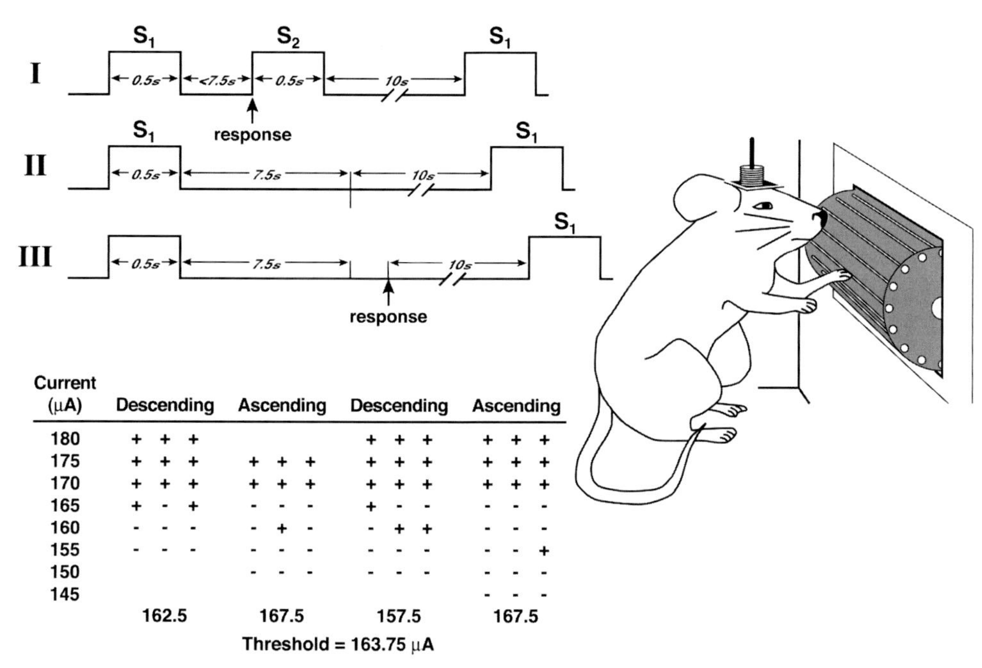

Current (μA)	Descending	Ascending	Descending	Ascending
180	+ + +		+ + +	+ + +
175	+ + +	+ + +	+ + +	+ + +
170	+ + +	+ + +	+ + +	+ + +
165	+ - +	- - -	+ - -	- - -
160	- - -	- + -	- + +	- - -
155	- - -	- - -	- - -	- - +
150		- - -	- - -	- - -
145				- - -
	162.5	167.5	157.5	167.5

Threshold = 163.75 μA

Figure 5 Intracranial self-stimulation threshold procedure. A rat is trained to turn a wheel (usually a quarter turn) to receive rewarding electrical stimulation directly in the brain. Panels I, II, and III illustrate the timing of events during three hypothetical discrete trials. Panel I shows a trial during which the rat responded within the 7.5 s following the delivery of the noncontingent stimulus (positive response). Panel II shows a trial during which the animal did not respond (negative response). Panel III shows a trial during which the animal responded during the intertrial interval (negative response). For demonstration purposes, the intertrial interval was set to 10 s. In reality, however, the interresponse interval has an average duration of 10 s, ranging from 7.5 to 12.5 s. The table at the bottom shows a hypothetical session and demonstrates how thresholds are defined for the four individual series. The threshold of the session is the mean of the four series' thresholds. *(Taken with permission from Markou A, Koob GF. Construct validity of a self-stimulation threshold paradigm: effects of reward and performance manipulations. Physiology and Behavior 1992;51:111–9 [31].)*

2.3 Conditioned place preference

Conditioned place preference or place conditioning is a non-operant procedure for assessing the reinforcing efficacy of drugs using a classical or Pavlovian conditioning procedure. In a simple version of the place preference paradigm, animals experience two distinct neutral environments that are paired spatially and temporally with distinct drug or nondrug states (Fig. 7). The animal is then given an opportunity to choose to enter and explore either environment, and the time spent in the drug-paired environment is considered an index of the reinforcing value of the drug. Animals exhibit a conditioned preference for an environment that is associated with drugs that function as positive reinforcers (i.e., spend more time in the drug-paired compartment compared with the

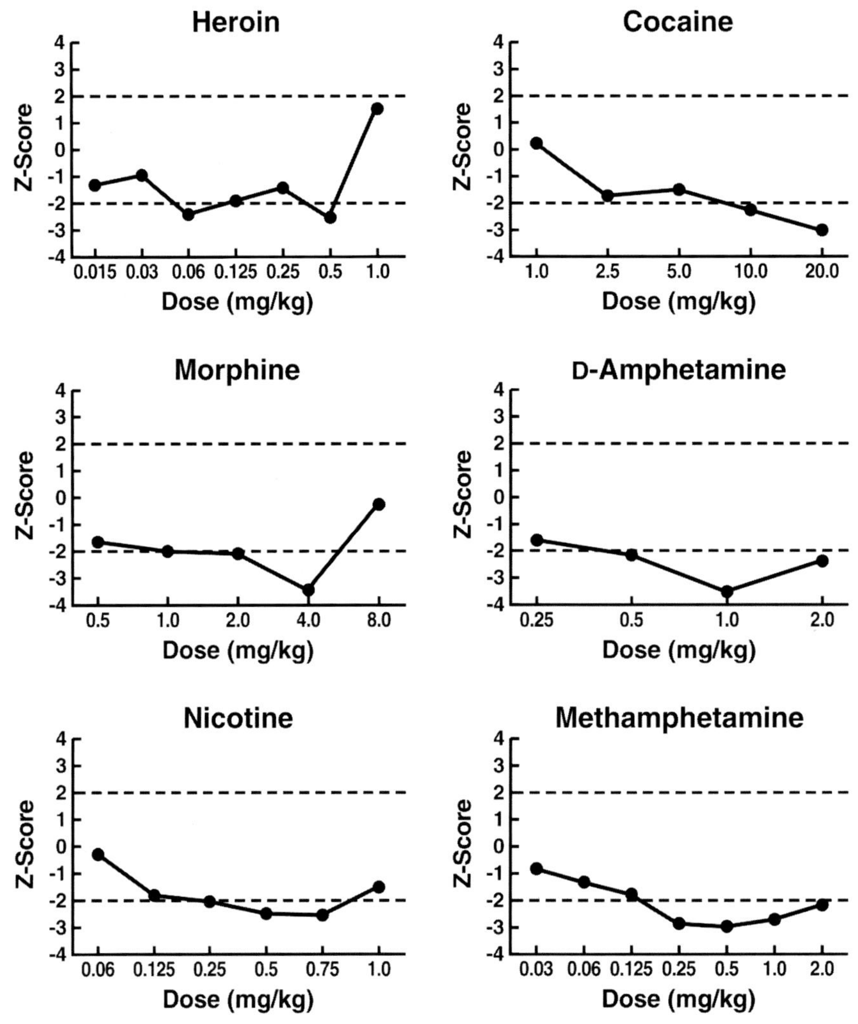

Figure 6 Changes in intracranial self-stimulation reward thresholds (Z-scores) in rats for various doses of heroin, morphine, nicotine, cocaine, D-amphetamine, and methamphetamine. A Z-score is based on the pre- and post-drug changes in threshold, and a Z-score of ± 2.0 indicates the 95% confidence limit based on the mean and standard deviation for all saline days. *(Taken with permission from Hubner CB, Kornetsky C. Heroin, 6-acetylmorphine and morphine effects on threshold for rewarding and aversive brain stimulation. Journal of Pharmacology and Experimental Therapeutics 1992;260:562–7 [233] (heroin, morphine); Huston-Lyons D, Kornetsky C. Effects of nicotine on the threshold for rewarding brain stimulation in rats. Pharmacology Biochemistry and Behavior 1992;41:755–9 [234] (nicotine); Izenwasser S, Kornetsky C. Brain-stimulation reward: a method for assessing the neurochemical baes of drug-induced euphoria. In: Watson RR, editor. Drugs of abuse and neurobiology. Boca Raton, FL: CRC Press; 1992. p. 1–21 [235] (cocaine); Kornetsky C. Brain-stimulation reward: a model for the neuronal bases for drug-induced euphoria. In: Brown RM, Friedman DP, Nimit Y, editors. Neuroscence methods in drug abuse research. NIDA research monograph, vol. 62. Rockville, MD: National Institute on Drug Abuse; 1985. p. 30–50 [236] (d-amphetamine); Sarkar M, Kornetsky C. Methamphetamine's action on brain-stimulation reward threshold and stereotypy. Experimental and Clinical Psychopharmacology 1995;3:112–7 [240] (methamphetamine).)*

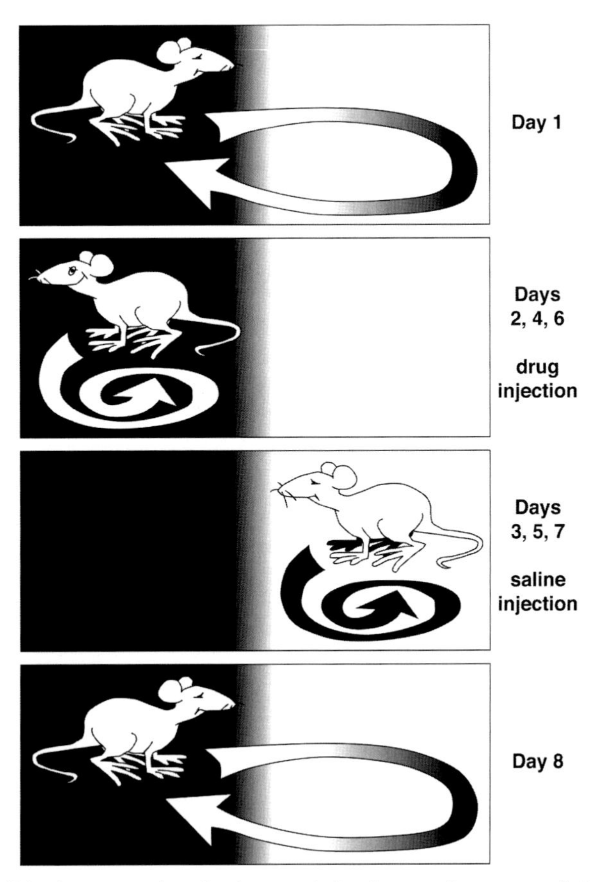

Figure 7 Place conditioning procedure in the rat. Animals experience two distinct neutral environments (here, black and white shaded) paired spatially and temporally with distinct unconditioned stimuli (here, drug on Days 2, 4, and 6 and saline on Days 3, 5, and 7). On Day 8, the rat is given the opportunity to enter either environment in a drug-free state, and the time spent in each environment is used as an index of the reinforcing value of each unconditioned stimulus. These time values are often compared with baseline preference for each environment (here, measured on Day 1). *(Taken with permission from Swerdlow NR, Gilbert D, Koob GF. Conditioned drug effects on spatial preference: critical evaluation. In: Boulton AA, Baker GB, Greenshaw AJ, editors. Psychopharmacology. Neuromethods, vol. 13. Clifton, NJ: Humana Press; 1989. p. 399–446 [241].)*

placebo-paired compartment) and avoid environments that are associated with aversive states (i.e., conditioned place aversion).

2.4 Drug discrimination

Drug discrimination procedures in animals have provided a powerful tool for identifying the relative similarity of the stimulus effects of drugs and, by comparison with known drugs of abuse, the formulation of hypotheses of the abuse potential of these drugs

[32]. Drug discrimination typically involves training an animal to produce a particular response in a given drug state for a food reinforcer and to produce a different response for the same food reinforcer in the placebo or nondrug state. The interoceptive cue state that is produced by the drug controls behavior as a discriminative stimulus or cue that informs the animal to make the appropriate response to gain reinforcement. The choice of the response that follows the administration of an unknown test compound can provide valuable information about the similarity of that drug's interoceptive cue properties to those of the training drug.

2.5 Genetic animal models of high alcohol drinking

Many genetically selected lines of rats that exhibit high versus low drinking phenotypes have been developed and include University of Chile A and B rats [33], Alko alcohol (AA) and Alko non alcohol (ANA) rats [34], University of Indiana alcohol-preferring (P) and -nonpreferring (NP) rats [35], University of Indiana high-alcohol-drinking (HAD) and low-alcohol-drinking (LAD) rats [36], and Sardinian alcohol-preferring (sP) and -nonpreferring (sNP) rats [37]. Some of these strains, such as P rats, voluntarily consume 6.5 g/kg alcohol per day in free-choice drinking and attain BALs in the 50–200 mg% range (Fig. 8).

Selectively Bred Alcohol-Preferring Rat Lines

Innate dependence-like phenotype

- **High alcohol preference/consumption**
- **Dimished sensitivity to sedative-hypnotic effects of alcohol**
- **Rapid functional tolerance**
- **Affective disruptions**

Preferring Line	Anxiety-like Behavior	Depressive-like Behavior
P	↑	↑
AA	↑	↑
HAD	no effect	no effect
sP	↑	↑
msP	↑	↑

Indiana **P** (preferring) / **NP** (non-preferring) [Wistar]
Alko Alcohol **AA** (preferring) [Wistar/Sprague Dawley]
Indiana High Alcohol Drinking **HAD** [N/NIH]
Sardinian **sP** (preferring) [Wistar]
Marchigian Sardinian **msP** (preferring) - derived from sP line [Wistar]

Figure 8 Phenotypes of selectively bred alcohol-preferring rat lines.

2.6 Drug taking in the presence of aversive consequences

Drug-taking or drug-seeking behavior that is impervious to environmental adversity, such as punishment or signals of punishment, has been hypothesized to capture elements of the compulsive nature of drug addiction that is associated with the *binge/intoxication* stage of the addiction cycle. From a DSM-5 perspective, this observable may fit well with "continued substance use despite knowledge of having a persistent physical or psychological problem." The presentation of an aversive stimulus suppresses cocaine-seeking behavior in rats with limited access and sucrose seeking in rats, but extended access to cocaine and sucrose produced differential effects [38]. Rats with extended access to cocaine did not suppress drug seeking in the presence of an aversive conditioned stimulus, but the conditioned aversive stimulus continued to suppress responding for sucrose.

The presentation of an aversive stimulus (e.g., a mild footshock) suppresses cocaine-seeking behavior in rats that have limited access (e.g., <3 h access to the drug per day). The aversive stimulus also suppresses sucrose seeking (i.e., a conventional reinforcer). This situation changes, however, when the animals are given extended access for 6 h, in which the animals' sucrose-seeking behavior continues to be suppressed by the aversive stimulus, but the rats' cocaine-seeking behavior does not diminish when confronted with the same aversive conditioned stimulus.

With oral alcohol self-administration, quinine adulteration (which is a bitter aversive solution when administered orally to rats) of the alcohol solution can be used as a punisher to suppress alcohol self-administration. The rats are maintained on an FR1 schedule until stable levels of alcohol self-administration are established before testing the effect of quinine on alcohol drinking. The alcohol solution is adulterated with increasing concentrations of quinine (0.005, 0.01, 0.025, and 0.05 g/L) between sessions (one concentration per session). This test measures the persistence of animals to consume alcohol despite the aversive bitter taste of quinine that is added to the alcohol solution and has been validated as a measure of compulsive intake [39–41].

For example, dependent rats exhibit more persistent alcohol consumption than nondependent rats as progressively greater concentrations of quinine are added to the alcohol solution (i.e., dependent rats continue to consume alcohol despite the aversive bitter taste of quinine). This is considered a measure of compulsive-like alcohol intake ([41]; Fig. 9).

2.7 Goal-directed versus habit learning

Two forms of learning have been defined in the habit domain: action-outcome (or goal-directed) learning and stimulus-response (or habit) learning [42]. When an operant response is goal-directed, it can be argued to be made with the intention of obtaining the goal, and the response for the goal is sensitive to devaluation. However, when the

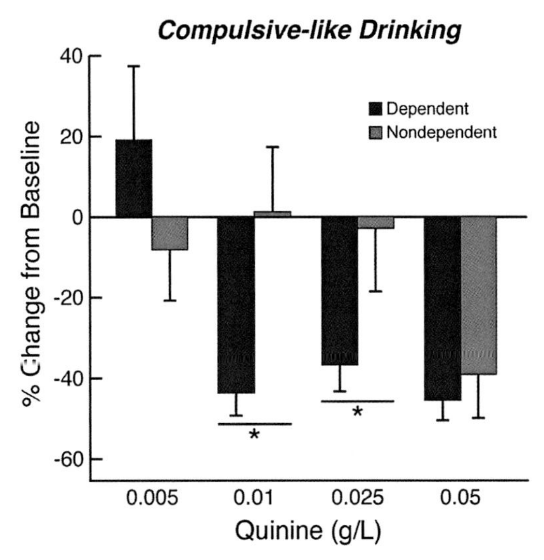

Figure 9 Compulsive-like drinking (i.e., persistent alcohol drinking despite the aversive bitter taste of quinine added to the alcohol solution) in dependent alcohol vapor-exposed rats during acute alcohol withdrawal. The data represent the percent change from baseline (i.e., lever presses for alcohol alone before quinine adulteration). *(Taken with permission from Vendruscolo LF, Barbier E, Schlosburg JE, Misra KK, Whitfield T Jr, Logrip ML, Rivier CL, Repunte-Canonigo V, Zorrilla EP, Sanna PP, Heilig M, Koob GF. Corticosteroid-dependent plasticity mediates compulsive alcohol drinking in rats. Journal of Neuroscience 2012;32:7563–71 [41].)*

operant response is acquired through associations with stimuli that are present during training, this reflects stimulus-response associations and is insensitive to devaluation [42,43].

One approach in rats is to use a drug seeking/taking chained schedule of intravenous cocaine self-administration and reward devaluation methods to examine whether drug seeking that is initially goal-directed becomes habitual after prolonged drug seeking and taking. For example, animals were trained to lever press one lever for intravenous cocaine infusions under a FR1 schedule of reinforcement. After reliable self-administration was established, responding for sucrose (10 μL of 20% sucrose via a liquid dipper) on a random-interval 60 s schedule was established. Cocaine self-administration under the FR1 schedule continued with simultaneous access to sucrose under a random-interval 60 s schedule until performance stabilized (three consecutive sessions >20 infusions, with <20% between-session variation; for details, see Ref. [44]). A chained schedule was then introduced for cocaine self-administration, in which every infusion cycle began with the insertion of a second lever, designated the drug-seeking lever, and the rats were moved from a random-interval 2 s/FR1 timeout 30 s schedule to a final random-interval 120 s/FR1 timeout 600 s chained cocaine seeking/taking schedule. Sucrose was made simultaneously available on the chained schedule throughout the latter part of training. After stable responding on the final chained schedule was achieved, drug taking was devalued by daily 2 h extinction sessions, but sucrose was still available. Cocaine seeking was then assessed in two identical tests: after devaluation and again after revaluation of the outcome of the cocaine-seeking lever (i.e., the drug-taking lever; for details, see Ref. [44]). Devaluation of the outcome of the drug-seeking link (i.e., the drug-taking link of the chained schedule) by extinction significantly decreased drug seeking, indicating that the behavior was goal-directed rather than habitual. With more prolonged drug experience, however, the animals transitioned to habitual cocaine seeking ([44]; Fig. 10). Thus, in these animals, cocaine seeking was insensitive to outcome devaluation. Similar results were observed in rats with oral alcohol and cocaine [45,46].

A within-subjects, goal-directed versus habitual instrumental lever-press task has also been developed in mice that is sensitive to neurobiological neuroadaptations in basal ganglia circuitry [47,48]. Here, mice readily perform the same action on a similar manipulandum for the same reward using a goal-directed versus habitual action strategy (Fig. 11). In this paradigm, schedules of reinforcement that are historically used to favor goal-directed or habitual control (random-ratio and random interval, respectively) are paired with a particular context. Devaluation reduces responding on the goal-directed random-ratio schedule but not on the habitual control schedule [47]. In this within-subjects procedure, it is not the animal that is habitual or goal-directed; it is the action control within a particular context that is habitual or goal-directed. However, habitual control either via extended training or random-interval schedules is known to recruit circuits of the dorsolateral striatum [48]. Each day, mice are concurrently trained to press a similar lever in the same location for the same reward (food pellets or a 20% sucrose

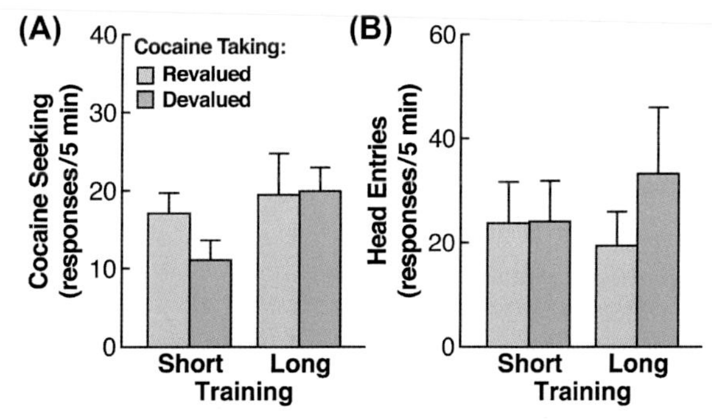

Figure 10 Effects of training duration on sensitivity to devaluation. Devaluation tests were performed after short training and again after extended training of the drug seeking/taking chained schedule. Cocaine-seeking responses and head entries into the sucrose magazine for each session, consisting of 12 drug infusions, were recorded. (A) Seeking responses for cocaine during the test. (B) Head entries into the sucrose magazine during the test. *(Taken with permission from Zapata A, Minney VL, Shippenberg TS. Shift from goal-directed to habitual cocaine seeking after prolonged experience in rats. Journal of Neuroscience 2010;30:15457–63 [44].)*

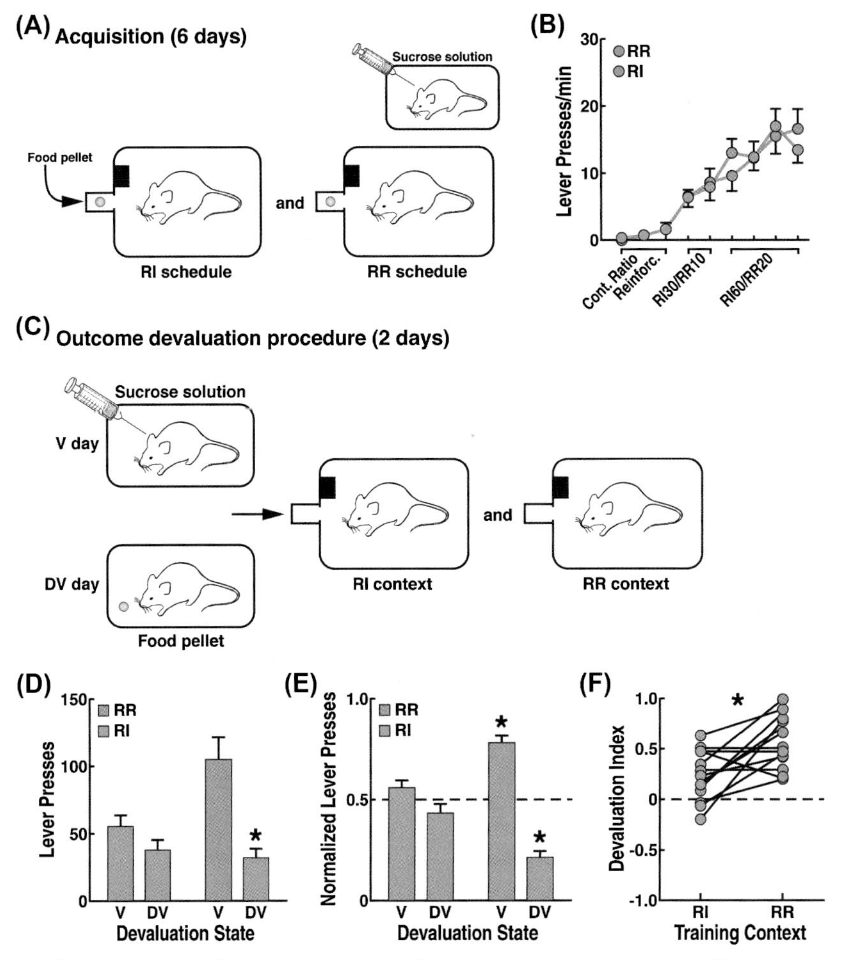

Figure 11 Within-subject shifting between goal-directed and habitual actions. (A) Acquisition schematic of lever pressing for a food outcome under random-interval (RI) and random-ratio (RR) schedules of reinforcement. The same mouse is placed in successive order in two operant chambers that are distinguished by contextual cues. There, the mouse is trained to press the same lever (e.g., left lever) for the same outcome (e.g., food pellets versus sucrose solution). The bias toward goal-directed actions is generated through the use of RR schedules of reinforcement, where the reinforcer is delivered following on average n lever press (2 days of $n = 10$, followed by 4 days of $n = 20$). In contrast, RI reinforcement schedules are used to bias toward the use of habitual actions, with the reinforcer delivered following the first lever press after, on average, an interval of t has passed (2 days of $t = 30$ s, followed by 4 days of $t = 60$ s). Each day after lever-press training, the other outcome (e.g., sucrose) is provided in the home cage. (B) Response rate for a control cohort under RI and RR schedules across acquisition. (C) Schematic of outcome devaluation procedure. On the valued (V) day, mice are fed (1 h) a control outcome (e.g., sucrose) that they have experienced in their home cage. On the devalued (DV) day, mice are prefed the outcome that is associated with the lever press (e.g., food pellet). Following prefeeding, the mice are placed into the RI and RR contexts, and lever presses are measured for 5 min in the absence of reinforcer delivery. (D) Lever pressing in V and DV states in RI (*gray*) and RR (*black*) contexts. (E) Distribution of lever presses between V and DV days in RI and RR training contexts. (F) Within-subject devaluation indices in previously trained RI and RR contexts, reflecting potential shifts in the magnitude of devaluation. Individual results and mean ± SEM are shown. $^*p < 0.05$. *(Taken with permission from Gremel CM, Chancey JH, Atwood BK, Luo G, Neve R, Ramakrishnan C, Deisseroth K, Lovinger DM, Costa RM. Endocannabinoid modulation of orbitostriatal circuits gates habit formation. Neuron June 15, 2016;90(6):1312–24 [47].)*

solution) under random-interval and random-ratio schedules. The other outcome is provided as a control in the home cage (i.e., it is not contingent on lever-press behavior). During training, these schedules produce largely similar rates of lever pressing. Before training commences, mice are food-restricted to 90% of their baseline body weight and then maintained at this lower weight for the duration of the experiments. The mice are trained to press a single lever (left or right) in a self-paced manner to achieve an outcome of either regular chow pellets (20 mg pellet per reinforcer or sucrose solution [20–30 mL of 20% solution] per reinforcer). The other outcome is provided later in their home cage and used as a control for general satiation in the outcome devaluation test. Each day, each mouse is trained on an FR1 schedule for 60 min in two separate operant chambers that are distinguished by contextual cues (i.e., black-and-white, vertical-striped, laminated paper in 3.2-mm-wide stripes on chamber walls or clear Plexiglas chamber walls). Upon the completion of training in one context, the mice are immediately trained in the remaining context. After acquiring lever-press behavior, the mice are then trained on random-interval and random-ratio schedules of reinforcement with schedules that are differentiated by context. Outcome devaluation testing then occurs across 2 consecutive days. On the valued day, the mice have *ad libitum* access to the home-cage outcome for 1 h before serial, brief, non-reinforced test sessions in the previous random-interval and random-ratio training contexts. On the devalued day, the mice are given 1 h *ad libitum* access to the outcome that was previously earned by lever pressing and then undergo serial, non-reinforced test sessions in each training context. Prefeeding occurs in a cage that is separate from the one where the mice were previously habituated, and the amount consumed is recorded.

2.8 Choice of drug versus alternative

In a choice experiment, there are two options for the animal: making a specific response (e.g., pressing a lever) for a drug or a different response (pressing another lever) for a nondrug reward, generally a palatable drink or food. Much of the rodent work in choice has involved choices between intravenous cocaine self-administration and a palatable sweet solution ([49]; Fig. 12). One example of the procedure that is used by Ahmed and colleagues can be found in Lenoir et al. [50]. Here, the chambers are equipped with two syringe pumps, one that delivers water or saccharin (or sucrose) solution into the drinking spout and one that delivers drug solution to the animal's catheter. Rats are allowed to choose between a cocaine-paired lever (lever C) and a saccharin-paired lever (lever S) in a discrete-trials choice procedure. Cocaine reward consists of one intravenous infusion of 0.25 mg cocaine delivered over 4 s. Saccharin reward consists of 20-s access to a drinking spout that delivers discrete volumes (0.02 mL) of a solution of sodium saccharin at a near optimal concentration of 0.2%; in 20 s, the rat can obtain a maximum of 15 vol, corresponding to 0.3 mL.

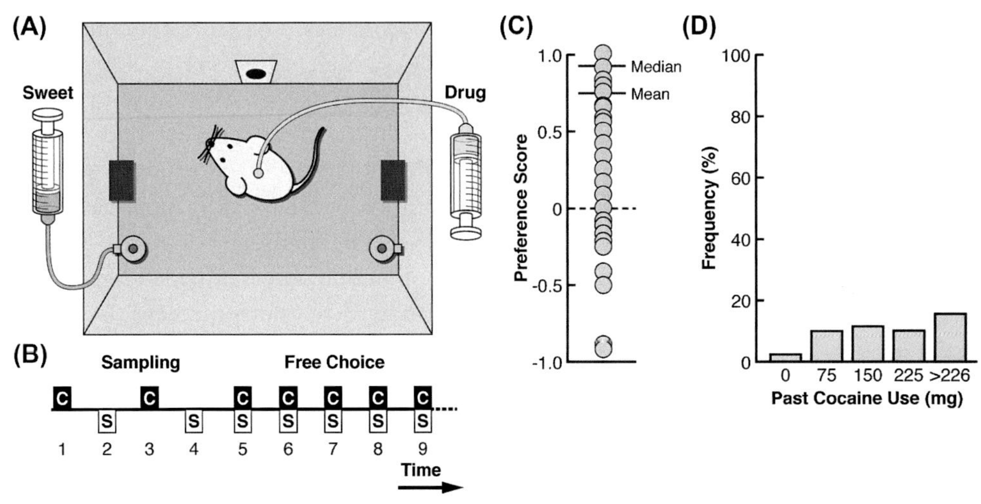

Figure 12 Shown on the left are the experimental set-up and design for choice studies. (A) Top view of an operant chamber showing a rat choosing between the cocaine-paired lever and the sweet water-paired lever. (B) Schematic diagram of the choice procedure. Each choice session consisted of at least 10 discrete trials, spaced by 10 min, and divided into two successive phases, sampling (four trials) followed by choice (six or more trials). S, saccharin-paired lever; C, cocaine-paired lever. Shown on the right is the impact of past cocaine consumption on cocaine choices. (C) Distribution of individual preferences regardless of past cocaine use. Only 16 individuals out of a total of 184 rats that were tested in the choice procedure preferred cocaine over sweet water (*closed circles*). (D) Histograms represent the frequency of cocaine-preferring individuals (i.e., cocaine choices >50% of completed trials) as a function of past cocaine consumption (i.e., total amount of self-administered cocaine before choice testing). *(Taken with permission from Ahmed, SH, Lenoir, M, Guillem, K. Neurobiology of addiction versus drug use driven by lack of choice. Current Opinion in Neurobiology 2013;23:581–7 [49].)*

Each choice session consists of 12 discrete trials, spaced by 10 min, and divided into two successive phases, sampling (four trials) and choice (eight trials). During sampling, each trial begins with the presentation of one single lever in this alternative order: C–S–C–S. Reward delivery is signaled by retraction of the lever and 40-s illumination of the cue light above this lever. If the rats fail to respond within 5 min, then the lever retracts, and no cue light or reward is delivered. Thus, during sampling, rats are allowed to separately associate each lever with its corresponding reward (lever C with cocaine, lever S with saccharin) before making their choice. During the choice procedure, each trial begins with the simultaneous presentation of both levers S and C. Rats have to select one of the two levers. During choice, reward delivery is signaled by the retraction of both levers and 40-s illumination of the cue light above the selected lever. If the rats fail to respond on either lever within 5 min, then both levers retract, and no cue light or reward is delivered.

When rats are allowed to choose between cocaine and water that is sweetened with noncaloric saccharin using this procedure, most of them choose almost exclusively the nondrug alternative, even after a long history of cocaine self-administration without choice. Even extended access to cocaine did not shift preference, suggesting that choice may produce a different readout of hedonic value of the drug than single reinforcer availability. Indeed, Ahmed and colleagues argued that the results with the choice paradigm show that the engagement of brain dopamine function by cocaine is insufficient alone to drive addiction-like behavior in rats [49]. Mice were allowed to choose between the optogenetic stimulation of dopaminergic neurons or consuming water that was sweetened with sucrose and chose sucrose over optogenetic stimulation when the dose of sucrose was sufficiently high [51]. However, drug preference over saccharin is more frequent with heroin than with cocaine after extended drug use, suggesting that the choice for heroin may be different than the choice for cocaine [52].

2.9 Summary of animal models of the binge/intoxication stage

The procedures outlined above have been shown to be reliable and have predictive validity in their ability to understand the neurobiological basis of the acute reinforcing effects of drugs of abuse and compulsive drug seeking that is associated with the *binge/intoxication* stage of the addiction cycle. One could reasonably argue that drug addiction mainly involves counteradaptive mechanisms that go far beyond the acute reinforcing actions of drugs. However, understanding the neurobiological mechanisms of the positive-reinforcing actions of drugs of abuse also provides a framework for understanding changes in the motivational effects of drugs that also result from counteradaptive mechanisms. A strength of the models that are outlined for the *binge/intoxication* stage is that any of the operant measures that are used as models for the reinforcing effects of drugs of abuse lend themselves to within-subjects designs, thus limiting the number of subjects required. Indeed, once an animal is trained, full dose-effect functions can be generated for different drugs, and the animal can be tested for weeks to months. Pharmacological manipulations can be conducted with standard reference compounds to validate any observed effects. Additionally, a rich literature on the experimental analysis of behavior is available for exploring the hypothetical constructs of drug action and modifying drug reinforcement by manipulating the history and contingencies of reinforcement.

One advantage of the ICSS paradigm as a model of the effects of drugs on motivation and reward is that the behavioral threshold measure that is provided by ICSS is easily quantifiable. Estimates of ICSS thresholds are very stable over periods of several months (for review, see Ref. [53]). Another considerable advantage of the ICSS technique is the high reliability with which it predicts the abuse liability of drugs. For example, there has never been a false positive with the discrete-trials threshold technique [29].

The advantages of place conditioning as a model for evaluating drugs of abuse include its high sensitivity to low doses of drugs, its potential utility in studying both positive- and negative-reinforcing effects, the fact that testing for drug reward occurs during drug-free conditions, and its allowance for precise control over the interaction between environmental cues and drug administration [54].

The animal models that involve responding in the face of punishment and the progressive-ratio schedule have both face validity and construct validity. Numerous human studies argue that individuals who meet the criteria for substance dependence will work harder to obtain drugs and as such present greater motivation for drug taking, and the behavioral repertoire narrows toward drug seeking and taking. Indeed, progressive-ratio studies in humans show similar patterns of responding as in animal models [55]. Clearly, responding in the face of punishment and progressive-ratio responding in rodents show individual differences that are reminiscent of the same degree of individual differences in the human population [56]. From a construct validity perspective, responding in the face of punishment and in progressive-ratio paradigms predicts a key role for midbrain dopamine systems in the reinforcing effects of cocaine [57,58]. Second-order schedules of a well-established cocaine "habit" revealed a key role for dorsal striatal mechanisms in the greater motivation to work for cocaine [59].

Measures of habit versus goal-directed responding are providing new insights into the neuroplasticity of compulsive-like behavior that develops late in the *binge/intoxication* stage. As noted above, in these paradigms, schedules of reinforcement are used to favor goal-directed or habitual control. Random-ratio and random-interval schedules, respectively, are paired with a particular context and, when combined with devaluation, reduce responding on the goal-directed schedule (random-ratio) but not on the habitual control schedule (random interval), giving a readout that is habitual or goal-directed. The development of habitual responding could be argued to "set the stage" for hedonic dysregulation in the *withdrawal/negative affect* stage because habit responding almost by definition parallels high drug intake.

In a choice experiment, there are two options for the animal: making a specific response (e.g., pressing a lever) for a drug dose or a different response (pressing another lever) for a nondrug reward, generally a palatable drink or food. Notably, in experiments with nondependent rats, the animals prefer the nondrug reinforcer over the drug reinforcer. However, at least with heroin, if the animals are made dependent, then there can be a shift in the preference to the drug in some animals. More work is needed in the choice procedure to argue that it indeed provides a better readout of hedonic value of the drug than single reinforcer availability. At a minimum, however, choice procedures allow more focus on measures of individual differences that may be of significance.

3. Animal models of the withdrawal/negative affect stage of the addiction cycle

Withdrawal from chronic drug administration is usually characterized by responses that are opposite to the acute initial actions of the drug, and the physical signs of withdrawal are often drug-specific. Many of the overt physical signs of drug withdrawal (e.g., alcohol, opioids, and nicotine) can be easily quantified and may provide a marker for the study of the neurobiological mechanisms of dependence. Standard rating scales have been developed for opioid, nicotine, and alcohol withdrawal [60–62]. Excellent indices of the onset of dependence have been developed in the classic sense [63], but unclear is that the physical signs of drug withdrawal (e.g., changes in heart rate and blood pressure, piloerection, tremor, or seizures) have major motivational significance for the neurobiology of addiction. Historically, the diagnostic criteria have focused on physical (somatic) signs of withdrawal while neglecting more motivational signs. The present argument is that motivational signs and not physical signs of withdrawal are a critical aspect of the addiction process.

Motivational measures of withdrawal have more validity for understanding the counteradaptive neurobiological mechanisms that drive addiction (see Chapter 1). Such motivational measures have been shown to be extremely sensitive to drug withdrawal and powerful tools for exploring the neurobiological bases of the motivational aspects of drug dependence. Indeed, somatic measures could be argued to be basically of little use for understanding negative reinforcement, drug seeking, or craving that is associated with acute and protracted abstinence. The somatic signs of withdrawal occur at higher doses and at later time points than motivational signs [64].

Animal models of the motivational effects of drug withdrawal have included operant schedules, conditioned place aversion, ICSS, the elevated plus maze, drug discrimination, and some of the same motivational measures of drug seeking that are used to characterize the *binge/intoxication* stage but where the valence has changed (place aversions versus place preference; elevation versus lowering of reward thresholds). Each of these models can address a different theoretical construct that is associated with a given motivational aspect of withdrawal; some reflect more general malaise; still others reflect more specific components of the withdrawal syndrome.

3.1 Brain stimulation reward (intracranial self-stimulation)

Withdrawal from the chronic administration of virtually all major drugs of abuse elevates ICSS thresholds (i.e., decreases reward; [64–71]; Fig. 13).

3.2 Conditioned place aversion

The aversive stimulus effects of withdrawal can be measured with a variant of conditioned place conditioning, termed conditioned place aversion. Place conditioning has been used as a measure of withdrawal and a measure of conditioned withdrawal (see Section 4.6 below). One method with opioid dependence is to precipitate withdrawal by administering as low dose of a competitive opioid receptor antagonist, such as naloxone [72,73]. Although naloxone itself will produce a conditioned place aversion in

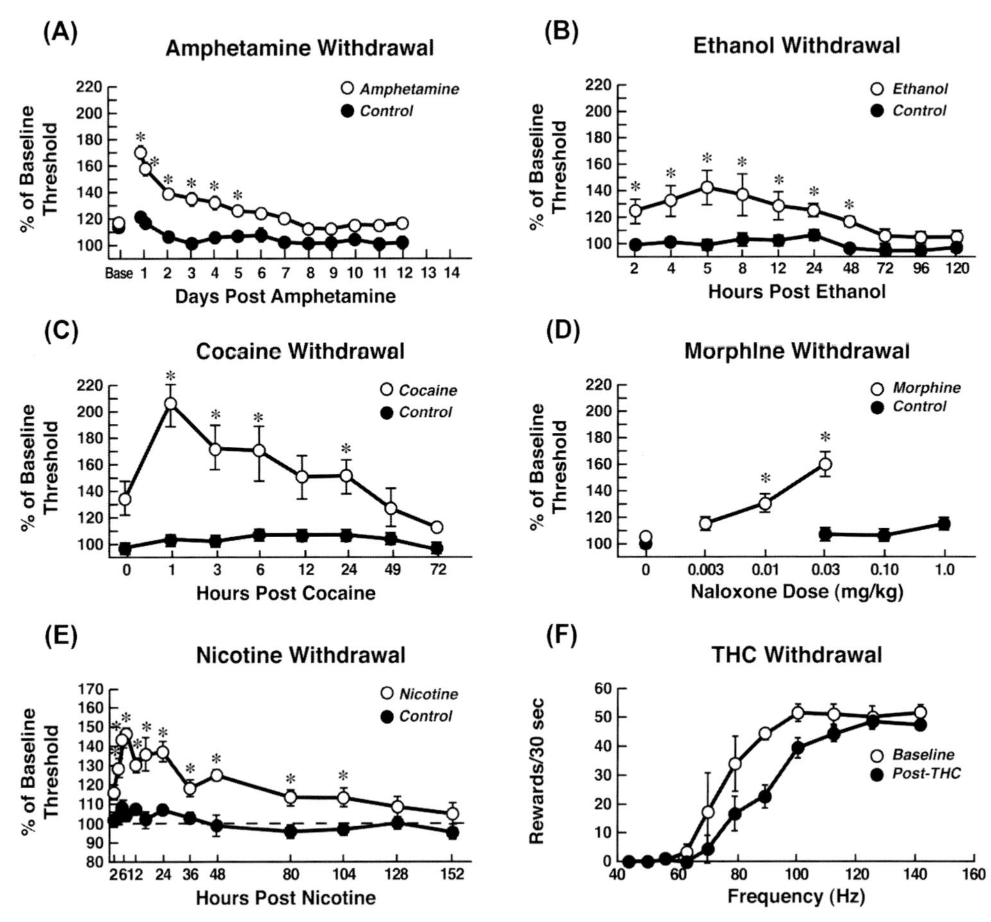

Figure 13 (A) Mean intracranial self-stimulation (ICSS) reward thresholds in rats during amphetamine withdrawal (10 mg/kg/day for 6 days). The data are expressed as a percentage of the mean of the last five baseline values prior to drug treatment. $*p < 0.05$, significant difference from saline control group. (B) Mean ICSS thresholds in rats during alcohol withdrawal (blood alcohol levels reached 197.29 mg%). Elevations of thresholds were time-dependent. $*p < 0.05$, significant differences from the control group. (C) Mean ICSS thresholds in rats during cocaine withdrawal 24 h following the cessation of cocaine self-administration. $*p < 0.05$, significant differences from the control group. (D) Mean ICSS thresholds in rats during naloxone-precipitated morphine withdrawal. The minimum dose of naloxone that elevated ICSS thresholds in the morphine group was 0.01 mg/kg. $*p < 0.05$, significant difference from control group. (E) Mean ICSS thresholds in rats during spontaneous nicotine withdrawal following the surgical removal of osmotic minipumps that delivered nicotine hydrogen tartrate (9 mg/kg/day) or saline. $*p < 0.05$, significant difference from control group. (F) Lower brain stimulation reward during withdrawal from acute 1.0 mg/kg dose of Δ^9-tetrahydrocannabinol (THC). Withdrawal significantly shifted the reward function to the right (indicating diminished reward). Note that because different equipment systems and threshold procedures were used in the collection of the above data, direct comparisons among the magnitude of effects that are induced by these drugs cannot be made. *((A) Taken with permission from Paterson NE, Myers C, Markou A. Effects of repeated*

nondependent rats, the threshold dose that is required to produce a place aversion decreases significantly in dependent rats [72]. Place aversions have also been observed with precipitated nicotine withdrawal [74] and acute spontaneous alcohol withdrawal [75].

3.3 Disrupted operant responding and drug discrimination

The response-disruptive effects of drug withdrawal using operant schedules have also been associated with the "amotivational" state of withdrawal [76–78]. Here, rats are trained to self-administer food on typically an FR schedule and then subjected to precipitated opioid withdrawal. This produces a marked suppression of lever pressing for food and has been used to explore the neural substrates for precipitated withdrawal ([78]; see Section 4.6 below).

Drug discrimination has also been used to characterize both specific and nonspecific aspects of withdrawal. For alcohol withdrawal, animals have been trained to discriminate the anxiogenic substance pentylenetetrazol from saline [79], and generalization to the pentylenetetrazol cue during withdrawal suggests an anxiogenic-like component of the withdrawal syndrome. Opioid-dependent animals have been trained to discriminate an opioid receptor antagonist from saline [80], and generalization to an opioid receptor antagonist provides a more general nonspecific measure of opioid withdrawal intensity and time course [81,82].

3.4 Response bias probabilistic reward measures

In human studies, investigators in the field of mood disorders have focused on the objective characterization of anhedonic-like symptoms using quantitative behavioral and neurobiological markers [83]. Such studies have demonstrated that non-depressed subjects who are given a choice between two correct responses but for an unequal frequency of reward exhibit a response bias (i.e., a systematic preference to identify the stimulus that

withdrawal from continuous amphetamine administration on brain reward function in rats. Psychopharmacology 2000;152:440–6 [70], (B) Taken with permission from Schulteis G, Markou A, Cole M, Koob G. Decreased brain reward produced by ethanol withdrawal. Proceedings of the National Academy of Sciences USA 1995;92:5880–4 [71], (C) Taken with permission from Markou A, Koob GF. Post-cocaine anhedonia: an animal model of cocaine withdrawal. Neuropsychopharmacology 1991;4:17–26 [69], (D) Taken with permission from Schulteis G, Markou A, Gold LH, Stinus L, Koob GF. Relative sensitivity to naloxone of multiple indices of opiate withdrawal: a quantitative dose-response analysis. Journal of Pharmacology and Experimental Therapeutics 1994;271:1391–8 [64], (E) Data adapted with permission from Epping-Jordan MP, Watkins SS, Koob GF, Markou A. Dramatic decreases in brain reward function during nicotine withdrawal. Nature 1998;393:76–9 [65], (F) Taken with permission from Gardner EL, Vorel SR. Cannabinoid transmission and reward-related events. Neurobiology of Disease 1998;5:502–33 [66].)

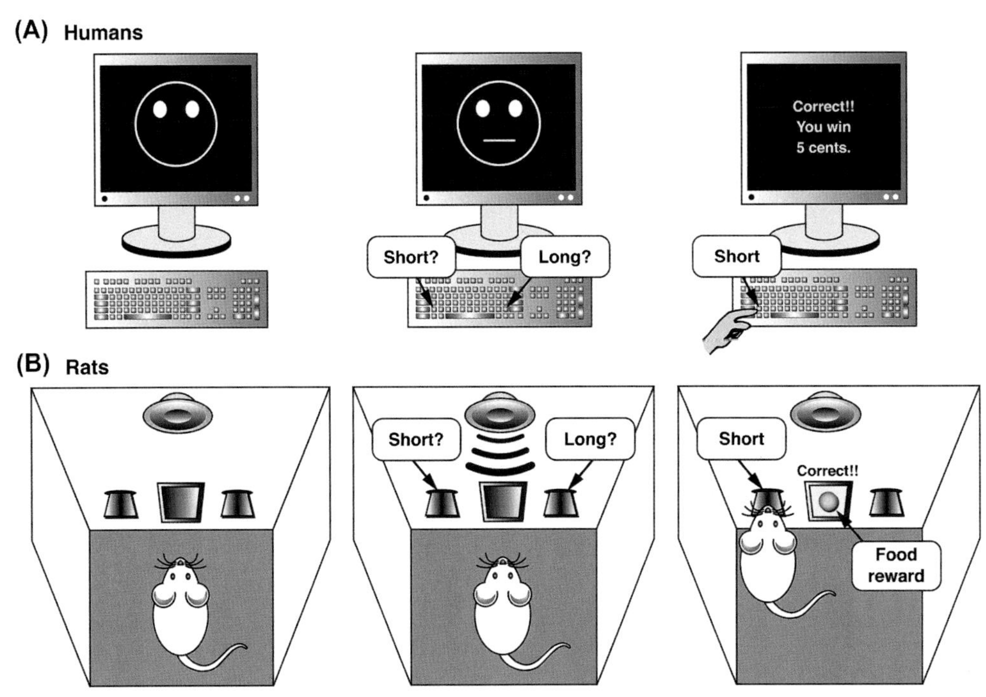

Figure 14 Schematic representation of the human and rat response bias probabilistic reward task. (A) In each trial, human participants were asked to choose whether a short (11.5 mm) or long (13 mm) mouth (briefly flashed for 100 ms) had been presented on a mouthless schematic face by pressing a key (e.g., *z* for short and / for long). In each of the three blocks (100 trials/block), the mouth stimuli were pseudorandomly presented in an equal number. For some of the correct trials, the participant received monetary reinforcement (5 cents). Unbeknownst to the participants, the reinforcement schedule was designed to favor 1 mouth length (i.e., rich) over the other (i.e., lean) in a 3:1 ratio. Only 40 correct trials were rewarded in each block (30 rich and 10 lean). The participants were instructed that the goal of the task was to win as much money as possible and that not all correct responses would receive a reward feedback. Response bias, the main variable of interest, was calculated as $\log b = \frac{1}{2} \log [(\text{rich}_{\text{correct}} \times \text{lean}_{\text{incorrect}})/(\text{rich}_{\text{incorrect}} \times \text{lean}_{\text{correct}})]$. As evident from the formula, a high response bias emerges when participants tend to correctly identify the rich stimulus and misclassify the lean stimulus. Discriminability, which is the degree to which the participant can distinguish the two target stimuli and is a measure of task difficulty, was used as a control variable and was calculated as $\log d = \frac{1}{2} \log [(\text{rich}_{\text{correct}} \times \text{lean}_{\text{correct}})/(\text{rich}_{\text{incorrect}} \times \text{lean}_{\text{incorrect}})]$. These formulas include the addition of 0.5 to each cell to allow for estimation in cases with a zero cell. Accuracy (percent hit rate) and reaction time in response to the rich and lean stimuli represented additional secondary behavioral variables. (B) Rats were food restricted and trained to discriminate between two tones that varied in duration (5 kHz or 60 dB and 0.5 or 2 s) by pressing one of the two levers that were associated with each tone. Tone durations and lever sides were counterbalanced across subjects, and tones were presented in a random order over 100 trials. Each trial was initiated with the presentation of a tone, after which the levers were extended and the rats had a 5-s limited hold period to respond. In each trial, the correct identification of tones resulted in the delivery of a single 45-mg food pellet. Both levers retracted after a correct, incorrect, or omitted response, followed by a variable intertrial interval (5–8 s). The rats were trained daily until they achieved ≥70% accuracy

is paired with the more frequent reward). Subjects with elevated depressive symptoms (i.e., high Beck Depression Inventory scores) failed to show a response bias. Impairments in reward responsiveness predicted higher anhedonic-like symptoms 1 month later, after controlling for general negative affectivity [84]. Here, reward responsiveness was assessed using a Response Bias Probabilistic Reward Task that measured the degree of response bias when choosing between two positively reinforced stimuli that differed in reinforcement frequency [83]. Lower response bias for the more frequently reinforced stimulus reflects lower reward responsiveness. Similar response biases have been observed in rats. Food-restricted rats were trained to discriminate between two tones that varied in duration (5 kHz or 60 dB and 0.5 or 2 s) using a lever press response on two levers that were associated with each tone. Rats that successfully discriminated the tones were then trained with tone durations of 0.7 and 1.8 s for 2 days and tone durations of 0.9 and 1.6 s for 2 days. During a subsequent test session, the ambiguous tone durations (i.e., 0.9 and 1.6 s) were reinforced for 60% and 20% of correct responses over 100 trials, which was identical to the 3:1 reinforcement ratio that is used in the human response bias probabilistic reward task (Figs. 14 and 15). Spontaneous withdrawal from nicotine blunted reward responsiveness similarly in humans and rats using an analogous task that was developed for both species [85].

3.5 Economic demand functions as a model of motivation for drug

Economic model descriptions of human behavior have been argued to be useful biomarkers of addiction severity [86,87]. Evidence indicates that economic demand correlates with lifetime years of drug use and, also for some drugs, drug dependence and craving [88]. As such, economic models provide a promising approach for a cross-species addiction biomarker [89] and provide a structured quantitative method that is mathematically identical across species [90].

A unique approach that has informed the neurobiology of addiction involves a mathematical approach for rapidly modeling economic demand in rats that are trained to self-

for 5 consecutive days. Rats that successfully discriminated the tones were then trained with tone durations of 0.7 and 1.8 s for 2 days and tone durations of 0.9 and 1.6 s for 2 days. During a subsequent test session, the ambiguous tone durations (i.e., 0.9 and 1.6 s) were reinforced for 60% and 20% of correct responses (counterbalanced across subjects) over 100 trials, which is identical to the 3:1 reinforcement ratio that is used in the human response bias probabilistic reward task. Response bias, the primary variable, and the three secondary behavioral variables (discriminability, accuracy, and reaction time) were computed using identical formulas as for the human experimental data. *(Taken with permission from Pergadia ML, Der-Avakian A, D'Souza MS, Madden PA, Heath AC, Shiffman S, Markou A, Pizzagalli DA. Association between nicotine withdrawal and reward responsiveness in humans and rats. JAMA Psychiatry November 2014;71(11):1238–45 [85].)*

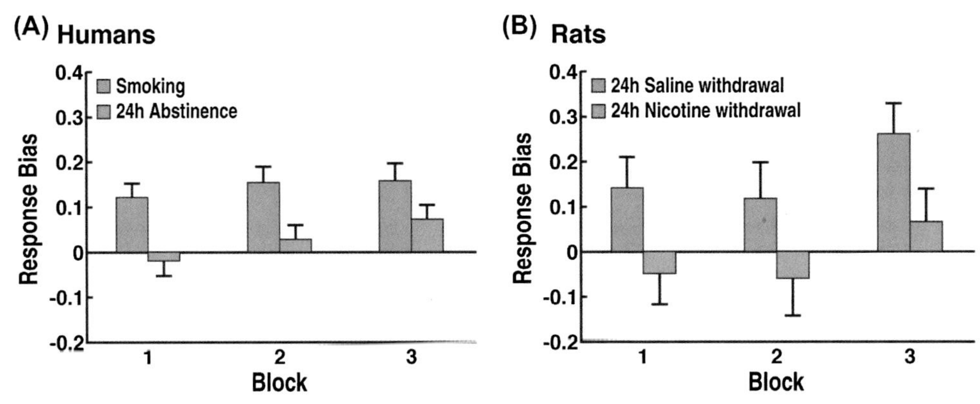

Figure 15 Withdrawal of nicotine and blunted reward responsiveness in humans and rats. (A) Human participants ($n = 31$) developed a response bias toward the more frequently rewarded (rich) stimulus when smoking at their usual rate. By contrast, 24-h abstinence from chronic tobacco smoking significantly decreased response bias. (B) Control rats that were administered saline developed a response bias toward the more frequently rewarded (rich) stimulus. By contrast, withdrawal from chronic nicotine administration significantly decreased response bias. $^{a}p < 0.05$. *(Taken with permission from Pergadia ML, Der-Avakian A, D'Souza MS, Madden PA, Heath AC, Shiffman S, Markou A, Pizzagalli DA. Association between nicotine withdrawal and reward responsiveness in humans and rats. JAMA Psychiatry November 2014;71(11):1238–45 [85].)*

administer cocaine within a single session [88]. Here, the authors showed that economic demand predicted a broad spectrum of these addiction-related behaviors, including the likelihood of reinstatement after extinction, drug seeking during abstinence, and compulsive (punished) drug taking [91]. They then validated the approach by showing that the model predicted greater compulsive (punished) drug taking in animals with extended access [91].

In this model, the animals were implanted with intravenous catheters and trained to intravenously self-administer cocaine on an FR1 schedule in daily 1 h sessions. The rats were then trained in the within-session threshold procedure for drug dose [88,92]. During a paradigm duration of 110 min, the rats received access to decreasing doses of cocaine in successive 10-min intervals on a quarter logarithmic scale (383.5, 215.6, 121.3, 68.2, 38.3, 21.6, 12.1, 6.8, 3.8, 2.2, and 1.2 μg per infusion) by decreasing the pump infusion duration (for details, see Ref. [93]). The animals were tested daily in the threshold procedure for a minimum of six sessions and until the last three sessions produced an α value (described below) that was within the range of ±25% of the mean of those days (for details, see Ref. [93]).

This economic demand approach yields two distinct measures of cocaine-self administration: *Q0* is a measure of drug consumption at null effort, and α (demand elasticity) is

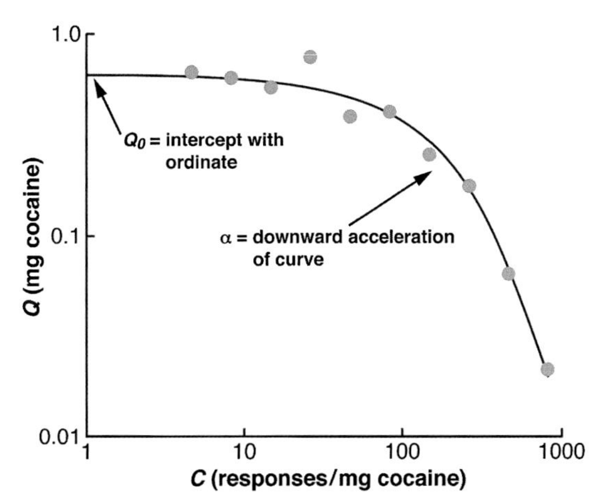

Figure 16 Example of a demand curve. Data points indicate cocaine consumption during the within-session threshold procedure from a single animal during a single session. The best-fit exponential demand curve was added. Q_0 is the theoretical demand (Q) at null cost, and α is a measure of how quickly consumption falls with increasing price. *(Taken with permission from Bentzley BS, Jhou TC, Aston-Jones G. Economic demand predicts addiction-like behavior and therapeutic efficacy of oxytocin in the rat. Proceedings of the National Academy of Sciences of the United States of America 2014;111(32): 11822–7 [91].)*

an inverse measure of motivation. The measures of cocaine demand are determined using a within-session threshold procedure [92] coupled to a method of demand curve analysis [88]. Using this approach, the demand for cocaine (consumption) was measured across increasing cocaine prices (i.e., lever responses/mg cocaine) by successively decreasing cocaine infusion doses in 10-min bins, noted above [92]. An exponential demand equation [90] was then fit to each animal's results [88] to determine the baseline economic demand for cocaine. The demand parameters *Q0* and α were derived from the resulting demand curves and describe drug consumption at null cost (*Q0*; free consumption) and the rate of consumption decline with increasing price (α; demand elasticity), respectively ([90,91]; Fig. 16). The authors emphasized that α (also referred to as "essential value") tracks inversely with motivation for drug. Thus, animals with high demand elasticity (high α) exhibit rapid reductions of drug consumption with increasing drug price. They further indicated that α is a measure of demand elasticity that is normalized to free drug consumption (*Q0*). They hypothesized that the normalized nature of α enables the motivation for drug to be tracked independently of changes in drug tolerance or sensitivity because α represents an intrinsic property of the drug's motivational efficacy, independent of the animal's preferred drug consumption at null cost, *Q0* [90,91]. The authors argued that the economic demand model supports the hypothesis that excessive motivation plays an important role in addiction (see Chapter 1) and that their results provide a structured, graded continuum by which to quantifiably measure a shift from recreational to compulsive-like drug use.

3.6 Measures of anxiety-like responses

A common response to acute withdrawal and protracted abstinence from all major drugs of abuse is the manifestation of anxiety-like responses, such as the fear of something bad that is about to happen, panic, irritability, hypervigilance, sweating, increased heart rate, higher blood pressure, and distractibility. Notice in this chapter and throughout the book, we often use the word "-like" to describe behaviors in animals, such as "anxiety-like behavior." The word "anxiety" or other such terms are used by people to describe their own subjective internal states, what they themselves are thinking or feeling. When studying animals, however, we cannot ask the rat to report what it is experiencing. Its behavioral or physiological state can be observed by researchers, and such states can bear a striking resemblance to states that are observed in and reported by people (face validity). Nevertheless, scientists and researchers are discouraged from using anthropomorphisms to describe human-like characteristics in nonhuman organisms.

In tests of anxiety-like behavior, the dependent variable is either a passive response to a novel and/or aversive stimulus, such as the open field or elevated plus maze and an active response. In passive response paradigms, the animal is simply placed in the apparatus, and its behavior is observed and recorded. In an active response paradigm, the response to an aversive stimulus, such as in the defensive burying test, is measured.

The ethologically based exploratory models of anxiety, such as the elevated plus maze, measure the ways in which animals, typically rats and mice, respond to a novel approach avoidance situation by measuring the relative exploration of two distinct environments, a lit and exposed runway versus a dark and walled runway that intersect in the form of a plus sign. Both runways are elevated above the floor. No motivational constraints are necessary, and the animal is free to remain in the darkened arm or venture out onto the open arms. High anxiety-like behavior is reflected by a lower percent time on the open arms and is very sensitive to treatments that produce disinhibition (such as sedative/hypnotic drugs) and stress and drug withdrawal.

In the defensive burying test, rodents have a natural defense reaction to unfamiliar and potentially dangerous objects by spreading bedding material over the object, leading to coverage of the object. One of the best-known procedures employs a metal prod that protrudes into the cage and on which, at first contact, a mild electric shock is delivered; once touched, it elicits burying of the prod with bedding material

[94]. The total time spent burying the prod, the total number of burying acts, and the height of the bedding material that is deposited over the prod all serve as validated measures of emotionality in this test. All of these measures have been validated to reflect emotionality-like behavior in this test [94]. Withdrawal from the repeated administration of cocaine, opioids, alcohol, and cannabinoids produces an anxiogenic-like response in the elevated plus maze and defensive burying test (for review, see Ref. [95]).

3.7 Drug self-administration with extended access and in dependent animals

A progressive increase in the frequency and intensity of drug use is one of the major behavioral phenomena that characterize the development of addiction and has face validity with several DSM-5 criteria for addiction, including "Need for markedly increased amounts of substance to achieve intoxication or desired effect," "Important social, occupational, or recreational activities given up or reduced because of substance use," "A great deal of time spent in activities necessary to obtain substance, to use substance, or recover from its effects," and "Substance in larger amounts or over a longer period than the person intended." A framework with which to model the transition from drug use to drug addiction can be found in recent animal models of prolonged access to drug self-administration and drug self-administration in dependent animals during withdrawal (Fig. 17).

When rats were allowed access to intravenous cocaine self-administration for 1 or 6 h/day, dramatic increases in self-administration were observed in animals with extended access, termed the escalation of intake [96]. Escalation has now been observed with extended access to all major drugs of abuse, including methamphetamine [97], heroin [98], nicotine [99], and alcohol [22,24].

The compulsive use of alcohol derives from multiple sources of reinforcement, and animal models have been developed not only for the acute positive reinforcing effects of alcohol but also for the negative reinforcing effects that are associated with removal of the aversive effects of alcohol withdrawal. One widely adapted model for measuring the source of motivation that is hypothesized to be driven by negative reinforcement measures self-administration in dependent rats and postdependent rats [100,101]. When the rats were tested repeatedly following the induction of alcohol dependence, they exhibited reliable increases in alcohol self-administration during withdrawal, in which the amount of intake approximately doubled, and the animals maintained BALs of 100–150 mg% for up to 12 h of withdrawal [101–104].

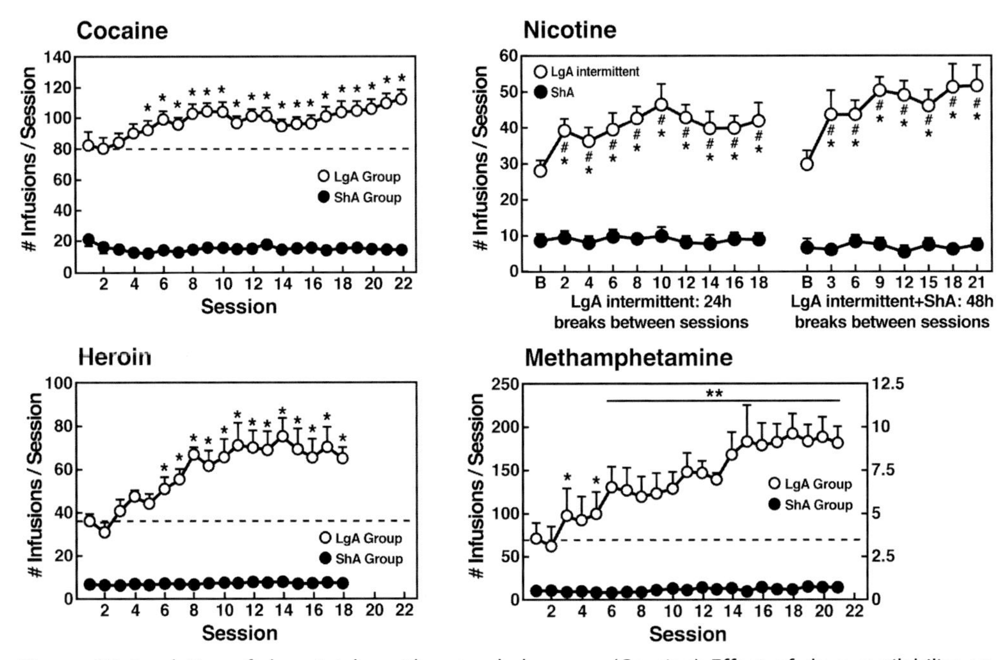

Figure 17 Escalation of drug intake with extended access. (Cocaine) Effect of drug availability on cocaine intake. In long-access (LgA) rats with 6 h access ($n = 12$) but not short-access (ShA) rats with 1 h access ($n = 12$), the mean total cocaine intake began to increase significantly from session 5 ($p < 0.05$; sessions 5 to 22 compared with session 1) and continued to increase thereafter ($p < 0.05$; session 5 compared with sessions 8–10, 12, 13, and 17–22). (Heroin) Effect of drug availability on total intravenous heroin self-infusions. During the escalation phase, the rats had access to heroin (40 μg per infusion) for 1 h per session (ShA rats, $n = 5–6$) or 11 h per session (LgA rats, $n = 5–6$). Regular 1-h (ShA rats) or 11-h (LgA rats) sessions of heroin self-administration were performed 6 days per week. The *dotted line* indicates the mean ± SEM number of heroin self-infusions in LgA rats during the first 11-h session. *$p < 0.05$, different from the first session. (Methamphetamine) Effect of extended access to intravenous methamphetamine on self-administration as a function of daily sessions in rats that were trained to self-administer 0.05, 0.1, and 0.2 mg/kg/infusion of intravenous methamphetamine during the 6-h session. ShA, 1-h session (each unit dose, $n = 6$). LgA, 6-h session (0.05 mg/kg/infusion, $n = 4$; 0.1 mg/kg/infusion, $n = 6$; 0.2 mg/kg/infusion, $n = 5$). *$p < 0.05$, **$p < 0.01$, compared with day 1. (Nicotine) Nicotine intake in rats that self-administered nicotine under a fixed-ratio (FR) 1 schedule in either 21 h (long access [LgA]) or 1 h (short access [ShA]) sessions. LgA rats increased their nicotine intake on an intermittent schedule with 24–48 h breaks between sessions, whereas LgA rats on a daily schedule did not. The left shows the total number of nicotine infusions per session when the intermittent schedule included 24 h breaks between sessions. The right shows the total number of nicotine infusions per session when the intermittent schedule included 48 h breaks between sessions. #$p < 0.05$, compared with baseline; *$p < 0.05$, compared with daily self-administration group ($n = 10$ per group). With all drugs, escalation is defined as a significant increase in drug intake within-subjects in extended-access groups, with no significant changes within-subjects in limited-access groups. *((Cocaine) Taken with permission from Ahmed SH, Koob GF. Transition from moderate*

In the alcohol-liquid diet procedure, alcohol is typically the sole source of calories that are available to rats (e.g. Ref. [105]), thereby forcing the rats to consume alcohol. Typically, the rats are provided a palatable liquid diet that contains 5–8.7% v/v alcohol as their sole source of calories, sufficient to produce dependence and maintain BALs of 100–130 mg% during the dark (active drinking) cycle [106–108]. High responders during withdrawal from an alcohol liquid diet will reach BALs of approximately 80–100 mg% [107,109].

Reliable alcohol self-administration in dependent animals using alcohol vapor exposure has been extensively characterized in rats, in which animals reach BALs in the 100–150 mg% range ([110,111]; Fig. 18). In a typical experiment, rats are separated into two groups that are exposed to either chronic intermittent alcohol vapor (alcohol-exposed group) or air vapor (alcohol-naive group) for a period of 28 days. The animals undergo cycles of 14 h ON (BALs range between 150 and 250 mg%) and 10 h OFF, during which behavioral testing occurs (i.e., 6–8 h after vapor is turned off when brain and blood alcohol levels are negligible). Control animals are treated similarly but exposed to chronic air vapor that does not contain alcohol. Tail blood samples are collected from all of the rats every third day just prior to the termination of alcohol vapor exposure (6 a.m.). Evaporated alcohol values (mL/h) are adjusted as necessary to maintain BALs in the 150–250 mg% (mg/dL) range [109].

Intermittent exposure to chronic alcohol using alcohol vapor chambers (14 h ON/10 h OFF) produces more rapid escalation of alcohol intake and higher amounts of intake than continuous 24 h exposure [100,112], and BALs are reliably >140 mg% after a 30 min session of self-administration in dependent animals [113]. Similarly, rats with a history of alcohol dependence exhibit an increase in alcohol self-administration, even weeks after acute withdrawal [110,114]. In both the liquid diet and alcohol vapor procedures, alcohol intake is directly related to the BAL range and the pattern of intermittent high-dose alcohol exposure [109]. Although the alcohol vapor model may have limited face validity, considering that alcohol is passively administered to animals, numerous studies have shown that it also has robust predictive validity for alcohol addiction ([11,101,115]; Fig. 19). High doses of alcohol solution are self-administered intragastrically after the animals are made dependent via passive intragastric infusion,

to excessive drug intake: change in hedonic set point. Science 1998;282:298–300 [96]. (Heroin) Taken with permission from Ahmed SH, Walker JR, Koob GF. Persistent increase in the motivation to take heroin in rats with a history of drug escalation. Neuropsychopharmacology 2000;22:413–21 [98]. (Methamphetamine) Taken with permission from Kitamura O, Wee S, Specio SE, Koob GF, Pulvirenti L. Escalation of methamphetamine self-administration in rats: a dose-effect function. Psychopharmacology 2006;186:48–53 [97]. (Nicotine) Taken with permission from Cohen A, Koob GF, George O. Robust escalation of nicotine intake with extended access to nicotine self-administration and intermittent periods of abstinence. Neuropsychopharmacology 2012;37:2153–60 [229].)

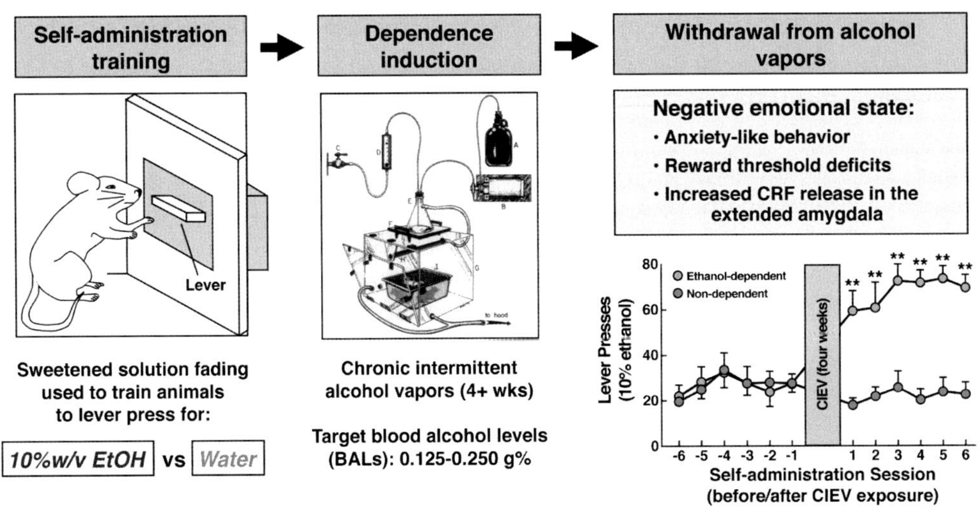

Figure 18 (Left) Schematic of a rat lever pressing for alcohol. Two levers are available, one for 10% alcohol, the other water. (Middle) Alcohol inhalation chamber. Alcohol vapor is produced by dripping 95% alcohol into 2000 mL Erlenmyer vacuum flasks that are at 50°C on a warming tray. Air is blown over the bottom of the flasks at 11 L/min. Concentrations of the alcohol vapor are adjusted by varying the rate at which alcohol is pumped into the flask. Alcohol vapor is introduced into sealed Plexiglas chambers through a stainless-steel manifold. The Plexiglas enclosure contains a standard laboratory animal cage that houses up to five rats. (Right) Alcohol self-administration in alcohol-dependent and nondependent animals, showing the induction of alcohol dependence and correlation of limited alcohol self-administration before and excessive drinking after dependence induction following chronic intermittent alcohol vapor exposure. *((Middle) Taken with permission from Rogers J, Wiener SG, Bloom FE. Long-term ethanol administration methods for rats: advantages of inhalation over intubation or liquid diets. Behavioral and Neural Biology 1979;27:466–86 [239]. (Right) Taken with permission from Edwards S, Guerrero M, Ghoneim OM, Roberts E, Koob GF. Evidence that vasopressin V1b receptors mediate the transition to excessive drinking in ethanol-dependent rats. Addiction Biology 2011;17:76–85 [232].)*

and rats will self-infuse 4–7 g/kg alcohol per day [116]. Here, BALs average 0.12 g%, measured 30 min after the start of a bout in which rats infuse 1.5 g/kg alcohol per 30 min.

Similar procedures have been developed for mice and produce reliable increases in withdrawal-induced drinking in dependent animals. Withdrawal-induced drinking usually involves C57BL/6 mice that are exposed to intermittent durations of alcohol vapor (three cycles of 16 h of alcohol vapor and 8 h of air) and then tested in a 2 h limited-access alcohol preference drinking test during the circadian dark period [117–119]. Intermittent alcohol vapor exposure significantly increased 15% (v/v) alcohol intake by

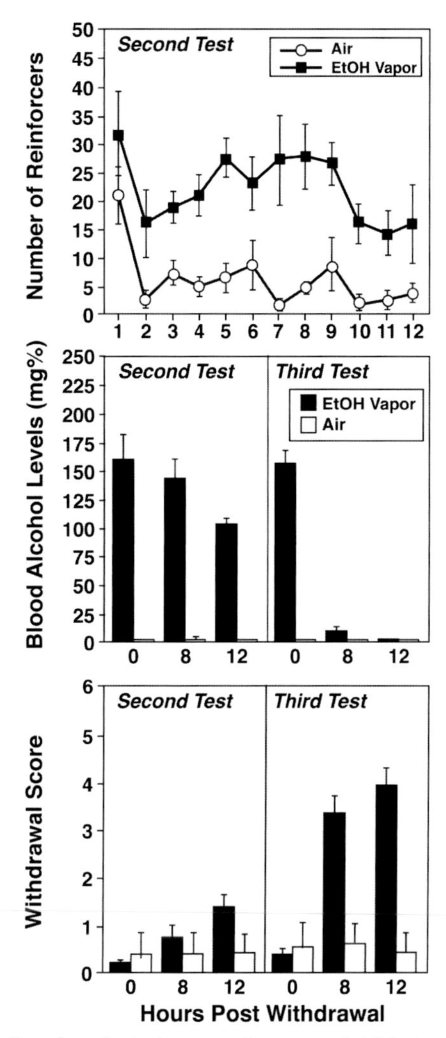

Figure 19 Operant responding for alcohol across the second 12-h test period in air-exposed and alcohol vapor-exposed rats (top). Blood alcohol levels (middle) and alcohol withdrawal severity (bottom) that were obtained during test 2 (while rats were allowed access to alcohol in the operant boxes) and test 3 (while in home cages) are shown. *(Taken with permission from Roberts AJ, Cole M, Koob GF. Intra-amygdala muscimol decreases operant ethanol self-administration in dependent rats, Alcoholism Clinical and Experimental Research 1996;20:1289–98 [101].)*

30–50% in the post-vapor period, usually after multiple cycles and usually after 24 h of withdrawal [118]. Similar results have been reported using an operant response in mice in 60 min test sessions for 10% (w/v) alcohol with intermittent vapor exposure of 14 h ON/ 10 h OFF [120].

3.8 Motivational changes associated with increased drug intake during extended access or dependence

The hypothesis that compulsive drug use is accompanied by a chronic perturbation of brain reward homeostasis has been tested in an animal model of the escalation of drug intake with prolonged access combined with measures of brain stimulation reward thresholds. Animals that were implanted with intravenous catheters and allowed differential access to intravenous self-administration of cocaine or heroin exhibited increases in drug self-administration from day to day in the long-access group but not in the short-access group. The differential exposure to drug self-administration had dramatic effects on reward thresholds that progressively increased in long-access rats but not in short-access or control rats across successive self-administration sessions [121−123].

For example, elevations of baseline reward thresholds temporally preceded and were highly correlated with the escalation of cocaine intake. Post-session elevations of reward thresholds failed to return to baseline levels before the onset of each subsequent self-administration session, thereby deviating progressively more from control levels. The progressive elevation of reward thresholds was associated with the dramatic escalation of cocaine intake that was observed previously. After escalation occurred, an acute cocaine challenge facilitated brain reward responsiveness to the same degree as previously but resulted in higher absolute brain reward thresholds in long-access rats compared with short-access rats [121]. Similar results were observed with extended access to heroin [123] and methamphetamine [122].

Another reflection of the change in motivation that is associated with dependence is a measure of reinforcement efficacy, measured by changes in progressive-ratio responding. In the progressive-ratio procedure, rats are allowed to reach baseline responding for cocaine under an FR1 schedule of reinforcement. For a progressive-ratio schedule, the response requirement (i.e., the number of lever presses that are required to receive a drug injection, or "ratio") increases using an exponential function, such as $5^{(0.2 \cdot \text{infusion number})} - 5$, yielding response requirements of 1, 2, 4, 6, 9, 12, 15, 20, 25, 32, 40, 50, 62, 77, 95, 118, 146, 178, 219, 268, etc. [124]. Sessions on this schedule are terminated when more than three-times the animal's longest baseline interresponse time elapses since the last self-administered cocaine injection [125]. Animals normally respond for 11−15 injections of cocaine, and the breakpoint is defined as the highest ratio completed in a session. The dependent measures in progressive-ratio experiments are the total number of injections obtained per session and the breakpoint. Extended access to drugs that results in the escalation of intake is also associated with a higher breakpoint for cocaine on a progressive-ratio schedule, suggesting greater motivation to seek cocaine or greater efficacy of cocaine reward [126,127]. Similar results have been observed with methamphetamine and withdrawal-induced drinking in rats that were made dependent with alcohol vapor ([128]; Fig. 20).

3.9 Summary of animal models of the withdrawal/negative affect stage

Motivational measures of drug withdrawal have significant face validity for the motivational measures of drug withdrawal in humans. Such symptoms as dysphoria, elements of anhedonia, loss of motivation, anxiety, and irritability are reflected in the animal models described above. Moreover, ICSS threshold procedures have high predictive validity for changes in reward valence. The disruption of operant responding during drug abstinence reflects, at a minimum, general malaise, and drug discrimination allows a powerful and sensitive comparison to other drug states. As such, both procedures provide a basis for further testing. Place aversion is hypothesized to reflect an aversive unconditioned stimulus. For alcohol, acamprosate and naltrexone had higher efficacy in rats that exhibited the escalation of intake either via dependence induction or prolonged access [24]. Escalation with all drugs of abuse is also associated with an increase in breakpoint on a progressive-ratio schedule, suggesting greater motivation to seek cocaine or enhanced efficacy of cocaine reward.

Reward responsiveness that is assessed using a response bias probabilistic reward task measures the degree of response bias when choosing between two positively reinforced stimuli that differ in reinforcement frequency. The value of this test is that it has direct face validity with human studies and as such can be used for the cross validation of neurobiological measures using brain imaging in humans (and rodents) and pharmacological probes in both humans and rodents.

The economic demand model is another measure that supports the hypothesis that excessive motivation plays an important role in addiction. Perhaps more importantly, it provides a structured, graded continuum with which to quantifiably measure a shift from recreational to compulsive-like drug use.

As increasingly more data are generated that establish the neurobiological bases for negative emotional states in animals that correspond to the neurobiological bases for such negative emotional states in humans, these measures will gain construct validity [129].

The use of multiple dependent variables for the study of the motivational effects of withdrawal may provide a powerful means of assessing overlapping neurobiological substrates and lay a heuristic groundwork for the counteradaptive mechanisms that are hypothesized to drive addiction. Finally, the reinforcing value of drugs may change with dependence. The neurobiological basis for such a change in humans is only beginning to be investigated [130], but much evidence has been generated that shows that drug dependence itself can produce an aversive or negative motivational state that is

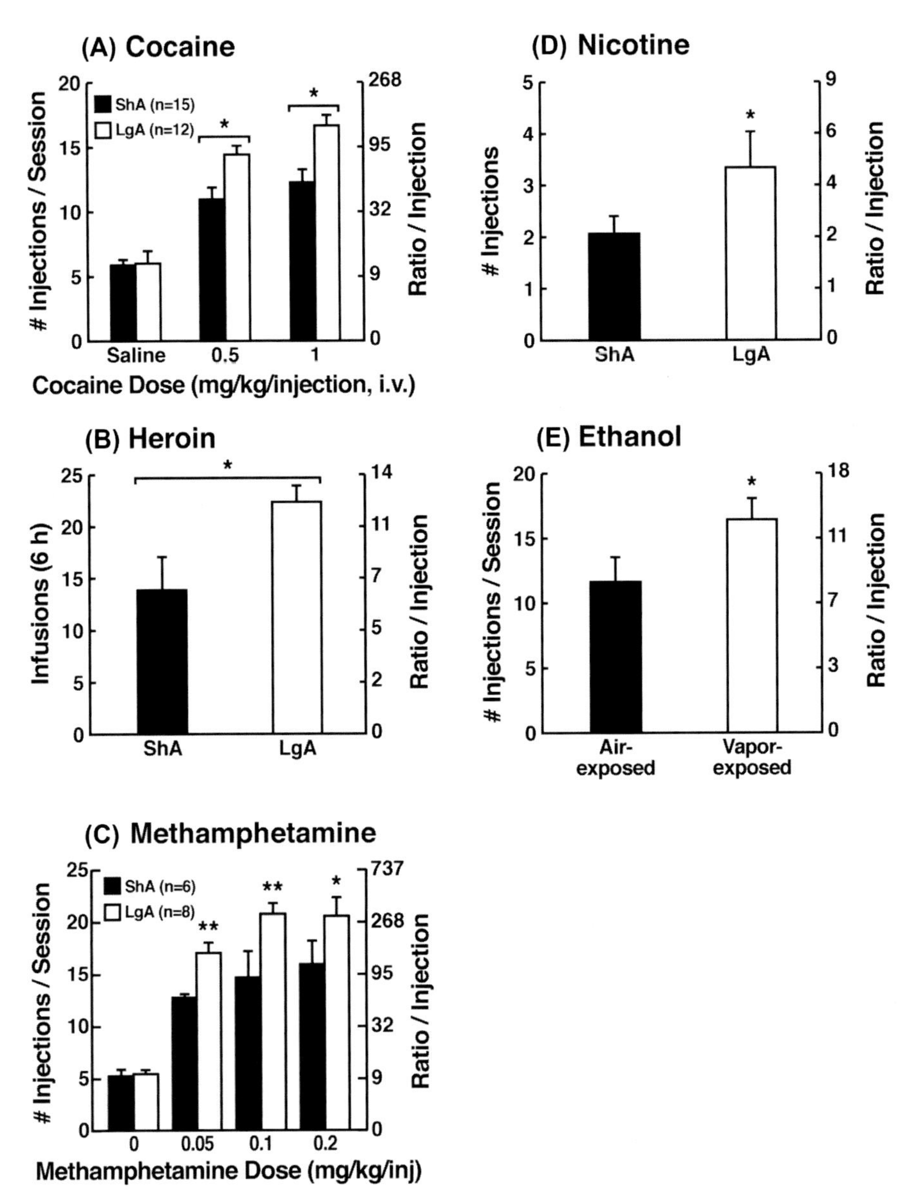

Figure 20 (A) Dose-response function of cocaine by rats responding under a progressive-ratio schedule. Test sessions under a progressive-ratio schedule ended when rats did not achieve reinforcement within 1 h. The data are expressed as the number of injections per session on the left axis and ratio per injection on the right axis. *$p < 0.05$, compared with short-access (ShA) rats at each dose of cocaine. (B) Responding for heroin under a progressive-ratio schedule of reinforcement in short-access (ShA) and long-access (LgA) rats. *$p < 0.05$, LgA significantly different from ShA. (C) Dose-response for methamphetamine under a progressive-ratio schedule. Test sessions under a progressive-ratio schedule ended when rats did not achieve reinforcement within 1 h. *$p < 0.05$, **$p < 0.01$, LgA significantly different from ShA. (D) Breakpoints on a progressive-ratio schedule in

manifested in changes in several behavioral measures, such as response disruption, changes in reward thresholds, and conditioned place aversion.

4. Animal models of the preoccupation/anticipation (craving) stage of the addiction cycle

A defining characteristic of addiction is its chronic, relapsing nature. Animal models of relapse fall into three categories of a broad-based conditioning construct: (*i*) drug-induced reinstatement, (*ii*) cue-induced reinstatement, and (*iii*) stress-induced reinstatement. The general conceptual framework for the conditioning construct is that cues, either internal or external, become associated with the reinforcing actions of a drug or drug abstinence by means of classical conditioning and then can elicit drug use.

4.1 Drug-induced reinstatement

The persistence of drug-seeking behavior in the absence of response-contingent drug availability can be measured with extinction procedures, but extinction is also a key element of most animal models of relapse. When noncontingent drug injections are administered after extinction, they produce an increase in responding on the lever that previously delivered the drug, and this responding is termed drug-primed reinstatement ([131]; Fig. 21).

4.2 Cue-induced reinstatement

Cues that are paired with drug self-administration can reliably and robustly reinstate responding after extinction [132]. Here, animals are trained to self-administer cocaine

LgA rats that self-administered nicotine with 48 h abstinence between sessions. LgA rats on an intermittent schedule reached significantly higher breakpoints than LgA rats that self-administered nicotine daily. The data are expressed as mean $\pm$ SEM. $*p < 0.05$. $n = 9$ rats per group. (E) Mean ($\pm$SEM) breakpoints for alcohol while in nondependent and alcohol-dependent states. $**p < 0.01$, main effect of vapor exposure on alcohol self-administration. *((A) Taken with permission from Wee S, Mandyam CD, Lekic DM, Koob GF. 1-Noradrenergic system role in increased motivation for cocaine intake in rats with prolonged access. European Neuropsychopharmacology 2008;18:303–11 [127], (B) Modified with permission from Barbier E, Vendruscolo LF, Schlosburg JE, Edwards S, Juergens N, Park PE, Misra KK, Cheng K, Rice KC, Schank J, Schulteis G, Koob GF, Heilig M. The NK1 receptor antagonist L822429 reduces heroin reinforcement. Neuropsychopharmacology 2013;38:976–84 [227], (C) Modified with permission from Wee S, Wang Z, Woolverton WL, Pulvirenti L, Koob GF. Effect of aripiprazole, a partial D2 receptor agonist, on increased rate of methamphetamine self-administration in rats with prolonged access. Neuropsychopharmacology 2007;32:2238–47 [242], (D) Taken with permission from Cohen A, Koob GF, George O. Robust escalation of nicotine intake with extended access to nicotine self-administration and intermittent periods of abstinence. Neuropsychopharmacology 2012;37:2153–60 [229], (E) Taken with permission from Walker BM, Koob GF. The γ-aminobutyric acid-B receptor agonist baclofen attenuates responding for ethanol in ethanol-dependent rats. Alcoholism Clinical and Experimental Research 2007; 31:11–8 [128].)*

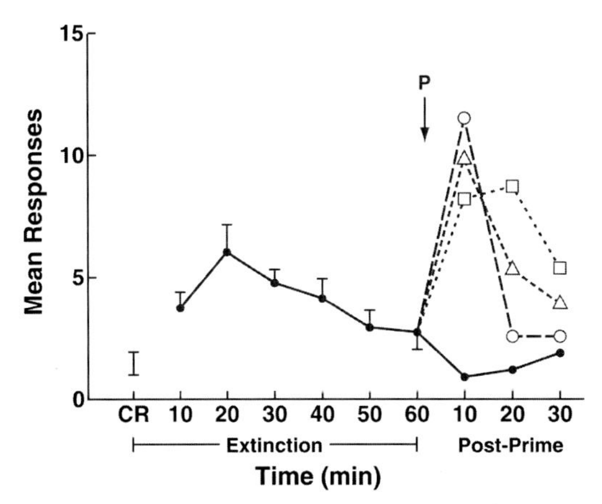

Figure 21 Mean number of responses in rats per 10 min during self-administration (continuous reinforcement [CR]), during extinction, and after cocaine priming injections of 2.0, 1.0, and 0.5 mg/kg or after a "dummy trial" (0.0 mg/kg). "P" indicates the point at which the priming injection was given. The mean values during extinction are based on eight determinations for each of five rats; the means after the priming infusions are based on two determinations per rat for each of five rats. *Closed circles*, 0.0 mg/kg; *open circles*, 0.5 mg/kg; *triangles*, 1.0 mg/kg; *squares*, 2.0 mg/kg. *(Taken with permission from de Wit H, Stewart J. Reinstatement of cocaine-reinforced responding in the rat. Psychopharmacology 1981;75:134–43 [230].)*

or other drugs of abuse via an operant response, usually lever pressing, with a cue that precedes and is contiguous with the delivery of the drug. After stable responding is achieved, the animal is subjected to extinction, in which lever pressing delivers no drug or cue. In reinstatement sessions, the animal is allowed to respond for the cue alone. The cue by itself is presented, and the animal is allowed to press the lever ([133]; Fig. 22). The cue then elicits vigorous lever pressing. This procedure is widely used to explore the neurobiological substrates of "relapse" [132–136]. Place conditioning can also be used as a model of cue-induced reinstatement. Here, conditioned place preference is induced by a drug, extinction is conducted, and then conditioned place preference is again induced by a priming injection [137].

4.3 Cue-induced reinstatement without extinction: "relapse"

Another model of drug seeking that has been termed a "relapse" model is one in which animals undergo cue-induced exposure following forced abstinence from chronic self-administration without extinction outside the testing cage [138]. This paradigm is based on the well-established "incubation effect," in which cocaine seeking that is induced by re-exposure to drug-associated cues progressively increases over the first 2 months after withdrawal from cocaine self-administration [139]. Thus, cues that are paired with drugs produce a robust increase in cocaine seeking after abstinence without extinction. This paradigm has more face validity for the human condition because individuals with

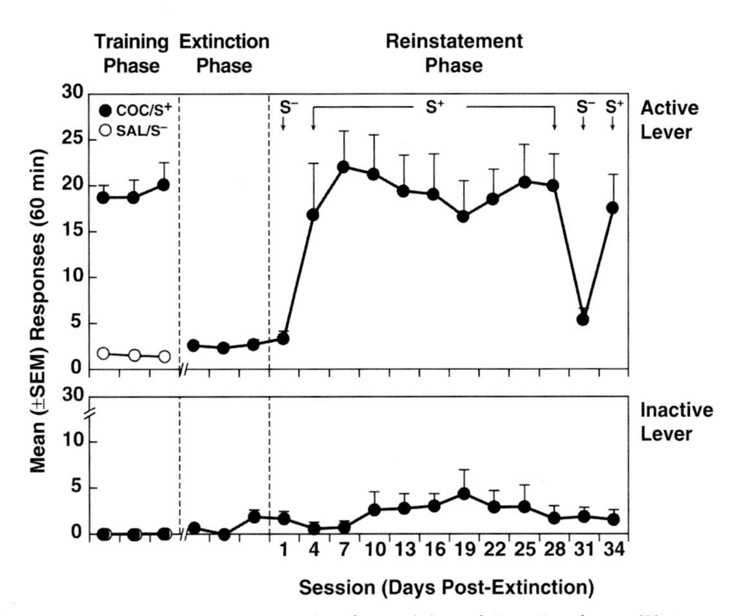

Figure 22 Lever-press responses at an active lever (A) and inactive lever (B) in rats during self-administration training, extinction, and reinstatement sessions. Training Phase: Cocaine-reinforced (●) and saline/nonreinforced (○) responses during the final 3 days of the self-administration phase in rats trained to associate discriminative stimuli with the availability of intravenous cocaine (S+) or saline (S−). Rats were designated for tests of resistance to the extinction of cocaine-seeking behavior induced by the cocaine S+ during the Reinstatement Phase. Extinction Phase: Extinction responses at criterion (<4 responses/session over 3 consecutive days). The number of days required to reach the criterion was 15.3 ± 3.9. Reinstatement Phase: Responses in the presence of the S+ and S−. Exposure to the S+ elicited significant recovery of responding in the absence of further drug availability, whereas responding in the presence of the S− remained at extinction levels. *(Taken with permission from Weiss F, Martin-Fardon R, Ciccocioppo R, Kerr TM, Smith DL, Ben-Shahar O. Enduring resistance to extinction of cocaine-seeking behavior induced by drug-related cues. Neuropsychopharmacology 2001; 25: 361−72 [133].)*

drug addiction rarely experience explicit extinction of drug seeking related to drug-paired cues.

4.4 Context-induced reinstatement: "renewal"

Drug-associated stimuli that signal the response-contingent availability of intravenous cocaine versus saline [140] also reliably elicit drug-seeking behavior in experimental animals, and responding for these stimuli is highly resistant to extinction [140−142]. Subsequent re-exposure after extinction to a cocaine discriminative stimulus but not a non-reward discriminative stimulus produces the strong recovery of responding at the previously active lever in the absence of any further drug availability. Cues that are associated with the availability of oral alcohol self-administration can also reinstate responding in the absence of the primary reinforcer [143−146].

4.5 Stress-induced reinstatement

In human studies, situations of stress are the most likely triggers of relapse to drug taking [147,148]. Animal models of stress-induced reinstatement show that stressors elicit the strong recovery of extinguished drug-seeking behavior in the absence of further drug availability ([149,150]; Fig. 23). The administration of acute intermittent footshock induced the reinstatement of cocaine-seeking behavior after prolonged extinction, and this was as effective as a priming injection of cocaine [149–151]. Such effects are observed even after a 4–6 week drug-free period [150] and appear to be drug-specific, in which food-seeking behavior is not reinstated [149]. Other stressors that are shown to be effective in reinstating drug seeking include food deprivation, restraint stress, tail pinch stress, swim stress, conditioned fear, social defeat stress, and administration of the α_2-adrenergic receptor antagonist yohimbine (an activator of the sympathetic nervous system; [152–159]).

4.6 Conditioned withdrawal

Motivational aspects of withdrawal can also be conditioned, and conditioned withdrawal has been repeatedly observed in opioid-dependent animals and humans. Here, most of the work has been done with opioids, alcohol, and nicotine. Opioid and alcohol withdrawal can be precipitated experimentally with low doses of an opioid receptor antagonist. Individuals who are addicted to opioids report withdrawal symptoms long after

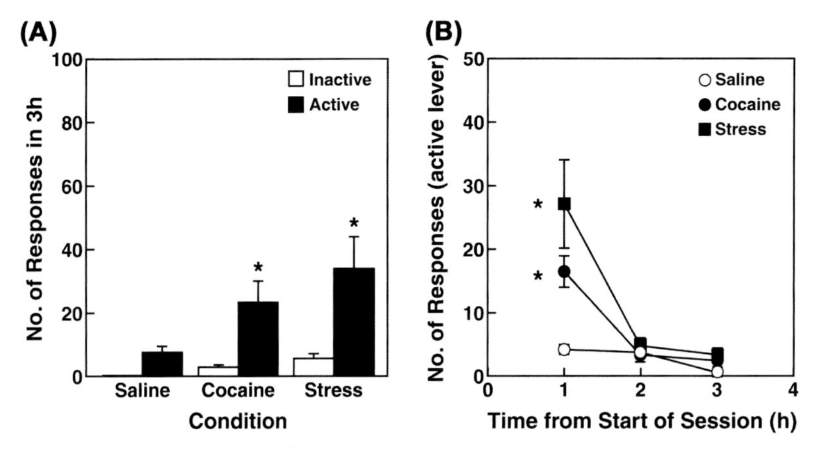

Figure 23 (A) Mean (±SEM) number of responses in rats on the previously inactive and active levers in the 3 h test for reinstatement after a noncontingent intravenous injection of saline, a noncontingent intravenous priming injection of cocaine (2.0 mg/kg), and intermittent footshock stress (10 min, 0.5 mA, 0.5 s on, mean off period of 40 s). (B) Mean (±SEM) number of responses on the previously active lever during each hour following the saline, cocaine, and footshock primes. *$p < 0.05$, significantly different from saline condition. *(Taken with permission from Erb S, Shaham Y, Stewart J. Stress reinstates cocaine-seeking behavior after prolonged extinction and a drug-free period. Psychopharmacology 1996;128:408–12 [150].)*

detoxification [160]. In an experimental setting, when patients with heroin addiction and who were maintained on methadone were repeatedly injected with a very low dose of an opioid receptor antagonist that was paired with a tone and peppermint odor. Ultimately, presentation of the tone and odor together with a vehicle injection elicited both subjective reports of aversive states and autonomic signs of withdrawal (e.g., elevations of respiration and heart rate; [161]). This phenomenon was termed conditioned withdrawal. There have been numerous reports of conditioned withdrawal-induced aversive states in humans [3,162].

Conditioned withdrawal-induced aversive states have also been modeled in animals. Conditioned place aversion was described above in the context of drug withdrawal (see Section 3.2). Here, one context is paired with spontaneous or pharmacologically precipitated withdrawal, and another context is paired with vehicle. Animals with chronic or repeated drug experience exhibit a strong aversion to the context that is paired with withdrawal, reflecting its negative motivational properties. Conditioned place aversion in animal models has been observed with alcohol and opioids [73,163–166]. The conditioned place aversion to opioids, once established, is long lasting without extinction, and conditioned withdrawal has been observed up to 16 weeks post-pairing ([167]; Fig. 24).

Conditioned withdrawal-induced aversive states have also been modeled in animals using operant self-administration. Here, drug cues that are conditioned to precipitated

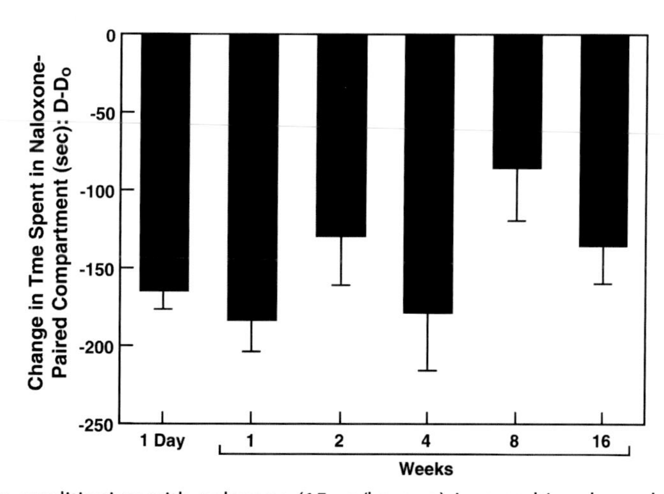

Figure 24 Place conditioning with naloxone (15 μg/kg, s. c.) in morphine-dependent rats. D0 indicates the time spent in the naloxone-paired compartment before conditioning. The rats were then evaluated 1 day and 1, 2, 4, 8, and 16 weeks later. The data are expressed as the mean ± SEM time spent in the naloxone compartment after conditioning minus the time spent in the naloxone compartment before conditioning. *(Data from Stinus L, Caille S, Koob GF. Opiate withdrawal-induced place aversion lasts for up to 16 weeks. Psychopharmacology 2000;149:115–20 [167].)*

withdrawal are able to elicit withdrawal and increase drug self-administration, similar to drug withdrawal [123,168,169]. When nonhuman primates or rats were challenged daily with repeated pairings of an opioid receptor antagonist and a light cue, presentation of the light cue and a vehicle injection resulted in a conditioned increase in responding for morphine [123,168]. In studies of heroin and nicotine in rats, presentation of the cue alone induced a reward deficit, measured by ICSS, suggesting a negative reinforcement process whereby drug use may increase to overcome reward deficits and avoid the onset of withdrawal [123,169].

4.7 Second-order schedules of reinforcement

Second-order schedules of reinforcement involve training animals to work for a previously neutral stimulus that ultimately predicts drug availability [170,171]. Second-order schedules maintain high rates of responding (e.g., up to thousands of responses per session in monkeys) and extended sequences of behavior before any drug administration. Thus, potentially disruptive, nonspecific, acute drug and treatment effects on response rates are minimized [172]. High response rates are maintained even for doses that decrease rates during a session on a regular FR schedule, indicating that performance on the second-order schedule is unaffected by the acute effects of the drug that disrupt operant responding ([170,171]; Fig. 25). In a typical paradigm in rats, the animals are trained on an FR1 schedule to lever press for an intravenous infusion of cocaine or heroin with an associated cue light that will become the conditioned stimulus. Once stable responding is established, the second-order schedule is initiated of the type FRx (FRy: S), where x is the response for the drug infusion, and y is the response for the conditioned stimulus. The rats are first trained to self-administer the drug under a schedule of continuous reinforcement. When rats acquire stable rates of drug self-administration, a second-order schedule with FR components with the initial values are an FR1 for the drug infusion and (FR1:S) for the conditioned stimulus. Subsequently, the rats are moved along the schedule in FR increments until a final schedule of fixed-interval 15 min (FR10:S) for cocaine or fixed-interval 15 min (FR5:S) for heroin [172]. Here, after every 10 responses, the conditioned stimulus is presented (this is often called the component or unit schedule); following the 10th response after completion of the FI 15 min, the reinforcer is delivered (e.g., the intravenous infusion of heroin or cocaine). Priming infusions are never given because they would influence measures of responding during the initial drug-free interval. The relative advantage of fixed-interval (FR) versus FR (FR) schedules is discussed in Everitt and Robbins [172]. The maintenance of performance in second-order schedules with drug-paired stimuli appears to be analogous to the maintenance and reinstatement of drug-seeking behavior in humans with the presentation of drug-paired stimuli [173,174].

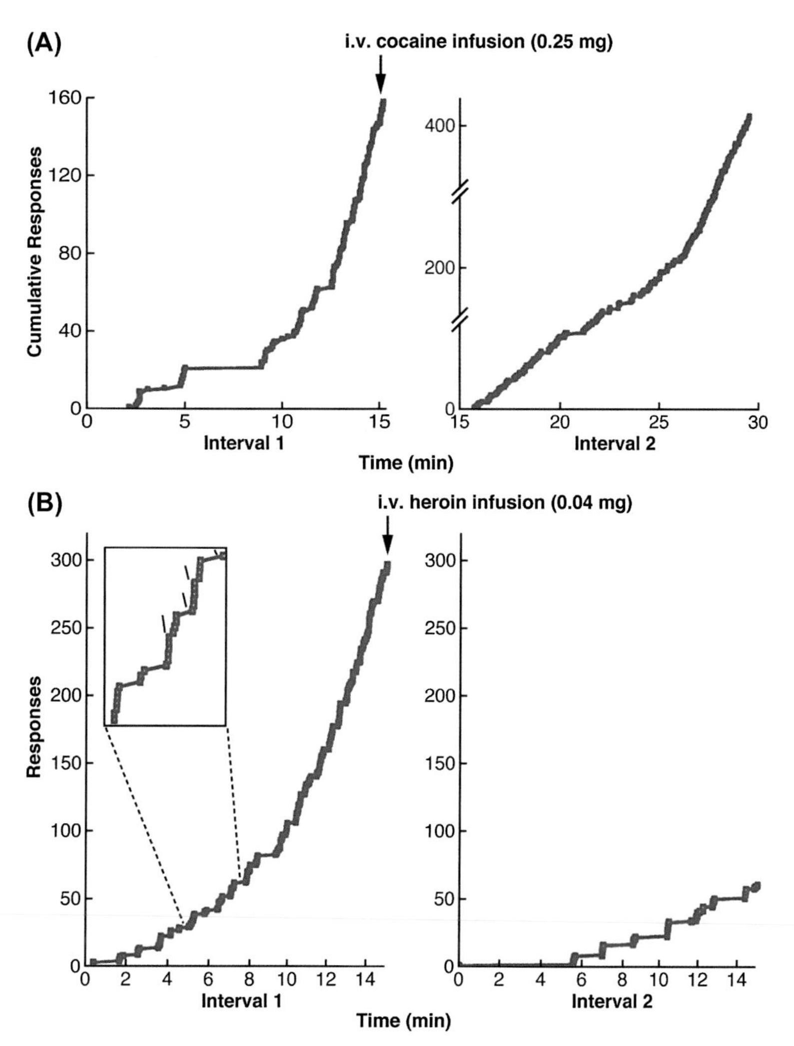

Figure 25 Representative records of responding by rats for (A) cocaine under an FI10 min (FR10:S) second-order schedule and (B) heroin under an FI10 min (FR5:S) second-order schedule. Only the first and second intervals are shown and thus the first interval (*left-hand panels*) represents responding for either drug in the drug-free state ("drug-seeking behavior"), whereas responding in the second intervals (*right-hand panels*) is affected by either cocaine or heroin that is self-administered at the end of the first interval. Note the change in scale in the *right-hand panel* of A, which shows the marked increase in responding during the second interval following cocaine self-administration. *(Data taken from Alderson HL, Robbins TW, Everitt BJ. The effects of excitotoxic lesions of the basolateral amygdala on the acquisition of heroin-seeking behaviour in rats. Psychopharmacology 2000;153:111–9 [226]; Pilla M, Perachon S, Sautel F, Garrido F, Mann A, Wermuth CG, Schwartz JC, Everitt BJ, Sokoloff P. Selective inhibition of cocaine-seeking behaviour by a partial dopamine D3 receptor agonist. Nature 1999;400:371–5 [237].)*

4.8 Alcohol deprivation effect

A robust and reliable feature of animal models of alcohol drinking is an increase in consumption after a period of deprivation. Termed the "alcohol deprivation effect," the increase in consumption has been observed in mice [175], rats [176–179], monkeys [180], and human social drinkers [181].

A representative paradigm can be found in Vengeliene et al. [182]. They examined the parameters of the alcohol deprivation effect in male Wistar rats and male C57BL/6N mice. The rats were given *ad libitum* access to tap water and 5%, 10%, and 20% (v/v) alcohol solutions. C57BL/6N mice received *ad libitum* access to tap water and up to 6% and 16% (v/v) alcohol solutions. After 8 weeks of continuous alcohol access, a 2-week deprivation period was introduced. All of the animals were given access to alcohol again, and then eight more deprivation periods were introduced in a random manner. Both the rats and mice initially exhibited a robust alcohol deprivation effect. During long-term alcohol consumption, however, C57BL/6N mice presented a decrease in alcohol intake. With repeated deprivation phases, the alcohol deprivation effect faded away ([182]; Fig. 26). Notably, the mice failed to show blockade of the alcohol deprivation effect with drugs that are usually effective in the rat model [182]. For an operant version of the alcohol deprivation effect in a representative study in rats, the animals were trained to self-administer alcohol in limited-access sessions, such as 30 min per day [183], and then various deprivation periods were introduced. An operant 30 min paradigm in rats produced a robust alcohol deprivation effect that appeared to be maximal

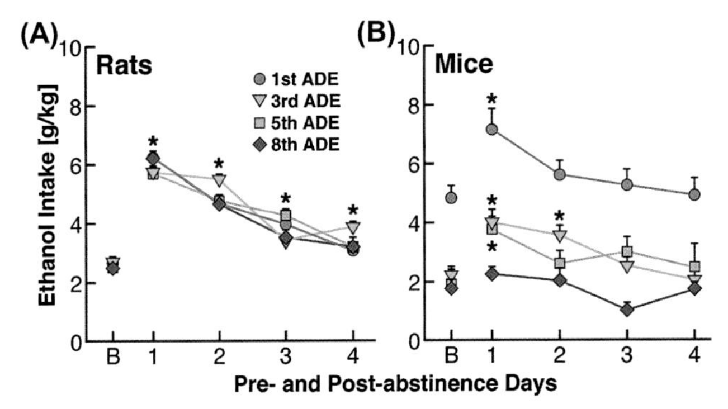

Figure 26 Total alcohol intake (in g/kg of body weight per day) before and after an alcohol deprivation period of 2 weeks in (A) male Wistar rats and (B) male C57BL/6N mice. The figure shows the first (1st ADE), third (3rd ADE), fifth (5th ADE), and eighth (8th ADE) abstinence cycle. The measurement of the last week of alcohol intake is given as pre-abstinence baseline drinking ("B"). The entire alcohol drinking procedure, including all eight deprivation phases, lasted 1 year. The data are expressed as mean $\pm$ SEM. *$p < 0.05$, significant difference from baseline drinking. *(Taken with permission from Vengeliene V, Bilbao A, Spanagel R. The alcohol deprivation effect model for studying relapse behavior: a comparison between rats and mice. Alcohol 2014;48:313–20 [182].)*

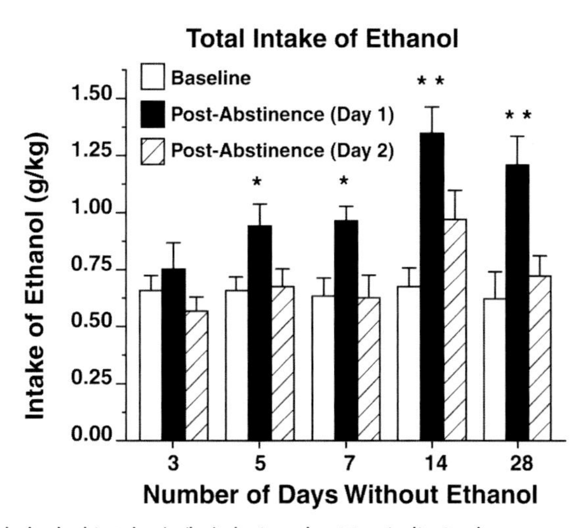

Figure 27 Mean total alcohol intake (g/kg) during the 30-min limited-access operant session. Alcohol intake increased as a function of the duration of alcohol deprivation. This increase was temporary and returned to baseline levels in 2–3 days. The data are expressed as mean ± SEM. $*p < 0.05$, significantly greater than baseline; $**p < 0.05$, significantly greater than baseline and groups deprived of alcohol for 3, 5, or 7 days. *(Taken with permission from Heyser CJ, Schulteis G, Koob GF. Increased ethanol self-administration after a period of imposed ethanol deprivation in rats trained in a limited access paradigm. Alcoholism Clinical and Experimental Research 1997;21:784–91 [183].)*

at 14–28 days ([183]; Fig. 27), consistent with the rat drinking model above. Blood alcohol levels ranged between 60 and 90 mg% post-deprivation, indicating a significant level of intoxication. The alcohol deprivation effect results in an increase in alcohol intake, an increase in alcohol preference, and an increase in the preference for higher concentrations of alcohol [184].

Parameters that facilitate the alcohol deprivation effect are the length of training (in rodents, usually ≥1 month), species, and strain. With repeated deprivations, rats develop compulsive-like drinking behavior, including insensitivity to taste adulteration with quinine, the loss of circadian drinking patterns during relapse-like drinking, and a shift toward drinking highly concentrated alcohol solutions to rapidly increase BALs [182].

4.9 Protracted abstinence

Relapse to drugs of abuse often occurs even after physical and motivational withdrawal signs have subsided, suggesting that neurochemical changes that occur during the development of dependence can persist beyond the overt signs of acute withdrawal. In humans with alcohol use disorder, numerous symptoms that can be characterized by negative emotional states persist long after acute physical withdrawal from alcohol, and these negative emotional states also contribute to the "craving" criteria for substance use disorder outlined in the DSM-5 [10]. Fatigue and tension have been reported to persist up to 5 weeks post-withdrawal [185]. Anxiety has been shown to persist up to 9 months [186], and anxiety and depression have been shown to persist for up to 2 years post-withdrawal in up to 20–25% of individuals with alcohol use disorder. These symptoms post-acute withdrawal tend to be affective in nature and subacute and often precede relapse [187,188]. A factor analysis of Marlatt's relapse taxonomy [189] found that negative emotion, including elements of anger, frustration, sadness, anxiety, and guilt, was a key factor in relapse [190], and the leading precipitant of relapse in a large-scale replication of Marlatt's taxonomy was negative affect [191]. In secondary analyses of participants in a 12-week clinical trial with alcohol dependence and not meeting criteria for any other DSM-IV [9] mood disorder, the association with relapse and a subclinical negative affective state was particularly strong [192]. This state has been termed "protracted abstinence" and has been defined in humans as a Hamilton Depression rating ≥ 8 with the following three items consistently reported by subjects: depressed mood, anxiety, and guilt [192].

Animal work has shown that prior dependence lowers the "dependence threshold," such that previously dependent animals that are made dependent again exhibit more severe physical and motivational withdrawal symptoms than groups that receive alcohol for the first time [193–196]. This supports the hypothesis that alcohol experience and particularly the development of dependence can lead to relatively permanent alterations of responsiveness to alcohol. However, relapse often occurs even after physical withdrawal signs have subsided, suggesting that neurochemical changes that occur during the development of dependence can persist beyond the final overt signs of withdrawal (i.e., "motivational withdrawal syndrome").

A history of dependence in male Wistar rats can produce a prolonged elevation of alcohol self-administration in daily 30-min sessions after acute withdrawal and detoxification [110,112,197,198]. This increase in alcohol self-administration is accompanied by higher BALs and persists for up to 8 weeks post-detoxification. The increase in self-administration is also accompanied by an increase in behavioral responsivity to stressors and greater responsivity to corticotropin-releasing factor receptor antagonists [198–200]. The persistent increase in alcohol self-administration has been hypothesized to involve an allostatic-like adjustment, such that the set point for alcohol reward is elevated [110,201]. These persistent alterations of alcohol self-administration and residual sensitivity to stressors can be arbitrarily defined as a state of "protracted abstinence." Protracted abstinence, defined as such in the rat, spans a period after acute physical withdrawal has disappeared when elevations of alcohol intake over baseline and greater behavioral responsivity to stress persist (2–8 weeks post-withdrawal from chronic alcohol).

Protracted abstinence has also been linked to elevations of brain reward thresholds, greater sensitivity to cues that are associated with withdrawal (conditioned place aversions to opioids), and increases in the sensitivity to anxiety-like behavior (alcohol) that have been shown to persist after acute withdrawal symptoms have subsided in animals with a history of dependence [167,200,202–204]. In summary, animals that are made dependent on alcohol will exhibit a residual increase in the motivation for alcohol 2–6 weeks post-withdrawal, a period without obvious physical signs of withdrawal but residual motivational signs when the animal is challenged [200]. Although little work has addressed a "protracted abstinence" syndrome with other drugs of abuse, the "incubation effect" that is described above for cocaine has some face validity for protracted abstinence and is more robust in animals with extended access. Additionally, others have shown that more robust stress-induced reinstatement is observed with extended access [205,206]. Indeed, although some studies have reported the stress (footshock)-induced reinstatement of cocaine seeking following self-administration in rats under specific conditions [207], Mantsch and colleagues reported the requirement of a history of self-administration under conditions of prolonged daily access [208].

4.10 Sleep disturbances

In humans, some of the problems that are most frequently reported by psychostimulant drug users are sleep disturbances [209]. Psychostimulant drugs have acute and chronic effects on sleep architecture. Cocaine users commonly report difficulties initiating and maintaining sleep, with a lower sleep efficiency and significant delay in sleep onset [210]. Chronic long-term effects of methamphetamine can lead to the dysregulation of sleep-wake states during active methamphetamine use, during withdrawal, and abstinence [211]. In humans, acute and chronic alcohol use has profound effects on sleep [212].

With acute alcohol administration, in the first half of the night, there is shorter sleep latency and increases in slow-wave sleep but a suppression of rapid-eye-movement sleep. In the second half of the night, sleep is characterized by poor-quality disrupted sleep [212]. In alcohol addiction, laboratory-based polysomnographic studies of abstinent alcohol-addicted individuals typically show a pattern of lower sleep efficiency that is consistent with self-reports of persistent sleep disturbances in alcohol-addicted individuals [212].

Perhaps more importantly, poor sleep quality, measured both subjectively (self-administered questionnaires) and objectively (polysomnographic sleep parameters), has been shown to predict relapse to drugs of abuse in addicted individuals [213,214]. A linear relationship has been found between sleep disruption and relapse risk following treatment (i.e., greater sleep dysfunction leads to a greater risk of relapse; [215,216]).

In an early animal study, acutely administered cocaine (orally or intraperitoneally) in rats reduced total sleep time, slow-wave sleep, and sleep latency and reduced rapid-eye-movement sleep during the first half of sleep recording [217]. The acute administration of cocaine and methamphetamine in rats also altered paradoxical sleep, slow-wave sleep, and sleep efficiency, and sleep rebound that was caused by paradoxical

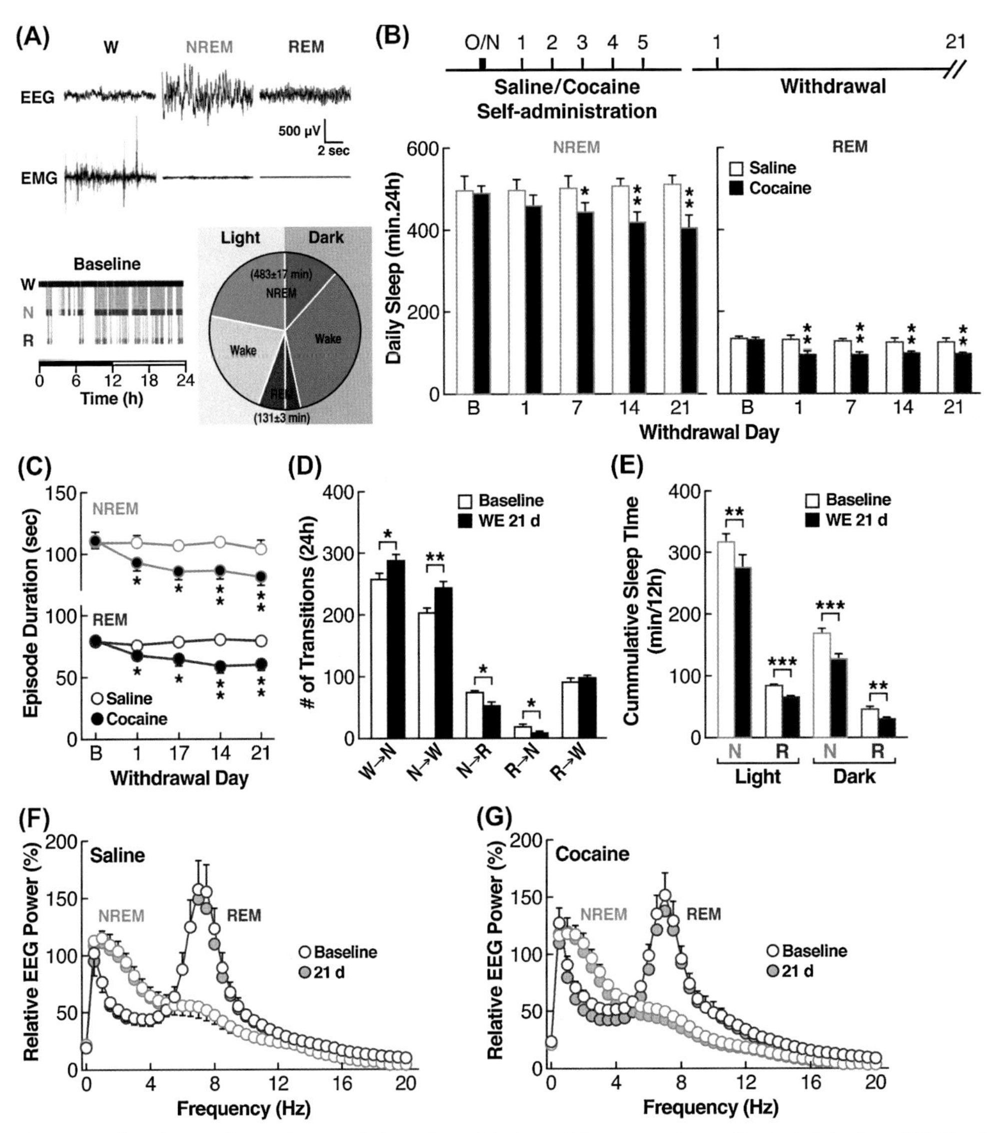

Figure 28 Persistent sleep abnormalities after withdrawal from cocaine self-administration. (A) (Top) Representative EEG and EMG recordings showing wakefulness (W), NREM, and REM sleep states. (Bottom left) Example hypnogram from the same rat showing wakefulness (W), NREM sleep (N), and REM sleep (R) states across the 12 h dark and 12 h light phases (*horizontal bar*) on a baseline recording day. (Bottom right) Summary pie chart showing the average proportion of time spent in wakefulness, NREM sleep, and REM sleep on baseline recording days (min/24 h), separated into the light and dark phases at midline. (B) (Top) Timeline of saline or cocaine self-administration training and withdrawal. (Bottom) Twenty-four hour total NREM and REM sleep time persistently decreased during the initial 21 days after withdrawal from cocaine self-administration (24 h/day recordings on baseline day and on withdrawal days 1, 7, 14, and 21). $*p < 0.05$, $**p < 0.01$, different from baseline. No

sleep deprivation [218]. In a more recent study in rats that were allowed to intravenously self-administer cocaine, prolonged withdrawal from cocaine self-administration produced a persistent reduction of non-rapid-eye-movement and rapid-eye-movement sleep and greater sleep fragmentation ([219]; Fig. 28). The rats were implanted with jugular catheters. Two gold-plated wire electroencephalographic (EEG) electrodes were implanted contralaterally through the skull over the parietal and frontal cortices. They were then trained to intravenously self-administer 0.75 mg/kg/injection cocaine and subjected to a 5 day self-administration procedure (2 h/session/day for 5 consecutive days on an FR1 schedule). For sleep measures, EEG and electromyographic (EMG) signals were amplified and processed using an analog-to-digital converter. The EEG signal was filtered, and then all signals were manually scored for sleep states in 10 s epochs. The authors identified wakefulness by desynchronized EEG and high EMG activity. Non-rapid-eye-movement sleep exhibited high-amplitude slow waves and lower EMG. Rapid-eye-movement sleep exhibited regular EEG theta activity and extremely low EMG activity (for details, see Ref. [219]).

4.11 Summary of animal models of the preoccupation/anticipation stage

Each of the models outlined above has some face validity with the human condition and ideally heuristic value for understanding the neurobiological bases for different aspects of the craving stage of the addiction cycle. The DSM-5 criteria that apply to the craving stage and loss of control over drug intake include "persistent desire or unsuccessful efforts to cut down or control drug or alcohol use or recover from its effects," "great deal of time is spent in activities necessary to obtain drug, use drug or alcohol, or recover from its effects," and "craving, or a strong desire or urge to use drug or alcohol."

significant changes were observed in the saline-exposed rats (compared with baseline). (C) Average episode durations of both NREM and REM sleep were reduced during the same 21 days after withdrawal from cocaine self-administration. $*p < 0.05$, $**p < 0.01$, different from baseline. No significant changes were observed in the saline-exposed rats (compared with baseline). (D) After 21 days of withdrawal (WD 21 d) from cocaine self-administration, there were more transitions between wakefulness and NREM sleep and fewer transitions to REM sleep. $*p < 0.05$, $**p < 0.01$, different from baseline. $n = 12$. (E) Cumulative NREM and REM sleep times decreased during both the light and dark phases by withdrawal day 21. $**p < 0.01$, $***p < 0.001$, different from baseline. $n = 12$. (F, G) Relative EEG power spectra of NREM and REM sleep episodes from saline-exposed (F) and cocaine-exposed (G) rats were compared between baseline and 21 days after the 5 day cocaine or saline self-administration. EEG signals were normalized to the average delta power (0–4 Hz for NREM) or theta power (5–10 Hz for REM) of the baseline condition (before cocaine exposure). No significant differences were detected in saline-exposed or cocaine-exposed rats between baseline and 21 days after self-administration. Error bars indicate SEM. *(Taken with permission from Chen B, Wang Y, Liu X, Liu Z, Dong Y, Huang YH. Sleep regulates incubation of cocaine craving. Journal of Neuroscience 2015;35: 13300–10 [219].)*

There is a general consensus that the animal reinstatement models have face validity [14,220]. However, predictive validity remains to be established. To date, there is little evidence of predictive validity from studies of the pharmacological treatments for drug relapse [14]. Very few clinical trials have tested medications that are effective in the reinstatement model, and very few anti-relapse medications have been tested in animal models of reinstatement. The main problem is that there are very few medications available for the treatment of craving and protracted abstinence. For example, there are no medications that are approved by the United States Food and Drug Administration for the treatment of cocaine use disorder or cannabis use disorder. From the perspective of functional equivalence or construct validity, there is some evidence of functional commonalities. For example, drug re-exposure or priming, stressors, and cues that are paired with drugs all produce reinstatement in animal models and promote relapse in humans. Much imaging data in humans during cue reactivity tasks are now available and provide fertile ground for parallel functional circuitry analyses in animal studies [130].

The second-order schedule paradigm has the advantage of assessing the motivational value of a drug infusion in the absence of acute effects of the self-administered drug that can influence performance or other processes that interfere with motivational functions.

Increasingly more interest is focusing on measures of protracted abstinence, conditioned withdrawal, and sleep disturbances. From a conceptual perspective, residual negative emotional states that generate hedonic deficits can be hypothesized to have an additive effect on cues, contexts, and acute stressors in driving relapse. Addiction can be heuristically divided into a three-stage cycle, but each stage of the cycle feeds into the next stage of the cycle to perpetuate the allostatic state (see Chapter 1; [1]). Protracted abstinence can be conceptualized as the ultimate vulnerability stage where genetics, environment, and residual allostatic deviations in reward thresholds and stress sensitivity contribute to the propensity to relapse.

Remaining to be accomplished is to show that both construct validity (functional equivalence) and predictive validity exist for these models. A further challenge for future studies will be to develop "endophenotypes" that cross species from animals to humans to allow further construct validity (functional equivalence) for studies of genetic and environmental vulnerability and the neurobiological mechanisms therein.

5. Animal models of vulnerability to addiction

5.1 Acquisition of drug seeking

Models of the vulnerability to addiction have historically involved acquisition studies, in which subjects that are naive to a particular drug learn a simple operant response to obtain an intravenous delivery of the drug in a limited-access situation. Individual differences in the response to psychostimulants and other drugs of abuse in general have been widely demonstrated in humans and laboratory animals. Although the importance of individual differences in humans is well accepted in clinical practice, it has generally been neglected in animal studies. One of most sensitive models for testing the vulnerability to drugs of abuse is to provide naive animals with very low doses of drugs in an acquisition paradigm, such that only the more sensitive individuals develop self-administration. The differences are hypothesized to reflect the differential reactivity of specific neurotransmitters ([221]; Fig. 29).

Such types of differential responses have been shown for cocaine, amphetamine, and heroin. The difference between animals within a group can be further exaggerated by dividing the group in half or by the median (i.e., 50%–50%) or to maximize phenotypic differences by comparing the lowest and highest interquartiles. Such models lead to at least two avenues of research: (*i*) investigating the biological and brain parameters that differentiate behavioral phenotypes and (*ii*) characterizing the vulnerable versus resistant phenotypes so that the knowledge or the measure of a behavioral characteristic predicts the type of response to the drug. These models have face validity because vulnerable subjects have a higher chance of developing addiction-like behavior, independent of the quantity of drug available (which appears to occur in the real world).

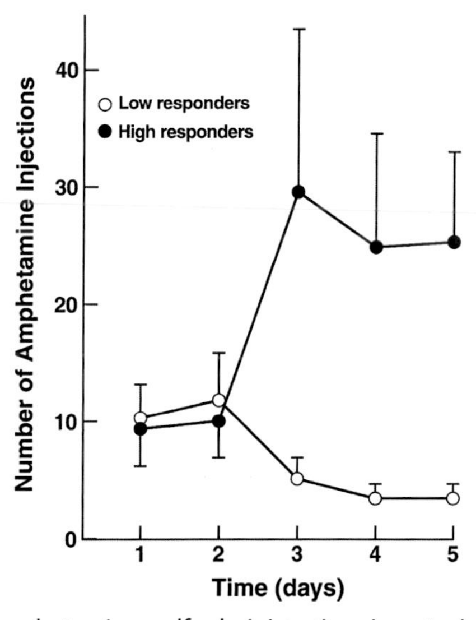

Figure 29 Acquisition of amphetamine self-administration in rats in high-responder and low-responder groups after repeated intraperitoneal administration of saline. After saline treatment, the groups ($n = 10$/group) differed in their acquisition of self-administration both in terms of total amphetamine administered over the 5 days and in terms of the number of injections over different days. *(Taken with permission from Piazza PV, Deminiere JM, Le Moal M, Simon H. Factors that predict individual vulnerability to amphetamine self-administration. Science 1989;245:1511–3 [221].)*

5.2 Vulnerability to addiction

Another approach to investigating individual differences in the vulnerability to addiction has been to characterize individual animals in terms of their propensity to exhibit responses that relate to the DSM-IV (now DSM-5) criteria for addiction. Three domains have been identified from the DSM-5 criteria that can be linked to animal models: (*i*) difficulty stopping or limiting drug use (items 3–4 in the DSM-IV), (*ii*) high motivation to seek and use the drug (items 5–6 in the DSM-IV), and (*iii*) maintenance of drug use despite negative consequences (item 7 in the DSM-IV). Using these three diagnostic domains, intravenous cocaine self-administration in Sprague-Dawley rats showed that the severity of cocaine use can be assessed in three animal models that represent these diagnostic domains: (*i*) drug seeking during periods when drug is unavailable, (*ii*) breakpoints on a progressive-ratio schedule of reinforcement, and (*iii*) persistence of self-administration despite punishment. Using such criteria, a certain percentage of animals can be ranked according to addiction-like behavior, and an "addiction score" can be computed, from 0 (no addiction-like behavior) to 3 (high addiction-like behavior; [56,59]). Animals with a score of 3 escalate drug intake more rapidly and to a much larger extent than animals with lower scores. They also present greater responses to contextual and drug-associated cues and exhibit much more reinstatement. This model is reliable; based on the link to DSM-IV and DSM-5 criteria, it also has some significant predictive validity, but see the Neurobiological Theories of Addiction chapter 3, Volume 1 for further discussion.

6. Summary of animal models of addiction

Most of the animal models discussed above have face validity for some component of the addiction cycle (compulsive use, withdrawal, or craving) and are reliable. For the positive reinforcing effects of drugs, drug self-administration, ICSS, and conditioned place preference, habit versus goal-directed responses have been shown to have predictive validity. Drug discrimination has predictive validity for abuse potential indirectly through generalization to the training drug. Animal models of withdrawal are focused on motivational constructs as opposed to the somatic signs of withdrawal, and now the response bias probabilistic reward measures and economic demand function models are being tested in both animals and humans. Animal models of conditioned drug effects are successful in predicting the potential for conditioned drug effects in humans. New challenges await in the domains of protracted abstinence and sleep. Predictive validity is more problematic for such concepts as craving, largely because of inadequate formulation of the concept of craving in humans [13,18,222]. Virtually all of the measures described above have reliability. Consistency and stability of the measures, small within-subject and between-subject variability, and reproducibility of the phenomena are characteristics of most of the measures that are employed in animal models of dependence.

Much remains to be explored about the predictive validity of these models. However, gaps in our knowledge may lie more in the human clinical laboratory domain than in the animal models domain [223–225]. The study of changes in the central nervous system that are associated with these models may provide insights into drug dependence and the etiology of psychopathologies that are associated (comorbid) with anxiety and affective disorders. Animal models of addiction provide a basis to continue such studies.

References

[1] Koob GF, Le Moal M. Drug abuse: hedonic homeostatic dysregulation. Science 1997;278:52–8.
[2] Koob GF, Sanna PP, Bloom FE. Neuroscience of addiction. Neuron 1998;21:467–76.
[3] Wikler A. Dynamics of drug dependence: implications of a conditioning theory for research and treatment. Archives of General Psychiatry 1973;28:611–6.
[4] Hebb DO. Organization of behavior: a neuropsychological theory. New York: Wiley; 1949.
[5] Richter CP. Animal behavior and internal drives. Quarterly Review of Biology 1927;2:307–43.
[6] Bindra D. A theory of intelligent behavior. New York: Wiley; 1976.
[7] Kling JW, Riggs LA. Woodworth and schlosberg's experimental psychology. 3rd ed. New York: Holt, Rinehart and Winston; 1971.
[8] Geyer MA, Markou A. Animal models of psychiatric disorders. In: Bloom FE, Kupfer DJ, editors. Psychopharmacology: the fourth generation of progress. New York: Raven Press; 1995. p. 787–98.
[9] American Psychiatric Association. Diagnostic and statistical manual of mental disorders. 4th ed. Washington, DC: American Psychiatric Press; 1994.
[10] American Psychiatric Association. Diagnostic and statistical manual of mental disorders. 5th ed. Washington, DC: American Psychiatric Publishing; 2013.
[11] Koob GF, Lloyd GK, Mason BJ. Development of pharmacotherapies for drug addiction: a Rosetta Stone approach. Nature Reviews Drug Discovery 2009;8:500–15.
[12] Ebel RL. Must all tests be valid? American Psychologist 1961;16:640–7.
[13] Sayette MA, Shiffman S, Tiffany ST, Niaura RS, Martin CS, Shadel WG. The measurement of drug craving. Addiction 2000;95(Suppl. 2):s189–210.
[14] Katz JL, Higgins ST. The validity of the reinstatement model of craving and relapse to drug use. Psychopharmacology 2003;168:21–30 [erratum: 168:244].
[15] McKinney WT. Models of mental disorders: a new comparative psychiatry. New York: Plenum; 1988.
[16] Geyer MA, Markou A. The role of preclinical models in the development of psychotropic drugs. In: Davis KL, Charney D, Coyle JT, Nemeroff C, editors. Neuropsychopharmacology: the fifth generation of progress. New York: Lippincott Williams and Wilkins; 2002. p. 445–55.
[17] Willner P. The validity of animal models of depression. Psychopharmacology 1984;83:1–16.
[18] Markou A, Weiss F, Gold LH, Caine SB, Schulteis G, Koob GF. Animal models of drug craving. Psychopharmacology 1993;112:163–82.
[19] Caine SB, Lintz R, Koob GF. Intravenous drug self-administration techniques in animals. In: Sahgal A, editor. Behavioural neuroscience: a practical approach, vol. 2. Oxford: IRL Press; 1993. p. 117–43.
[20] Caine SB, Koob GF. Modulation of cocaine self-administration in the rat through D-3 dopamine receptors. Science 1993;260:1814–6.
[21] Li TK. Pharmacogenetics of responses to alcohol and genes that influence alcohol drinking. Journal of Studies on Alcohol 2000;61:5–12.

[22] Wise RA. Voluntary ethanol intake in rats following exposure to ethanol on various schedules. Psychopharmacologia 1973;29:203–10.
[23] Simms JA, Bito-Onon JJ, Chatterjee S, Bartlett SE. Long-Evans rats acquire operant self-administration of 20% ethanol without sucrose fading. Neuropsychopharmacology 2010;35:1453–63.
[24] Simms JA, Steensland P, Medina B, Abernathy KE, Chandler LJ, Wise R, Bartlett SE. Intermittent access to 20% ethanol induces high ethanol consumption in Long-Evans and Wistar rats. Alcoholism Clinical and Experimental Research 2008;32:1816–23.
[25] Rhodes JS, Best K, Belknap JK, Finn DA, Crabbe JC. Evaluation of a simple model of ethanol drinking to intoxication in C57BL/6J mice. Physiology & Behavior 2005;84:53–63.
[26] Olds J, Milner P. Positive reinforcement produced by electrical stimulation of septal area and other regions of rat brain. Journal of Comparative and Physiological Psychology 1954;47:419–27.
[27] Gallistel CR. Self-stimulation. In: Deutsch JA, editor. The physiological basis of memory. 2nd ed. New York: Academic Press; 1983. p. 73–7.
[28] Kornetsky C, Bain G. Brain-stimulation reward: a model for drug induced euphoria. In: Adler MW, Cowan A, editors. Testing and evaluation of drugs of abuse. Modern methods in pharmacology, vol. 6. New York: Wiley-Liss; 1990. p. 211–31.
[29] Kornetsky C, Esposito RU. Euphorigenic drugs: effects on the reward pathways of the brain. Federation Proceedings 1979;38:2473–6.
[30] Campbell KA, Evans G, Gallistel CR. A microcomputer-based method for physiologically interpretable measurement of the rewarding efficacy of brain stimulation. Physiology & Behavior 1985;35:395–403.
[31] Markou A, Koob GF. Construct validity of a self-stimulation threshold paradigm: effects of reward and performance manipulations. Physiology & Behavior 1992;51:111–9.
[32] Holtzman SG. Discriminative stimulus effects of drugs: relationship to potential for abuse. In: Adler MW, Cowan A, editors. Testing and evaluation of drugs of abuse. Modern methods in pharmacology, vol. 6. New York: Wiley; 1990. p. 193–210.
[33] Mardones J, Segovia-Riquelme N. Thirty-two years of selection of rats by ethanol preference: UChA and UChB strains. Neurobehavioral Toxicology and Teratology 1983;5:171–8.
[34] Eriksson K. Genetic selection for voluntary alcohol consumption in the Albino rat. Science 1968;159:739–41.
[35] Lumeng L, Hawkins TD, Li TK. New strains of rats with alcohol preference and nonpreference. In: Thurman RG, Williamson JR, Drott HR, Chance B, editors. Alcohol and aldehyde metabolizing systems. Intermediary metabolism and neurochemistry, vol. III. New York: Academic Press; 1977. p. 537–44.
[36] Li TK, Lumeng L, Doolittle DP. Selective breeding for alcohol preference and associated responses. Behavior Genetics 1993;23:163–70.
[37] Fadda F, Mosca E, Colombo G, Gessa GL. Effect of spontaneous ingestion of ethanol on brain dopamine metabolism. Life Sciences 1989;44:281–7.
[38] Vanderschuren LJ, Everitt BJ. Drug seeking becomes compulsive after prolonged cocaine self-administration. Science 2004;305:1017–9.
[39] Leao R, Cruz F, Vendruscolo L, de Guglielmo G, Logrip M, Planeta C, Hope B, Koob G, George O. Chronic nicotine activates stress/reward-related brain regions and facilitates the transition to compulsive alcohol drinking. Journal of Neuroscience 2015;35:6241–53.
[40] Seif T, Chang SJ, Simms JA, Gibb SL, Dadgar J, Chen BT, Harvey BK, Ron D, Messing RO, Bonci A, Hopf FW. Cortical activation of accumbens hyperpolarization-active NMDARs mediates aversion-resistant alcohol intake. Nature Neuroscience August 2013;16(8):1094–100.
[41] Vendruscolo LF, Barbier E, Schlosburg JE, Misra KK, Whitfield Jr T, Logrip ML, Rivier CL, Repunte-Canonigo V, Zorrilla EP, Sanna PP, Heilig M, Koob GF. Corticosteroid-dependent plasticity mediates compulsive alcohol drinking in rats. Journal of Neuroscience 2012;32:7563–71.
[42] Everitt BJ, Robbins TW. Neural systems of reinforcement for drug addiction: from actions to habits to compulsion. Nature Neuroscience 2005;8:1481–9 [erratum: 9(7):979].

[43] Dickinson A, Nicholas DJ, Adams CD. The effect of the instrumental training ontengency on susceptibility to reinforcer devaluation. The Quarterly Journal of Experimental Psychology 1983;35: 35–51.
[44] Zapata A, Minney VL, Shippenberg TS. Shift from goal-directed to habitual cocaine seeking after prolonged experience in rats. Journal of Neuroscience 2010;30:15457–63.
[45] Dickinson A, Wood N, Smith JW. Alcohol seeking by rats: action or habit? Quarterly Journal of Experimental Psychology B Comparative and Physiological Psychology 2002;55B:331–48.
[46] Miles FJ, Everitt BJ, Dickinson A. Oral cocaine seeking by rats: action or habit? Behavioral Neuroscience 2003;117(5):927–38.
[47] Gremel CM, Chancey JH, Atwood BK, Luo G, Neve R, Ramakrishnan C, Deisseroth K, Lovinger DM, Costa RM. Endocannabinoid modulation of orbitostriatal circuits gates habit formation. Neuron June 15, 2016;90(6):1312–24.
[48] Gremel CM, Costa RM. Orbitofrontal and striatal circuits dynamically encode the shift between goal-directed and habitual actions. Nature Communications 2013;4:2264.
[49] Ahmed SH, Lenoir M, Guillem K. Neurobiology of addiction versus drug use driven by lack of choice. Current Opinion in Neurobiology 2013;23:581–7.
[50] Lenoir M, Serre F, Cantin L, Ahmed SH. Intense sweetness surpasses cocaine reward. PLoS One 2007;2:e698.
[51] Domingos AI, Vaynshteyn J, Voss HU, Ren X, Gradinaru V, Zang F, Deisseroth K, de Araujo IE, Friedman J. Leptin regulates the reward value of nutrient. Nature Neuroscience 2011;14:1562–8.
[52] Lenoir M, Cantin L, Vanhille N, Serre F, Ahmed SH. Extended heroin access increases heroin choices over a potent nondrug alternative. Neuropsychopharmacology 2013;38:1209–20.
[53] Stellar JR, Stellar E. The neurobiology of motivation and reward. New York: Springer-Verlag; 1985.
[54] Mucha RF, van der Kooy D, O'Shaughnessy M, Bucenieks P. Drug reinforcement studied hello helloby the use of place conditioning in rat. Brain Research 1982;243:91–105.
[55] Stoops WW. Reinforcing effects of stimulants in humans: sensitivity of progressive-ratio schedules. Experimental and Clinical Psychopharmacology 2008;16:503–12.
[56] Deroche-Gamonet V, Belin D, Piazza PV. Evidence for addiction-like behavior in the rat. Science 2004;305:1014–7.
[57] McGregor A, Roberts DC. Dopaminergic antagonism within the nucleus accumbens or the amygdala produces differential effects on intravenous cocaine self-administration under fixed and progressive ratio schedules of reinforcement. Brain Research 1993;624:245–52.
[58] Roberts DC, Loh EA, Vickers G. Self-administration of cocaine on a progressive ratio schedule in rats: dose-response relationship and effect of haloperidol pretreatment. Psychopharmacology 1989; 97:535–8.
[59] Belin D, Everitt BJ. Cocaine seeking habits depend upon dopamine-dependent serial connectivity linking the ventral with the dorsal striatum. Neuron 2008;57:432–41.
[60] Gellert VF, Holtzman SG. Development and maintenance of morphine tolerance and dependence in the rat by scheduled access to morphine drinking solutions. Journal of Pharmacology and Experimental Therapeutics 1978;205:536–46.
[61] Macey DJ, Schulteis G, Heinrichs SC, Koob GF. Time-dependent quantifiable withdrawal from ethanol in the rat: effect of method of dependence induction. Alcohol 1996;13:163–70.
[62] Malin DH, Lake JR, Newlin-Maultsby P, Roberts LK, Lanier JG, Carter VA, Cunningham JS, Wilson OB. Rodent model of nicotine abstinence syndrome. Pharmacology Biochemistry and Behavior 1992;43:779–84.
[63] Himmelsbach CK. Can the euphoric, analgetic, and physical dependence effects of drugs be separated? IV. With reference to physical dependence. Federation Proceedings 1943;2:201–3.
[64] Schulteis G, Markou A, Gold LH, Stinus L, Koob GF. Relative sensitivity to naloxone of multiple indices of opiate withdrawal: a quantitative dose-response analysis. Journal of Pharmacology and Experimental Therapeutics 1994;271:1391–8.
[65] Epping-Jordan MP, Watkins SS, Koob GF, Markou A. Dramatic decreases in brain reward function during nicotine withdrawal. Nature 1998;393:76–9.

[66] Gardner EL, Vorel SR. Cannabinoid transmission and reward-related events. Neurobiology of Disease 1998;5:502–33.
[67] Kokkinidis L, McCarter BD. Postcocaine depression and sensitization of brain-stimulation reward: analysis of reinforcement and performance effects. Pharmacology Biochemistry and Behavior 1990;36:463–71.
[68] Leith NJ, Barrett RJ. Amphetamine and the reward system: evidence for tolerance and post-drug depression. Psychopharmacologia 1976;46:19–25.
[69] Markou A, Koob GF. Post-cocaine anhedonia: an animal model of cocaine withdrawal. Neuropsychopharmacology 1991;4:17–26.
[70] Paterson NE, Myers C, Markou A. Effects of repeated withdrawal from continuous amphetamine administration on brain reward function in rats. Psychopharmacology 2000;152:440–6.
[71] Schulteis G, Markou A, Cole M, Koob G. Decreased brain reward produced by ethanol withdrawal. Proceedings of the National Academy of Sciences of the United States of America 1995;92:5880–4.
[72] Hand TH, Koob GF, Stinus L, Le Moal M. Aversive properties of opiate receptor blockade: evidence for exclusively central mediation in naive and morphine-dependent rats. Brain Research 1988;474: 364–8.
[73] Stinus L, Le Moal M, Koob GF. Nucleus accumbens and amygdala are possible substrates for the aversive stimulus effects of opiate withdrawal. Neuroscience 1990;37:767–73.
[74] Guillem K, Vouillac C, Koob GF, Cador M, Stinus L. Monoamine oxidase inhibition dramatically prolongs the duration of nicotine withdrawal-induced place aversion. Biological Psychiatry 2008;63: 158–63.
[75] Morse AC, Schulteis G, Holloway FA, Koob GF. Conditioned place aversion to the "hangover" phase of acute ethanol administration in the rat. Alcohol 2000;22:19–24.
[76] Denoble U, Begleiter H. Response suppression on a mixed schedule of reinforcement during alcohol withdrawal. Pharmacology Biochemistry and Behavior 1976;5:227–9.
[77] Gellert VF, Sparber SB. A comparison of the effects of naloxone upon body weight loss and suppression of fixed-ratio operant behavior in morphine-dependent rats. Journal of Pharmacology and Experimental Therapeutics 1977;201:44–54.
[78] Koob GF, Wall TL, Bloom FE. Nucleus accumbens as a substrate for the aversive stimulus effects of opiate withdrawal. Psychopharmacology 1989;98:530–4.
[79] Gauvin DV, Holloway FA. Cue dimensionality in the three-choice pentylenetetrazole-saline-chlordiazepoxide discrimination task. Behavioural Pharmacology 1991;2:417–28.
[80] Emmett-Oglesby MW, Mathis DA, Moon RT, Lal H. Animal models of drug withdrawal symptoms. Psychopharmacology 1990;101:292–309.
[81] France CP, Woods JH. Discriminative stimulus effects of naltrexone in morphine-treated rhesus monkeys. Journal of Pharmacology and Experimental Therapeutics 1989;250:937–43.
[82] Gellert VF, Holtzman SG. Discriminative stimulus effects of naltrexone in the morphine-dependent rat. Journal of Pharmacology and Experimental Therapeutics 1979;211:596–605.
[83] Pizzagalli DA, Jahn AL, O'Shea JP. Toward an objective characterization of an anhedonic phenotype: a signal-detection approach. Biological Psychiatry 2005;57:319–27.
[84] Pizzagalli DA, Iosifescu D, Hallett LA, Ratner KG, Fava M. Reduced hedonic capacity in major depressive disorder: evidence from a probabilistic reward task. Journal of Psychiatric Research 2008;43:76–87.
[85] Pergadia ML, Der-Avakian A, D'Souza MS, Madden PA, Heath AC, Shiffman S, Markou A, Pizzagalli DA. Association between nicotine withdrawal and reward responsiveness in humans and rats. JAMA Psychiatry November 2014;71(11):1238–45.
[86] Chase HW, Mackillop J, Hogarth L. Isolating behavioural economic indices of demand in relation to nicotine dependence. Psychopharmacology 2013;226(2):371–80.
[87] MacKillop J, Miranda Jr R, Monti PM, Ray LA, Murphy JG, Rohsenow DJ, McGeary JE, Swift RM, Tidey JW, Gwaltney CJ. Alcohol demand, delayed reward discounting, and craving in relation to drinking and alcohol use disorders. Journal of Abnormal Psychology 2010;119(1): 106–14.

[88] Bentzley BS, Fender KM, Aston-Jones G. The behavioral economics of drug selfadministration: a review and new analytical approach for within-session procedures. Psychopharmacology 2013; 226(1):113–25.
[89] Bickel WK, Johnson MW, Koffarnus MN, MacKillop J, Murphy JG. The behavioral economics of substance use disorders: reinforcement pathologies and their repair. Annual Review of Clinical Psychology 2014;10:641–77.
[90] Hursh SR, Silberberg A. Economic demand and essential value. Psychological Review 2008;115(1): 186–98.
[91] Bentzley BS, Jhou TC, Aston-Jones G. Economic demand predicts addiction-like behavior and therapeutic efficacy of oxytocin in the rat. Proceedings of the National Academy of Sciences of the United States of America 2014;111(32):11822–7.
[92] Oleson EB, Richardson JM, Roberts DCS. A novel IV cocaine self-administration procedure in rats: differential effects of dopamine, serotonin, and GABA drug pretreatments on cocaine consumption and maximal price paid. Psychopharmacology 2011;214(2):567–77.
[93] Bentzley BS, Aston-Jones G. Orexin-1 receptor signaling increases motivation for cocaine-associated cues. European Journal of Neuroscience 2015;41(9):1149–56.
[94] Andrews JS, Broekkamp CLE. Procedures to identify anxiolytic or anxiogenic agents. In: Sahgal A, editor. Behavioural neuroscience: a practical approach, vol. 2. Oxford: IRL Press; 1993. p. 37–54.
[95] Koob GF. A role for brain stress systems in addiction. Neuron 2008;59:11–34.
[96] Ahmed SH, Koob GF. Transition from moderate to excessive drug intake: change in hedonic set point. Science 1998;282:298–300.
[97] Kitamura O, Wee S, Specio SE, Koob GF, Pulvirenti L. Escalation of methamphetamine self-administration in rats: a dose-effect function. Psychopharmacology 2006;186:48–53.
[98] Ahmed SH, Walker JR, Koob GF. Persistent increase in the motivation to take heroin in rats with a history of drug escalation. Neuropsychopharmacology 2000;22:413–21.
[99] George O, Ghozland S, Azar MR, Cottone P, Zorrilla EP, Parsons LH, O'Dell LE, Richardson HN, Koob GF. CRF-CRF_1 system activation mediates withdrawal-induced increases in nicotine self-administration in nicotine-dependent rats. Proceedings of the National Academy of Sciences of the United States of America 2007;104:17198–203.
[100] O'Dell LE, Roberts AJ, Smith RT, Koob GF. Enhanced alcohol self-administration after intermittent versus continuous alcohol vapor exposure. Alcoholism Clinical and Experimental Research 2004;28:1676–82.
[101] Roberts AJ, Cole M, Koob GF. Intra-amygdala muscimol decreases operant ethanol self-administration in dependent rats. Alcoholism Clinical and Experimental Research 1996;20:1289–98.
[102] Aufrere G, Le Bourhis B, Beauge F. Ethanol intake after chronic intoxication by inhalation of ethanol vapour in rat: behavioural dependence. Alcohol 1997;14:247–53.
[103] Lallemand F, Soubrie PH, De Witte PH. Effects of CB1 cannabinoid receptor blockade on ethanol preference after chronic ethanol administration. Alcoholism Clinical and Experimental Research 2001;25:1317–23.
[104] Naassila M, Beauge FJ, Sebire N, Daoust M. Intracerebroventricular injection of antisense oligos to nNOS decreases rat ethanol intake. Pharmacology Biochemistry and Behavior 2000;67: 629–36.
[105] Moy SS, Knapp DJ, Criswell HE, Breese GR. Flumazenil blockade of anxiety following ethanol withdrawal in rats. Psychopharmacology 1997;131:354–60.
[106] Brown G, Jackson A, Stephens DN. Effects of repeated withdrawal from chronic ethanol on oral self-administration of ethanol on a progressive ratio schedule. Behavioural Pharmacology 1998;9: 149–61.
[107] Schulteis G, Hyytia P, Heinrichs SC, Koob GF. Effects of chronic ethanol exposure on oral self-administration of ethanol or saccharin by Wistar rats. Alcoholism Clinical and Experimental Research 1996;20:164–71.
[108] Valdez GR, Sabino V, Koob GF. Increased anxiety-like behavior and ethanol self-administration in dependent rats: reversal via corticotropin-releasing factor-2 receptor activation. Alcoholism Clinical and Experimental Research 2004;28:865–72.

[109] Gilpin NW, Smith AD, Cole M, Weiss F, Koob GF, Richardson HN. Operant behavior and alcohol levels in blood and brain of alcohol-dependent rats. Alcoholism Clinical and Experimental Research 2009;33(12):2113–23.
[110] Roberts AJ, Heyser CJ, Cole M, Griffin P, Koob GF. Excessive ethanol drinking following a history of dependence: animal model of allostasis. Neuropsychopharmacology 2000;22:581–94.
[111] Roberts AJ, Heyser CJ, Koob GF. Operant self-administration of sweetened versus unsweetened ethanol: effects on blood alcohol levels. Alcoholism Clinical and Experimental Research 1999;23: 1151–7.
[112] Rimondini R, Arlinde C, Sommer W, Heilig M. Long-lasting increase in voluntary ethanol consumption and transcriptional regulation in the rat brain after intermittent exposure to alcohol. Federation of American Societies for Experimental Biology Journal 2002;16:27–35.
[113] Richardson HN, Lee SY, O'Dell LE, Koob GF, Rivier CL. Alcohol self-administration acutely stimulates the hypothalamic-pituitary-adrenal axis, but alcohol dependence leads to a dampened neuroendocrine state. European Journal of Neuroscience 2008;28:1641–53.
[114] Rimondini R, Sommer W, Heilig M. A temporal threshold for induction of persistent alcohol preference: behavioral evidence in a rat model of intermittent intoxication. Journal of Studies on Alcohol 2003;64(4):445–9.
[115] Heilig M, Koob GF. A key role for corticotropin-releasing factor in alcohol dependence. Trends in Neurosciences 2007;30:399–406.
[116] Fidler TL, Clews TW, Cunningham CL. Reestablishing an intragastric ethanol self-infusion model in rats. Alcoholism Clinical and Experimental Research 2006;30:414–28.
[117] Becker HC, Lopez MF. Increased ethanol drinking after repeated chronic ethanol exposure and withdrawal experience in C57BL/6 mice. Alcoholism Clinical and Experimental Research 2004; 28:1829–38.
[118] Finn DA, Snelling C, Fretwell AM, Tanchuck MA, Underwood L, Cole M, Crabbe JC, Roberts AJ. Increased drinking during withdrawal from intermittent ethanol exposure is blocked by the CRF receptor antagonist D-Phe-CRF(12-41). Alcoholism Clinical and Experimental Research 2007;31: 939–49.
[119] Lopez MF, Becker HC. Effect of pattern and number of chronic ethanol exposures on subsequent voluntary ethanol intake in C57BL/6J mice. Psychopharmacology 2005;181:688–96.
[120] Chu K, Koob GF, Cole M, Zorrilla EP, Roberts AJ. Dependence-induced increases in ethanol self-administration in mice are blocked by the CRF1 receptor antagonist antalarmin and by CRF1 receptor knockout. Pharmacology Biochemistry and Behavior 2007;86:813–21.
[121] Ahmed SH, Kenny PJ, Koob GF, Markou A. Neurobiological evidence for hedonic allostasis associated with escalating cocaine use. Nature Neuroscience 2002;5:625–6.
[122] Jang CG, Whitfield T, Schulteis G, Koob GF, Wee S. A dysphoric-like state during early withdrawal from extended access to methamphetamine self-administration in rats. Psychopharmacology 2013; 225:753–63.
[123] Kenny PJ, Chen SA, Kitamura O, Markou A, Koob GF. Conditioned withdrawal drives heroin consumption and decreases reward sensitivity. Journal of Neuroscience 2006;26:5894–900.
[124] Roberts DCS, Richardson NR. Self-administration of psychomotor stimulants using progressive ratio schedules of reinforcement. In: Boulton AA, Baker GB, Wu PH, editors. Animal models of drug addiction. Neuromethods, vol. 24. Totowa, NJ: Human Press; 1992. p. 233–69.
[125] Bedford JA, Bailey LP, Wilson MC. Cocaine reinforced progressive ratio performance in the rhesus monkey. Pharmacology Biochemistry and Behavior 1978;9:631–8.
[126] Paterson NE, Markou A. Increased motivation for self-administered cocaine after escalated cocaine intake. NeuroReport 2003;14:2229–32.
[127] Wee S, Mandyam CD, Lekic DM, Koob GF. 1-Noradrenergic system role in increased motivation for cocaine intake in rats with prolonged access. European Neuropsychopharmacology 2008;18: 303–11.
[128] Walker BM, Koob GF. The γ-aminobutyric acid-B receptor agonist baclofen attenuates responding for ethanol in ethanol-dependent rats. Alcoholism Clinical and Experimental Research 2007;31: 11–8.

[129] Koob GF, Volkow ND. Neurocircuitry of addiction. Neuropsychopharmacology Reviews 2010;35: 217–38 [erratum: 35:1051].
[130] Koob GF, Volkow ND. Neurobiology of addiction: a neurocircuitry analysis. Lancet Psychiatry 2016;3:760–73.
[131] Stewart J, de Wit H. Reinstatement of drug-taking behavior as a method of assessing incentive motivational properties of drugs. In: Bozarth MA, editor. Methods of assessing the reinforcing properties of abused drugs. New York: Springer-Verlag; 1987. p. 211–27.
[132] Shaham Y, Shalev U, Lu L, de Wit H, Stewart J. The reinstatement model of drug relapse: history, methodology and major findings. Psychopharmacology 2003;168:3–20.
[133] Weiss F, Martin-Fardon R, Ciccocioppo R, Kerr TM, Smith DL, Ben-Shahar O. Enduring resistance to extinction of cocaine-seeking behavior induced by drug-related cues. Neuropsychopharmacology 2001;25:361–72.
[134] Epstein DH, Preston KL, Stewart J, Shaham Y. Toward a model of drug relapse: an assessment of the validity of the reinstatement procedure. Psychopharmacology 2006;189:1–16.
[135] Schmidt HD, Anderson SM, Famous KR, Kumaresan V, Pierce RC. Anatomy and pharmacology of cocaine priming-induced reinstatement of drug seeking. European Journal of Pharmacology 2005; 526:65–76.
[136] See RE. Neural substrates of cocaine-cue associations that trigger relapse. European Journal of Pharmacology 2005;526:140–6.
[137] Mueller D, Stewart J. Cocaine-induced conditioned place preference: reinstatement by priming injections of cocaine after extinction. Behavioural Brain Research 2000;115:39–47.
[138] Yahyavi-Firouz-Abadi N, See RE. Anti-relapse medications: preclinical models for drug addiction treatment. Pharmacology & Therapeutics 2009;124:235–47.
[139] Lu L, Grimm JW, Hope BT, Shaham Y. Incubation of cocaine craving after withdrawal: a review of preclinical data. Neuropharmacology 2004;47(Suppl. 1):214–26.
[140] Weiss F, Maldonado-Vlaar CS, Parsons LH, Kerr TM, Smith DL, Ben-Shahar O. Control of cocaine-seeking behavior by drug-associated stimuli in rats: effects on recovery of extinguished operant-responding and extracellular dopamine levels in amygdala and nucleus accumbens. Proceedings of the National Academy of Sciences of the United States of America 2000;97:4321–6.
[141] Hyytia P, Schulteis G, Koob GF. Intravenous heroin and ethanol self-administration by alcohol-preferring AA and alcohol-avoiding ANA rats. Psychopharmacology 1996;125:248–54.
[142] See RE, Grimm JW, Kruzich PJ, Rustay N. The importance of a compound stimulus in conditioned drug-seeking behavior following one week of extinction from self-administered cocaine in rats. Drug and Alcohol Dependence 1999;57:41–9.
[143] Ciccocioppo R, Angeletti S, Weiss F. Long-lasting resistance to extinction of response reinstatement induced by ethanol-related stimuli: role of genetic ethanol preference. Alcoholism Clinical and Experimental Research 2001;25:1414–9.
[144] Crombag HS, Bossert JM, Koya E, Shaham Y. Context-induced relapse to drug seeking: a review. Philosophical Transactions of the Royal Society of London B Biological Sciences 2008;363: 3233–43.
[145] Crombag HS, Grimm JW, Shaham Y. Effect of dopamine receptor antagonists on renewal of cocaine seeking by reexposure to drug-associated contextual cues. Neuropsychopharmacology 2002;27: 1006–15.
[146] Katner SN, Magalong JG, Weiss F. Reinstatement of alcohol-seeking behavior by drug-associated discriminative stimuli after prolonged extinction in the rat. Neuropsychopharmacology 1999;20: 471–9.
[147] Brown SA, Vik PW, Patterson TL, Grant I, Schuckit MA. Stress, vulnerability and adult alcohol relapse. Journal of Studies on Alcohol 1995;56:538–45.
[148] Kosten TR, Rounsaville BJ, Kleber HD. A 2.5-year follow-up of depression, life crises, and treatment effects on abstinence among opioid addicts. Archives of General Psychiatry 1986;43:733–8.
[149] Ahmed SH, Koob GF. Cocaine- but not food-seeking behavior is reinstated by stress after extinction. Psychopharmacology 1997;132:289–95.

[150] Erb S, Shaham Y, Stewart J. Stress reinstates cocaine-seeking behavior after prolonged extinction and a drug-free period. Psychopharmacology 1996;128:408—12.
[151] Deroche V, Marinelli M, Le Moal M, Piazza PV. Glucocorticoids and behavioral effects of psychostimulants: II Cocaine intravenous self-administration and reinstatement depend on glucocorticoid levels. Journal of Pharmacology and Experimental Therapeutics 1997;281:1401—7.
[152] Kreibich AS, Briand L, Cleck JN, Ecke L, Rice KC, Blendy JA. Stress-induced potentiation of cocaine reward: a role for CRF R1 and CREB. Neuropsychopharmacology 2009;34:2609—17.
[153] Le AD, Harding S, Juzytsch W, Funk D, Shaham Y. Role of alpha-2 adrenoceptors in stress-induced reinstatement of alcohol seeking and alcohol self-administration in rats. Psychopharmacology 2005; 179:366—73.
[154] Lu L, Shepard JD, Hall FS, Shaham Y. Effect of environmental stressors on opiate and psychostimulant reinforcement, reinstatement and discrimination in rats: a review. Neuroscience & Biobehavioral Reviews 2003;27:457—91.
[155] Redila VA, Chavkin C. Stress-induced reinstatement of cocaine seeking is mediated by the kappa opioid system. Psychopharmacology 2008;200:59—70.
[156] Ribeiro Do Couto B, Aguilar MA, Manzanedo C, Rodriguez-Arias M, Armario A, Minarro J. Social stress is as effective as physical stress in reinstating morphine-induced place preference in mice. Psychopharmacology 2006;185:459—70.
[157] Sanchez CJ, Sorg BA. Conditioned fear stimuli reinstate cocaine-induced conditioned place preference. Brain Research 2001;908:86—92.
[158] Shalev U, Highfield D, Yap J, Shaham Y. Stress and relapse to drug seeking in rats: studies on the generality of the effect. Psychopharmacology 2000;150:337—46.
[159] Shepard JD, Bossert JM, Liu SY, Shaham Y. The anxiogenic drug yohimbine reinstates methamphetamine seeking in a rat model of drug relapse. Biological Psychiatry 2004;55:1082—9.
[160] Wikler A. Neurophysiological aspects of the opiate and barbiturate abstinence syndromes. Research Publications - Association for Research in Nervous and Mental Disease 1953;32:269—86.
[161] O'Brien CP, Testa T, O'Brien TJ, Brady JP, Wells B. Conditioned narcotic withdrawal in humans. Science 1977;195:1000—2.
[162] Bradley BP, Phillips G, Green L, Gossop M. Circumstances surrounding the initial lapse to opiate use following detoxification. British Journal of Psychiatry 1989;154:354—9.
[163] Cunningham CL, Gremel CM, Groblewski PA. Drug-induced conditioned place preference and aversion in mice. Nature Protocols 2006;1:1662—770.
[164] Gracy KN, Dankiewicz LA, Koob GF. Opiate withdrawal-induced Fos immunoreactivity in the rat extended amygdala parallels the development of conditioned place aversion. Neuropsychopharmacology 2001;24:152—60.
[165] Heinrichs SC, Menzaghi F, Schulteis G, Koob GF, Stinus L. Suppression of corticotropin-releasing factor in the amygdala attenuates aversive consequences of morphine withdrawal. Behavioural Pharmacology 1995;6:74—80.
[166] Stinus L, Cador M, Zorrilla EP, Koob GF. Buprenorphine and a CRF1 antagonist block the acquisition of opiate withdrawal-induced conditioned place aversion in rats. Neuropsychopharmacology 2005;30:90—8.
[167] Stinus L, Caille S, Koob GF. Opiate withdrawal-induced place aversion lasts for up to 16 weeks. Psychopharmacology 2000;149:115—20.
[168] Goldberg SR, Morse WH, Goldberg DM. Behavior maintained under a second-order schedule by intramuscular injection of morphine or cocaine in rhesus monkeys. Journal of Pharmacology and Experimental Therapeutics 1976;199:278—86.
[169] Kenny PJ, Markou A. Conditioned nicotine withdrawal profoundly decreases the activity of brain reward systems. Journal of Neuroscience 2005;25:6208—12.
[170] Katz JL, Goldberg SR. Second-order schedules of drug injection. In: Bozarth MA, editor. Methods of assessing the reinforcing properties of abused drugs. New York: Springer-Verlag; 1987. p. 105—15.
[171] Katz JL, Goldberg SR. Second-order schedules of drug injection: implications for understanding reinforcing effects of abused drugs. Advances in Substance Abuse 1991;4:205—23.

[172] Everitt BJ, Robbins TW. Second-order schedules of drug reinforcement in rats and monkeys: measurement of reinforcing efficacy and drug-seeking behaviour. Psychopharmacology 2000;153: 17–30.
[173] Childress AR, McLellan AT, Ehrman R, O'Brien CP. Classically conditioned responses in opioid and cocaine dependence: a role in relapse? In: Ray BA, editor. Learning factors in substance abuse. NIDA research monograph, vol. 84. Rockville, MD: National Institute on Drug Abuse; 1988. p. 25–43.
[174] McLellan AT, Childress AR, Ehrman R, O'Brien CP, Pashko S. Extinguishing conditioned responses during opiate dependence treatment turning laboratory findings into clinical procedures. Journal of Substance Abuse Treatment 1986;3:33–40.
[175] Salimov RM, Salimova NB. The alcohol-deprivation effect in hybrid mice. Drug and Alcohol Dependence 1993;32:187–91.
[176] Le Magnen J. Etude de quelques facteurs associe a des modification de la consommation spontanee d'alcool ethylique par le rat [Study of some factors associated with modifications of spontaneous ingestion of ethyl alcohol by the rat]. Journal de Radiologie, d'Electrologie, et de Medecine Nucleaire 1960;52:873–84.
[177] Sinclair JD, Senter RJ. Increased preference for ethanol in rats following alcohol deprivation. Psychonomic Science 1967;8:11–2.
[178] Sinclair JD, Senter RJ. Development of an alcohol-deprivation effect in rats. The Quarterly Journal of Social Affairs 1968;29:863–7.
[179] Spanagel R, Hölter SM, Allingham K, Landgraf R, Zieglgänsberger W. Acamprosate and alcohol: I. Effects on alcohol intake following alcohol deprivation in the rat. European Journal of Pharmacology 1996;305:39–44.
[180] Kornet M, Goosen C, Van Ree JM. Effect of naltrexone on alcohol consumption during chronic alcohol drinking and after a period of imposed abstinence in free-choice drinking rhesus monkeys. Psychopharmacology 1991;104:367–76.
[181] Burish TG, Maisto SA, Cooper AM, Sobell MB. Effects of voluntary short-term abstinence from alcohol on subsequent drinking patterns of college students. Journal of Studies on Alcohol 1981; 42:1013–20.
[182] Vengeliene V, Bilbao A, Spanagel R. The alcohol deprivation effect model for studying relapse behavior: a comparison between rats and mice. Alcohol 2014;48(3):313–20.
[183] Heyser CJ, Schulteis G, Koob GF. Increased ethanol self-administration after a period of imposed ethanol deprivation in rats trained in a limited access paradigm. Alcoholism Clinical and Experimental Research 1997;21:784–91.
[184] Heyser CJ, Schulteis G, Koob GF. The alcohol deprivation effect: experimental conditions, applications, and treatment. In: Hannigan JH, Spear LP, Spear NE, Goodlett CR, editors. Alcohol and alcoholism: effects on brain and development. Mahwah, NJ: Lawrence Erlbaum; 1999. p. 161–76.
[185] Alling C, Balldin J, Bokstrom K, Gottfries CG, Karlsson I, Langstrom G. Studies on duration of a late recovery period after chronic abuse of ethanol: a cross-sectional study of biochemical and psychiatric indicators. Acta Psychiatrica Scandinavica 1982;66:384–97.
[186] Roelofs SM. Hyperventilation, anxiety, craving for alcohol: a subacute alcohol withdrawal syndrome. Alcohol 1985;2:501–5.
[187] Annis HM, Sklar SM, Moser AE. Gender in relation to relapse crisis situations, coping, and outcome among treated alcoholics. Addictive Behaviors 1998;23:127–31.
[188] Hershon HI. Alcohol withdrawal symptoms and drinking behavior. Journal of Studies on Alcohol 1977;38:953–71.
[189] Marlatt G, Gordon J. Determinants of relapse: implications for the maintenance of behavioral change. In: Davidson P, Davidson S, editors. Behavioral medicine: changing health lifestyles. New York: Brunner/Mazel; 1980. p. 410–52.
[190] Zywiak WH, Connors GJ, Maisto SA, Westerberg VS. Relapse research and the Reasons for Drinking Questionnaire: a factor analysis of Marlatt's relapse taxonomy. Addiction 1996;91(Suppl. l): s121–30.

[191] Lowman C, Allen J, Stout RL. Replication and extension of Marlatt's taxonomy of relapse precipitants: overview of procedures and results. Addiction 1996;91(Suppl. l):s51–71.
[192] Mason BJ, Ritvo EC, Morgan RO, Salvato FR, Goldberg G, Welch B, Mantero-Atienza E. A double-blind, placebo-controlled pilot study to evaluate the efficacy and safety of oral nalmefene HCl for alcohol dependence. Alcoholism Clinical and Experimental Research 1994;18:1162–7.
[193] Becker HC. Positive relationship between the number of prior ethanol withdrawal episodes and the severity of subsequent withdrawal seizures. Psychopharmacology 1994;116:26–32.
[194] Becker HC, Hale RL. Ethanol-induced locomotor stimulation in C57BL/6 mice following RO15-4513 administration. Psychopharmacology 1989;99:333–6.
[195] Branchey M, Rauscher G, Kissin B. Modifications in the response to alcohol following the establishment of physical dependence. Psychopharmacologia 1971;22:314–22.
[196] Breese GR, Overstreet DH, Knapp DJ, Navarro M. Prior multiple ethanol withdrawals enhance stress-induced anxiety-like behavior: inhibition by CRF1 and benzodiazepine-receptor antagonists and a 5-HT1a-receptor agonist. Neuropsychopharmacology 2005;30:1662–9.
[197] Rimondini R, Sommer WH, Dall'Olio R, Heilig M. Long-lasting tolerance to alcohol following a history of dependence. Addiction Biology 2008;13:26–30.
[198] Sommer WH, Rimondini R, Hansson AC, Hipskind PA, Gehlert DR, Barr CS, Heilig MA. Upregulation of voluntary alcohol intake, behavioral sensitivity to stress, and amygdala crhr1 expression following a history of dependence. Biological Psychiatry 2008;63:139–45.
[199] Gehlert DR, Cippitelli A, Thorsell A, Le AD, Hipskind PA, Hamdouchi C, Lu J, Hembre EJ, Cramer J, Song M, McKinzie D, Morin M, Ciccocioppo R, Heilig M. 3-(4-Chloro-2-morpholin-4-yl-thiazol-5-yl)-8-(1-ethylpropyl)-2,6-dimethyl-imidazo[1,2-b]pyridazine: a novel brain-penetrant, orally available corticotropin-releasing factor receptor 1 antagonist with efficacy in animal models of alcoholism. Journal of Neuroscience 2007;27:2718–26.
[200] Valdez GR, Zorrilla EP, Roberts AJ, Koob GF. Antagonism of corticotropin-releasing factor attenuates the enhanced responsiveness to stress observed during protracted ethanol abstinence. Alcohol 2003;29:55–60.
[201] Koob GF, Le Moal M. Drug addiction, dysregulation of reward, and allostasis. Neuropsychopharmacology 2001;24:97–129.
[202] Niikura K, Zhou Y, Ho A, Kreek MJ. Proopiomelanocortin (POMC) expression and conditioned place aversion during protracted withdrawal from chronic intermittent escalating-dose heroin in POMC-EGFP promoter transgenic mice. Neuroscience 2013;236:220–32.
[203] Skjei KL, Markou A. Effects of repeated withdrawal episodes, nicotine dose, and duration of nicotine exposure on the severity and duration of nicotine withdrawal in rats. Psychopharmacology 2003;168:280–92.
[204] Valdez GR, Roberts AJ, Chan K, Davis H, Brennan M, Zorrilla EP, Koob GF. Increased ethanol self-administration and anxiety-like behavior during acute withdrawal and protracted abstinence: regulation by corticotropin-releasing factor. Alcoholism Clinical and Experimental Research 2002;26:1494–501.
[205] Mantsch JR, Baker DA, Francis DM, Katz ES, Hoks MA, Serge JP. Stressor- and corticotropin releasing factor-induced reinstatement and active stress-related behavioral responses are augmented following long-access cocaine self-administration by rats. Psychopharmacology 2008;195:591–603.
[206] Mantsch JR, Baker DA, Serge JP, Hoks MA, Francis DM, Katz ES. Surgical adrenalectomy with diurnal corticosterone replacement slows escalation and prevents the augmentation of cocaine-induced reinstatement in rats self-administering cocaine under long-access conditions. Neuropsychopharmacology 2008;33:814–26.
[207] Shelton KL, Beardsley PM. Interaction of extinguished cocaine-conditioned stimuli and footshock on reinstatement in rats. International Journal of Comparative Psychology 2005;18:154–66.
[208] Mantsch JR, Vranjkovic O, Twining RC, Gasser PJ, McReynolds JR, Blacktop JM. Neurobiological mechanisms that contribute to stress-related cocaine use. Neuropharmacology 2014;76(Pt B):383–94.
[209] McKetin R, Kozel N, Douglas J, Ali R, Vicknasingam B, Lund J, Li JH. The rise of methamphetamine in Southeast and East Asia. Drug and Alcohol Review 2008;27(3):220–8.

[210] Pluddemann A, Parry CDH. Methamphetamine use and associated problems among adolescents in the Western Cape province of South Africa: a need for focused interventions. Cape Town: South African Medical Research Council; 2012. p. 1–3.
[211] Lipinska G, Timol R, Thomas KG. The implications of sleep disruption for cognitive and affective processing in methamphetamine abuse. Medical Hypotheses 2015;85(6):914–21.
[212] Colrain IM, Nicholas CL, Baker FC. Alcohol and the sleeping brain. Handbook of Clinical Neurology 2014;125:415–31.
[213] Brower KJ. Insomnia, alcoholism and relapse. Sleep Medicine Reviews 2003;7:523–39.
[214] Gillin JC. Are sleep disturbances risk factors for anxiety, depressive and addictive disorders? Acta Psychiatrica Scandinavica 1998;98:39–43.
[215] Brower KJ, Perron BE. Sleep disturbance as a universal risk factor for relapse in addictions to psychoactive substances. Medical Hypotheses 2010;74:928–33.
[216] Gossop MR, Bradley BP, Brewis RK. Amphetamine withdrawal and sleep disturbance. Drug and Alcohol Dependence 1982;10:177–83.
[217] Hill SY, Mendelson WB, Bernstein DA. Cocaine effects on sleep parameters in the rat. Psychopharmacology 1977;51(2):125–7.
[218] Martins RC, Andersen ML, Shih MC, Tufik S. Effects of cocaine, methamphetamine and modafinil challenge on sleep rebound after paradoxical sleep deprivation in rats. Brazilian Journal of Medical and Biological Research 2008;41:68–77.
[219] Chen B, Wang Y, Liu X, Liu Z, Dong Y, Huang YH. Sleep regulates incubation of cocaine craving. Journal of Neuroscience 2015;35(39):13300–10.
[220] Epstein DH, Preston KL. The reinstatement model and relapse prevention: a clinical perspective. Psychopharmacology 2003;168:31–41.
[221] Piazza PV, Deminiere JM, Le Moal M, Simon H. Factors that predict individual vulnerability to amphetamine self-administration. Science 1989;245:1511–3.
[222] Tiffany ST, Carter BL, Singleton EG. Challenges in the manipulation, assessment and interpretation of craving relevant variables. Addiction 2000;95(Suppl. 2):s177–87.
[223] Koob GF, Mason BJ. Existing and future drugs for the treatment of the dark side of addiction. Annual Review of Pharmacology and Toxicology 2016;56:299–322.
[224] Kwako L, Koob GF. Neuroclinical framework for the role of stress in addiction. Chronic Stress 2017;1:1–14.
[225] Kwako LE, Momenan R, Litten RZ, Koob GF, Goldman D. Addictions neuroclinical assessment: a neuroscience-based framework for addictive disorders. Biological Psychiatry 2016;80:179–89.
[226] Alderson HL, Robbins TW, Everitt BJ. The effects of excitotoxic lesions of the basolateral amygdala on the acquisition of heroin-seeking behaviour in rats. Psychopharmacology 2000;153:111–9.
[227] Barbier E, Vendruscolo LF, Schlosburg JE, Edwards S, Juergens N, Park PE, Misra KK, Cheng K, Rice KC, Schank J, Schulteis G, Koob GF, Heilig M. The NK1 receptor antagonist L822429 reduces heroin reinforcement. Neuropsychopharmacology 2013;38:976–84.
[228] Bauco P, Wise RA. Synergistic effects of cocaine with lateral hypothalamic brain stimulation reward: lack of tolerance or sensitization. Journal of Pharmacology and Experimental Therapeutics 1997;283:1160–7.
[229] Cohen A, Koob GF, George O. Robust escalation of nicotine intake with extended access to nicotine self-administration and intermittent periods of abstinence. Neuropsychopharmacology 2012;37:2153–60.
[230] de Wit H, Stewart J. Reinstatement of cocaine-reinforced responding in the rat. Psychopharmacology 1981;75:134–43.
[231] Depoortere RY, Li DH, Lane JD, Emmett-Oglesby MW. Parameters of self-administration of cocaine in rats under a progressive-ratio schedule. Pharmacology Biochemistry and Behavior 1993;45:539–48.
[232] Edwards S, Guerrero M, Ghoneim OM, Roberts E, Koob GF. Evidence that vasopressin V1b receptors mediate the transition to excessive drinking in ethanol-dependent rats. Addiction Biology 2011;17:76–85.

[233] Hubner CB, Kornetsky C. Heroin, 6-acetylmorphine and morphine effects on threshold for rewarding and aversive brain stimulation. Journal of Pharmacology and Experimental Therapeutics 1992;260:562–7.

[234] Huston-Lyons D, Kornetsky C. Effects of nicotine on the threshold for rewarding brain stimulation in rats. Pharmacology Biochemistry and Behavior 1992;41:755–9.

[235] Izenwasser S, Kornetsky C. Brain-stimulation reward: a method for assessing the neurochemical baes of drug-induced euphoria. In: Watson RR, editor. Drugs of abuse and neurobiology. Boca Raton, FL: CRC Press; 1992. p. 1–21.

[236] Kornetsky C. Brain-stimulation reward: a model for the neuronal bases for drug-induced euphoria. In: Brown RM, Friedman DP, Nimit Y, editors. Neuroscence methods in drug abuse research. NIDA research monograph, vol. 62. Rockville, MD: National Institute on Drug Abuse; 1985. p. 30–50.

[237] Pilla M, Perachon S, Sautel F, Garrido F, Mann A, Wermuth CG, Schwartz JC, Everitt BJ, Sokoloff P. Selective inhibition of cocaine-seeking behaviour by a partial dopamine D3 receptor agonist. Nature 1999;400:371–5.

[238] Rassnick S, Pulvirenti L, Koob GF. SDZ 205,152, a novel dopamine receptor agonist, reduces oral ethanol self-administration in rats. Alcohol 1993;10:127–32.

[239] Rogers J, Wiener SG, Bloom FE. Long-term ethanol administration methods for rats: advantages of inhalation over intubation or liquid diets. Behavioral and Neural Biology 1979;27:466–86.

[240] Sarkar M, Kornetsky C. Methamphetamine's action on brain-stimulation reward threshold and stereotypy. Experimental and Clinical Psychopharmacology 1995;3:112–7.

[241] Swerdlow NR, Gilbert D, Koob GF. Conditioned drug effects on spatial preference: critical evaluation. In: Boulton AA, Baker GB, Greenshaw AJ, editors. Psychopharmacology. Neuromethods, vol. 13. Clifton, NJ: Humana Press; 1989. p. 399–446.

[242] Wee S, Wang Z, Woolverton WL, Pulvirenti L, Koob GF. Effect of aripiprazole, a partial D2 receptor agonist, on increased rate of methamphetamine self-administration in rats with prolonged access. Neuropsychopharmacology 2007;32:2238–47.

CHAPTER 3

Neurobiological theories of addiction

Contents

Introduction to Addiction
ISBN 978-0-12-816863-9, https://doi.org/10.1016/B978-0-12-816863-9.00003-0

1. Introduction

The purpose of the present chapter is to outline and summarize the different neurobiological theories of addiction that correspond to overall neurocircuitry theories and theories with a focus on each of the three stages of the addiction cycle: *binge/intoxication*, *withdrawal/negative affect*, and *preoccupation/anticipation*—independent of any particular class of drugs. A historical and level of analysis (neurocircuitry, cellular, and molecular) overlay is provided that forms an undercurrent within the reviews herein, and an attempt is made to represent the latest versions of the authors' perspectives on each theory. Within each section, we first summarize the major tenets of each theory, followed by our thoughts thereof in *italics*. Representative diagrams and summaries from all of the original sources are provided for the reader's use. Finally, an attempt is made to provide a generic circuit of addiction with the main frameworks of reward, behavioral output, and craving. The generic circuit is not all-inclusive and undoubtedly has left out specific brain regions that may be of importance in addiction. However, it is hoped that it will provide a heuristic framework for future work.

2. Overall neurocircuitry theories of addiction

2.1 Mesolimbic dopamine reward hypothesis of addiction or "two-neuron" theory of reward

Wise RA. Action of drugs of abuse on brain reward systems. Pharmacology Biochemistry and Behavior 1980;13(Suppl. 1):213–23 [1].

One of the original theories of the actions of drugs of abuse on the brain reward system conceptualized an action on a critical dopaminergic synapse where all reward sites were argued to be afferent to a critical dopaminergic synapse. Two components of the brain reward system were outlined: (*i*) high-frequency-sensitive, fast-conducting myelinated fibers of the medial forebrain bundle and (*ii*) ventral tegmental dopamine system that is hypothesized to synapse directly on the dopamine link. At that time, Wise did not specify the mesolimbic dopamine system. Rather, most references to specific dopamine pathways were to the ventral tegmentum or tegmental-striatal projections [1]. Later discussions came to focus on the mesolimbic component *per se* (see below).

Data that were generated to support this theory at the time included evidence that amphetamine and cocaine act directly on the dopamine synapse in terminal areas of the mesolimbic dopamine system and evidence that opioids act at dopamine cell bodies or the dopamine synapse. Alcohol, barbiturates, and benzodiazepines were speculated to act via the naloxone-reversible inhibition of noradrenergic function that disinhibited rather than directly excited the dopamine reward link ([1]; Fig. 1). Dopamine receptor antagonists were shown to block amphetamine and cocaine reward [2–4] and food reward [5,6]. All major drugs of abuse were shown to facilitate brain stimulation reward and by extrapolation were linked to their activation of the dopamine system. Wise wrote, "It is attractive to consider the possibility that opiate reward, like brain stimulation, food and stimulant reward, ultimately activates a common dopaminergic substrate" [1].

This theory had a profound effect on the neurobiology of drug abuse and has guided research in this area more than any other theory. Dopamine receptor blockade was argued to interfere with all

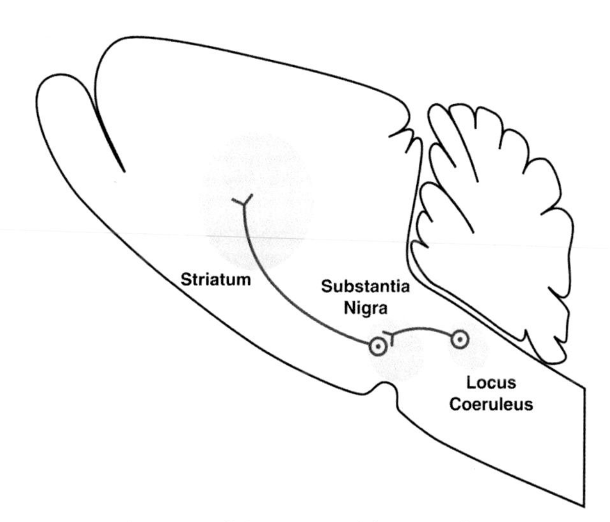

Figure 1 ***Early neurocircuitry diagram of drug reward for opioids [1].*** Suggested sites of potential interaction of opiates with brain reward circuitry. Opiate receptor fields are shaded in the region of the striatal dopamine terminal field, the tegmental dopamine cell region, and the region of the locus coeruleus, which is thought to inhibit reward circuitry, perhaps by an inhibitory synapse on the dopamine cells themselves. Opiates inhibit locus coeruleus firing; their actions in the tegmentum and striatum are not yet understood, and may be either pre- or postsynaptic in either region. Thus, opiates may act on, or either afferent or efferent to, the dopamine cells implicated in reward function. *(Taken with permission from Wise RA. Action of drugs of abuse on brain reward systems. Pharmacology Biochemistry and Behavior 1980;13(Suppl. 1):213–23 [1].)*

rewards (tested up until the publication of [1]), and the developed model had as its central element the dopamine neuron and its efferents. Additionally, at least one afferent link was added to illustrate the ways in which target neurons for brain stimulation reward and psychomotor stimulant reward were linked, thus yielding the now famous "two-neuron" theory of reward. Wise argued that the specific anatomy of the medial forebrain bundle connection was unknown (and to some extent, it is still unknown) and that drugs of abuse may enter the system to activate the dopamine system at different sites, a hypothesis that has persisted to this day ([1]; Fig. 2). Thus, Wise wrote, "While the rewarding and reward-facilitating effects of opiates, benzodiazepines, alcohol, and barbiturates may be mediated at the level of the dopamine neuron, it seems more likely that these agents interact with the dopamine link in reward circuitry through its afferents, either by exciting dopaminergic activity directly or by causing disinhibition" [1].

In summary, the focal point of drug reward—and by extrapolation reward per se—had become inextricably linked to the activation of midbrain dopamine neurons and remains so today.

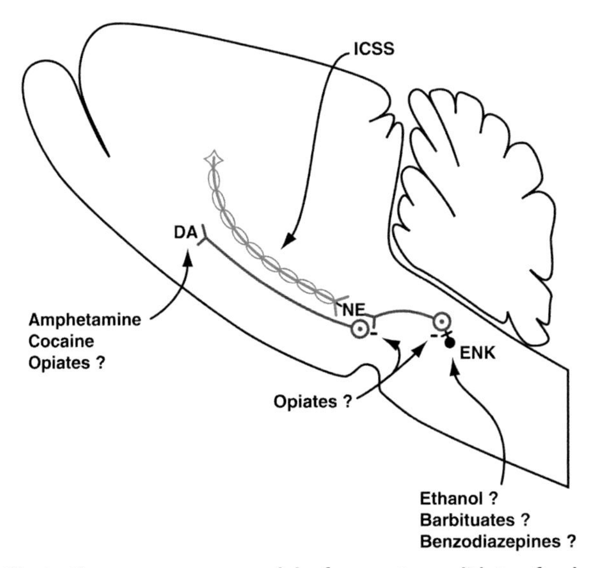

Figure 2 ***Diagram illustrating a summary model of current candidates for brain reward circuitry related to drugs of abuse and brain stimulation reward [1].*** The dopamine neuron is thought to be at least one synapse efferent to the directly activated fiber system in brain stimulation reward, which is shown as myelinated. Amphetamine and cocaine are known to act at the dopamine link, presumably in the synapse though perhaps at tegmental autoreceptors. Opiates might act at any level of the diagrammed model. Alcohol, barbiturates, and benzodiazepines are speculated to link through inputs to an opiate receptor to a noradrenergic inhibitory control over the dopamine cells; current evidence for this particular site of anxiolytic action is suggestive at best, but some disinhibitory links with the reward system must be taken as a serious possibility in current models of reward circuitry. ENK, enkephalin; DA, dopamine; NE, norepinephrine; ICSS, intracranial self-stimulation. *(Taken with permission from Wise RA. Action of drugs of abuse on brain reward systems. Pharmacology Biochemistry and Behavior 1980;13(Suppl. 1):213–23 [1].)*

2.2 Psychomotor stimulant theory of addiction

Wise RA, Bozarth MA. A psychomotor stimulant theory of addiction. Psychological Review 1987; 94:469–92 [7].

The thesis of this conceptual framework was based on the premise that addiction was synonymous with operant reinforcement and specified that independent psychomotor stimulant properties were predictors of whether a drug will be reinforcing in an operant situation. In short, "the crux of the theory is that the reinforcing effects of drugs, and thus their addiction liability, can be predicted from their ability to induce psychomotor activation" [7]. Building on the Glickman and Schiff [8] formulation that approach behaviors and positive reinforcement can each be elicited by activation of the medial forebrain bundle, Wise and Bozarth [7] argued that all drugs that are positive reinforcers should elicit forward locomotion.

Much evidence was marshaled to support this hypothesis from the domain of behavioral pharmacology. Wise and Bozarth argued that all drugs that have addiction potential, such as amphetamine, cocaine, nicotine, caffeine, opioids, barbiturates, alcohol, benzodiazepines, cannabis, and phencyclidine, have psychomotor stimulant actions, and these psychostimulant actions are attributable to the activation of central dopaminergic systems [7]. Wise and Bozarth further argued that even drugs of abuse with central nervous system depression as a dominant effect also have stimulant properties that are mediated by the same brain mechanisms that mediate the psychostimulant properties and by extrapolation the addictive properties of drugs of abuse with psychomotor stimulation as the dominant effect. The increase in locomotor activity that is caused by all of these drugs was hypothesized to be attributable to activation of the mesolimbic dopamine system.

The final conceptual argument of the psychomotor theory of addiction was that a common biological mechanism played homologous roles in psychomotor stimulation and positive reinforcement. Forward locomotion was proposed to be the unconditioned response to all positive reinforcers and that the medial forebrain bundle mediated such a response. The midbrain dopamine systems were thought to be critical for the reinforcing effects of brain stimulation reward [9,10], and the mesolimbic dopamine system was also thought to be critical not only for psychomotor stimulation that is produced by all drugs of abuse but also for their reinforcing actions. Most of the data at that time were generated with psychomotor stimulants and opioids [3,11,12].

Subsequent studies largely discredited the psychomotor theory of addiction, but the theory had a major effect. More than any other theory, the focus was placed on the mesolimbic dopamine system as critical for the positive reinforcing effects of drugs of abuse. Evidence against the theory today includes observations that very little psychomotor activation, if any, is observed in rats that receive alcohol or cannabis at doses that are regularly self-administered. Additionally, locomotor activation and self-administration that are produced by opioids, alcohol, and phencyclidine in rats and mice can be observed in the absence of the mesolimbic dopamine system [13–18], suggesting that these drugs have reinforcing effects that are independent of the mesolimbic dopamine system. Despite powerful arguments to suggest that dopamine is only one part of reward circuitry, there are persistent arguments in the literature for a central role for the mesolimbic dopamine system in the positive reinforcement of drugs and other arguments that positive reinforcement is directly equated with addiction.

In summary, the psychomotor theory of addiction, in its literal form, has been largely discredited. However, it has had the major impact of further propagating the theory that the mesolimbic dopamine system is the critical substrate for the acute positive reinforcing effects of drugs of abuse.

2.3 Mesolimbic dopamine reward hypothesis of addiction

Wise RA, Bozarth MA. A psychomotor stimulant theory of addiction. Psychological Review 1987; 94:469–92 [19].

A subsequent version of addiction neurocircuitry with a focus on the mesolimbic dopamine system moved the focus from forward locomotion to "neuroadaptations associated with the learning of the drug seeking habit" [19]. The argument put forth by Wise [19] was that the memory of early drug experiences is "stamped in" by the same reinforcement process that stamps in ordinary habits (non-drug habits) via weaker incentives. The theoretical framework was moved to a neuroadaptational perspective, in which brain neuroadaptations are again designated to be within the domain of the mesolimbic dopamine system but are argued to be the "neuroadaptations of habit formation."

The mesolimbic dopamine system via its cortical inputs (glutamatergic afferents) and nucleus accumbens output (γ-aminobutyric acid [GABA]ergic efferents) was hypothesized to comprise a major portion of the endogenous circuitry through which "the pleasures of the flesh come to shape the habits of an animal" ([19]; Fig. 3). This neurocircuit strikingly resembles the neurocircuitry that was outlined much earlier by Koob ([205]; see Section 2.4. *Drugs of abuse: anatomy, pharmacology, and function of reward pathways*) but with a critical dopaminergic link in series with reward. Wise argued that the mesolimbic dopamine system is activated transsynaptically by the normal pleasures of life or is activated directly by drugs of abuse or electrical brain stimulation reward.

There was relatively little that is new with this conceptualization, with the exception that the "forward locomotion" proposal of Wise and Bozarth [7] was replaced by augmentation of the "consolidation—by stamping in—the still active memory traces of the exteroceptive (reward-associated) and interoceptive (response feedback) stimuli that led to the behavior that preceded activation of the system."

The segue of the function of the mesolimbic dopamine system, from forward locomotion and pure reward to memory consolidation, appears to depend on critical observations that were made in primates by Schultz and colleagues that the mesolimbic dopamine system is more activated by the distant sensory message that predicts reward rather than the receipt of reward itself [20–24]. Monkeys that were trained to respond to juice that was delivered to their mouth exhibited the activation of midbrain dopamine neurons initially in response to the juice itself. With repeated testing, the neuronal response became less associated with the juice and more associated with stimuli that predicted stimulus presentation and could be reinstated by the presentation of a novel incentive, such as an apple slice [20].

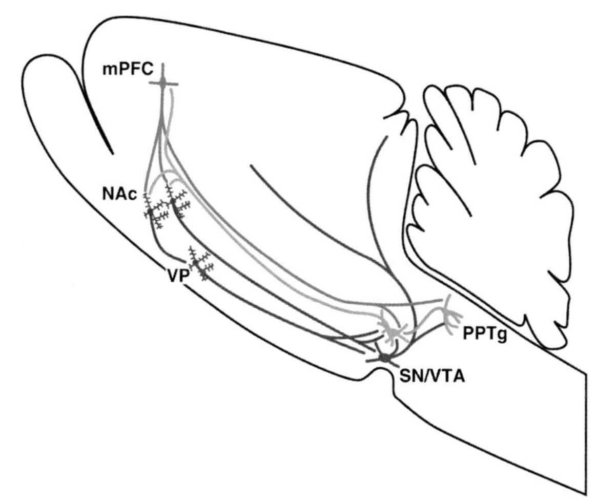

Figure 3 ***Selected elements and neurocircuitry of brain reward related to drugs of abuse and brain stimulation reward [19].*** The mesolimbic dopamine system is in gold. Amphetamine and cocaine are rewarding because they act at the dopamine transporter to elevate nucleus accumbens (NAc) dopamine levels. Nicotine is rewarding because of actions on nicotinic cholinergic receptors, expressed at both the cell bodies and the terminals of the mesolimbic system, resulting in elevated dopamine release in the NAc. Dopamine in NAc inhibits output neurons of the NAc. The normal cholinergic input to these receptors in the VTA is from the pedunculopontine tegmental nucleus (PPTg) and latero-dorsal pontine tegmental nucleus. These nuclei send branching projections to several basal forebrain targets (not shown). Rewarding electrical stimulation of the lateral hypothalamus is thought to be rewarding because it activates fibers to the PPTg. Excitatory amino acid (glutamate) projections of the medial prefrontal cortex (mPFC) are in *blue*. Projections from this and other cortical areas that receive mesolimbic dopamine input (amygdala, hippocampus) also project to the NAc. The amygdala also projects to the substantia nigra and ventral tegmental area (SN/VTA). Phencyclidine is rewarding because it blocks NMDA-type glutamate receptors in the NAc and mPFC. The blockade of NMDA receptors in the NAc reduces the excitatory input to GABAergic output neurons. Electrical stimulation of the mPFC is rewarding because it causes glutamate release in the VTA and dopamine release in the NAc. Two subsets of GABAergic projection neurons exit the NAc: one projects to the ventral pallidum (VP) and the other to the SN/VTA. GABAergic neurons in the VP also project to the SN/VTA. Most of the GABAergic projection to the SN synapses again on GABAergic neurons; these, in turn, project to the PPTg, deep layers of the superior colliculus, and dorsomedial thalamus. Heroin and morphine have two rewarding actions: inhibition of GABAergic cells that normally hold the mesolimbic dopamine system under inhibitory control (thus morphine disinhibits the dopamine system) and inhibition of output neurons in the NAc. Alcohol and cannabis act by unknown mechanisms to increase the firing of the mesolimbic dopamine system and are apparently rewarding for that reason. The habit-forming effects of barbiturates and benzodiazepines appear to be triggered at one or more of the GABAergic links in the circuitry, not necessarily through feedback links to the dopamine system. Caffeine appears to be rewarding through some independent circuitry. *(Taken with permission from Wise RA. Brain reward circuitry: insights from unsensed incentives. Neuron 2002;36:229–40 [19].)*

The authors concluded that these data provided "evidence for the involvement of dopamine neurons in arousing, motivational and behavioral activating processes that determine behavioral reactivity without encoding specific information about the behavioral reaction" [20]. Wise concluded that these data suggest that it was the receipt of reward predictors (i.e., the promise of reward) that produces the most arousal. Therefore, these reward predictors are conditioned rewards and not primary rewards and thus are rewards only because of prior learning. Activation of the midbrain dopamine system serves to establish response habits that are followed by its activation that is caused by either the normal pleasures of life or directly by intravenous drugs or electrical brain stimulation (i.e., a form of consolidation, stamping in of the still-active memory traces of the stimuli that led to the behavior that preceded activation of the system). This neo-dopamine reward theory strikingly resembles the argument that was put forth years earlier of a role for midbrain dopamine systems in incentive motivation and the activation that is associated with the presentation of incentives [25–27] with the addition of a new role for dopamine to "stamp in" the memory of the association between previously neutral stimuli and the incentive (for an earlier version of the dopamine learning hypothesis, see Ref. [28]).

In summary, from the perspective of drug abuse, Wise moved his dopamine theory of primary drug reward to a more important role for midbrain dopamine to establish response habits that are followed by its activation that is caused by drugs of abuse. As such, Wise moved the role of dopamine more from a primary reward function to what was later to become pathological habit theories (see below).

2.4 Drugs of abuse: anatomy, pharmacology, and function of reward pathways

Koob GF. Drugs of abuse: anatomy, pharmacology, and function of reward pathways. Trends in Pharmacological Sciences 1992b;13:177–84 [29].

In one of the earliest attempts to explore the interactions between drugs of abuse and reward pathways, a midbrain-forebrain-extrapyramidal circuit with a focus on the nucleus accumbens was proposed ([29]; Fig. 4). Three neurochemical systems were hypothesized to be involved in the initial reinforcing (or rewarding) actions of drugs of abuse: dopamine, opioid, and γ-aminobutyric acid (GABA). For indirect sympathomimetics, such as cocaine and amphetamines, the mesolimbic dopamine system was proposed to be critical and supported by numerous neuropharmacological studies. For example, neurotoxin-specific lesions of the mesolimbic dopamine system blocked the self-administration of cocaine and amphetamine [30,31] but not the self-administration of heroin [16] or alcohol [17]. For opioids, opioid receptors were hypothesized to be involved in a critical first step in the reinforcing actions of opioid drugs, with a predominant role for μ opioid receptors and both pre- and postsynaptic sites in the mesolimbic dopamine system in the nucleus accumbens and ventral tegmental area. For alcohol, $GABA_A$ receptors were hypothesized to be an initial site of the reinforcing actions of alcohol, with a predominant role for $GABA_A$ receptors in the nucleus accumbens and amygdala [32].

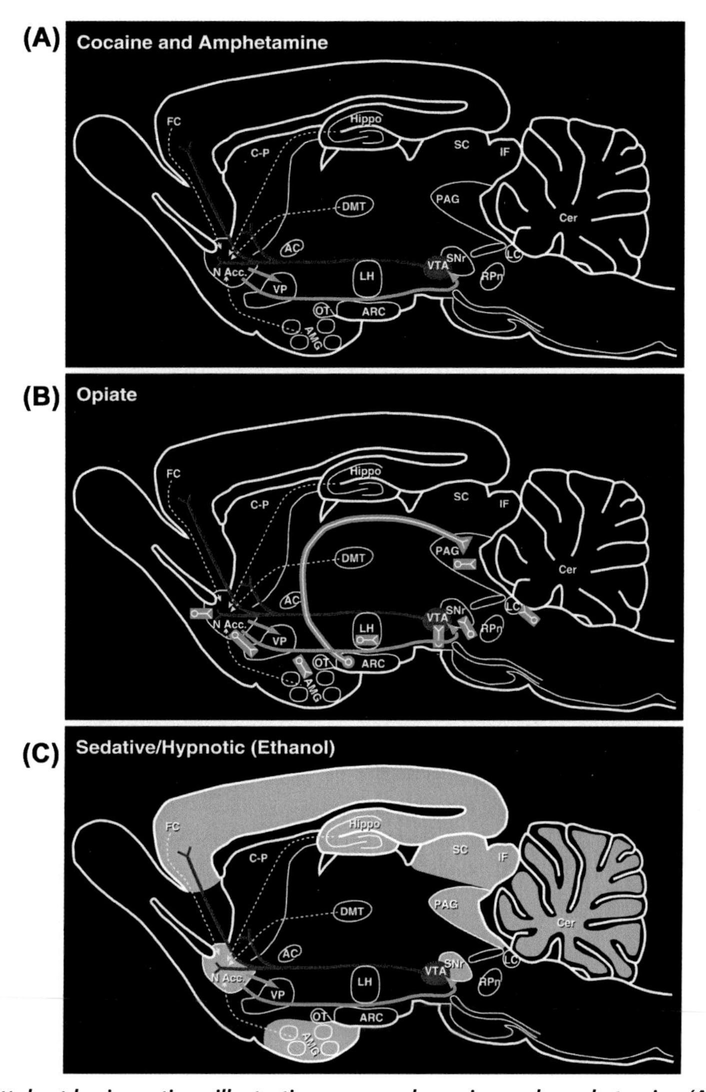

Figure 4 ***Sagittal rat brain sections illustrating proposed cocaine and amphetamine (A), opioid peptide (B), and sedative/hypnotic (C) drug reward neurocircuits.*** (A) The cocaine and amphetamine reward circuit includes a limbic-extrapyramidal motor interface. *Yellow* indicates limbic afferents to the nucleus accumbens, and *orange* represents efferents from the nucleus accumbens thought to be involved in psychomotor stimulant reward. *Red* indicates projections of the mesocorticolimbic dopamine system thought to be a critical substrate for psychomotor stimulant reward. This system originates in the A10 cell group of the ventral tegmental area and projects to the nucleus accumbens, olfactory tubercle, and ventral striatal domains of the caudate putamen. (B) These opioid peptide systems (*green*) include local enkephalin circuits (short segments) and the hypothalamic midbrain β-endorphin circuit (long segment). The system is superimposed on the neural reward circuit shown in (A). (C) Approximate distribution of $GABA_A$ receptor complexes (*blue*) determined by the relative distribution of both [^{3}H]flumazenil binding and expression of the α, β, and γ subunits of the $GABA_A$ receptor [478–480]. This distribution is superimposed on the neural reward circuit shown in (A). AC, anterior commissure; AMG, amygdala; ARC, arcuate nucleus; Cer, cerebellum; C-P, caudate putamen; DMT, dorsomedial thalamus; FC, frontal cortex; Hippo, hippocampus; IF, inferior colliculus; LC, locus coeruleus; LH, lateral hypothalamus; N Acc, nucleus accumbens; OT, olfactory tubercle; PAG, periaqueductal gray; RPn, reticular pontine nucleus; SC, superior colliculus; SNr, substantia nigra pars reticulata; VP, ventral pallidum; VTA, ventral tegmental area. *(Taken with permission from Koob GF. Drugs of abuse: anatomy, pharmacology, and function of reward pathways. Trends in Pharmacological Sciences 1992b;13:177–84 [29].)*

Based on this synthesis, Koob [29] proposed an early neurobiological circuit of drug reward. The starting point for the reward circuit was the medial forebrain bundle, which is composed of myelinated fibers that connect the olfactory tubercle and nucleus accumbens with the hypothalamus and ventral tegmental area [33] and ascending monoamine pathways (e.g., the mesocorticolimbic dopamine system). Drug reward was hypothesized to depend on dopamine release in the nucleus accumbens for cocaine and amphetamine, opioid receptor activation in the ventral tegmental area and nucleus accumbens for opioids, and $GABA_A$ receptors in the amygdala for alcohol.

Even at this time, the independence of both opioid and alcohol reward from a critical role for dopamine was noted and was used as a basis to argue for multiple independent neurochemical elements in drug reward. Koob further argued that the nucleus accumbens was strategically situated to receive important limbic information from the amygdala, frontal cortex, and hippocampus that could be converted to motivational action via its connections with the extrapyramidal motor system. Finally, the nucleus accumbens was suggested to not be a homogeneous structure, in which the "shell" part (medial and ventral) may be part of an extended amygdala system (see below), and the core part resembled more the corpus striatum [34,35].

In summary, an early neurobiological circuit for drug reward included a midbrain-forebrain-extrapyramidal circuit with a focus on the nucleus accumbens and multiple parallel neurochemical sites of action for different drugs of abuse.

2.5 Development and maintenance of drug addiction: reconciliation … sort of

Wise RA, Koob GF. The development and maintenance of drug addiction. Neuropsychopharmacology 2014;39:254–62 [36].

Invited to do the first "Perspectives" piece for the journal *Neuropsychopharmacology* in 2014, Wise and Koob "dusted off "their long-term discussion about the relative importance of two forms of reinforcement—positive reinforcement and negative reinforcement—in addiction. Areas of agreement were that addiction begins with the formation of habits through positive reinforcement and that drug-opposite physiological responses often establish the conditions for negative reinforcement as tolerance appears to develop to positive reinforcement. The work of Roy Wise has focused on positive-reinforcement mechanisms that are important for establishing drug-seeking habits and reinstating them quickly after periods of abstinence, whereas George Koob's work has focused on the negative-reinforcement mechanisms that become most obvious in the late stages of sustained addiction. However, Koob and Wise continue to hold different views with regard to (*i*) the point between early and late at which the diagnosis of "addiction" should be invoked, (*ii*) the relative importance of positive and negative reinforcement that leads to this transition, and (*iii*) the degree to which the specifics of negative reinforcement can be generalized across the range of addictive agents. Wise continues to see addiction as the result of the self-administration of drugs that more strongly elevate dopamine levels than natural rewards (e.g., food or sex) and the subsequent "stamping" in of the memory traces that are associated with such self-administration [37]. Koob, in contrast, focuses on neuroadaptations that are caused by the drug itself and argues that drug addiction involves not just the recruitment or sensitization of reward but the sensitization of anti-reward that drives negative reinforcement (Fig. 5).

So how are these views similar and different? They both agree that "positive reinforcement leads to the initial repetition of drug taking that becomes habitual and eventually compulsive." Wise does not question that positive reinforcement is necessarily limited by opponent processes—positive reinforcement involves a positive feedback system, and positive feedback mechanisms, if not limited by some form of opponent process, necessarily self-destruct—nor does he question that negative reinforcement comes to contribute very significantly to the compulsive, chronic use of addictive drugs [36]. Where they continue to differ is in the domains of semantics: "dependent" versus "addicted" and the types of animal models they continue to study. In the end, these differences likely only reflect which part of the proverbial elephant of addiction on which we focus.

2.6 Neurocircuitry/neurobiology of addiction: a neurocircuitry analysis

Koob GF, Volkow ND. Neurocircuitry of addiction. Neuropsychopharmacology Reviews 2010;35: 217–38 [erratum: 35:1051] [38].

Koob GF, Volkow ND. Neurobiology of addiction: a neurocircuitry analysis. Lancet Psychiatry 2016;3:760–73 [39].

Drug addiction is a chronically relapsing disorder that has been characterized by (*i*) compulsion to seek and take the drug, (*ii*) the loss of control in limiting intake, and (*iii*) the emergence of a negative emotional state (e.g., dysphoria, anxiety, and irritability), reflecting a motivational withdrawal syndrome, when access to the drug is prevented. Drug addiction had been conceptualized as a disorder that involves elements of both impulsivity and compulsivity that yield a composite addiction cycle that is composed of three stages: *binge/intoxication*, *withdrawal/negative affect*, and *preoccupation/anticipation* (craving; [40]). Subsequently, drug addiction was argued to represent the dramatic dysregulation of motivational circuits that is caused by a combination of exaggerated incentive salience and habit formation, reward deficits, stress surfeits, and compromised executive function [38]. Animal and human imaging studies have revealed discrete circuits that mediate the three stages of the addiction cycle, with key elements of the ventral tegmental area and ventral striatum as a focal point for the *binge/intoxication* stage, a key role for the extended amygdala in the *withdrawal/negative affect* stage, and a key role for a widely distributed network (orbitofrontal cortex–dorsal striatum, prefrontal cortex, basolateral amygdala, hippocampus, and insula in craving and cingulate gyrus, dorsolateral prefrontal, and inferior frontal cortices in disrupted inhibitory control) in the *preoccupation/anticipation* stage. The transition to addiction was hypothesized to involve neuroplasticity in all of these structures. The rewarding effects of drugs of abuse, development of incentive salience, and development of drug-seeking habits in the *binge/intoxication* stage involve changes in dopamine and opioid peptides in the basal ganglia. The increases in negative emotional states and

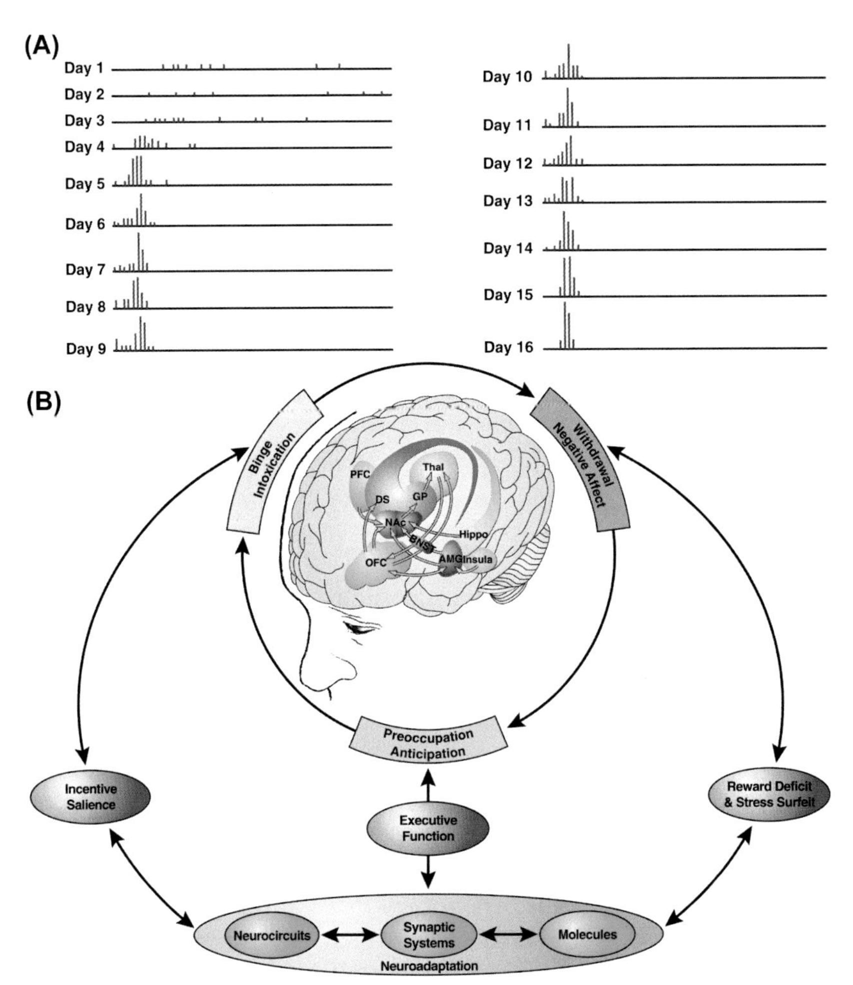

Figure 5 (A) The Roy Wise perspective. Interresponse interval histograms from a rat lever-pressing for intravenous cocaine in 4-h daily sessions at a dose per injection of 1 mg/kg. Of interest is the decrease in variability across days of testing. The narrowing of the distribution of interresponse times over days offers an objective measure of the steady progression to compulsive cocaine taking that develops even with limited daily access to the drug. Wise offers the ratio of mean to standard deviation (M/SD) interresponse time as an objective measure of the subjective label "compulsive." (B) The George Koob perspective. Diagram showing the neurocircuitry of addiction divided heuristically into the three stages of the addiction cycle: *binge/intoxication, blue*; *withdrawal/negative affect, red*; *preoccupation/anticipation, green*. The neurocircuits involved are also color-coded, with the basal ganglia, including the nucleus accumbens (NAc), dorsal striatum (DS), globus pallidum (GP), and thalamus (Thal) as key elements of the *binge/intoxication* stage; the extended amygdala, including the central nucleus of the amygdala (AMG), bed nucleus of the stria terminals (BNST), and a transition area in the shell of the nucleus accumbens (NAc) as key elements of the *withdrawal/negative affect* stage; and the frontal cortex and allocortex, including the prefrontal cortex (PFC), orbitofrontal cortex (OFC), hippocampus (Hippo), and insula (Insula) as key elements of the *preoccupation/anticipation* stage. Molecular, synaptic, and neurocircuitry neuroadaptations combine to render the four key elements of the transition to addiction: increased incentive salience (Koob's shortcut translation of Wise's "dominance for the cues that guide and motivate drug seeking over the cues that guide and motivate the seeking of the more natural pleasures of life"), decreased reward, increased stress, and decreased executive function. *((A and B) Taken with permission from Wise RA, Koob GF. The development and maintenance of drug addiction. Neuropsychopharmacology 2014;39:254–62 [36].)*

dysphoric and stress-like responses in the *withdrawal/negative affect* stage involve decreases in the function of the dopamine component of the reward system and the recruitment of brain stress neurotransmitters, such as corticotropin-releasing factor (CRF) and dynorphin, in the neurocircuitry of the extended amygdala. The craving and deficits in executive function in the *preoccupation/anticipation* stage involve the dysregulation of key afferent projections from the prefrontal cortex and insula, including glutamate, to the basal ganglia and extended amygdala. Molecular genetic studies have identified transduction and transcription factors that act in neurocircuitry that is associated with the development and maintenance of addiction that might mediate initial vulnerability, maintenance, and relapse that are associated with addiction.

The conceptual framework for the three-stage addiction cycle and the subsequent neurobiological circuitry derived originally from the social psychology of impulse control disorders and the development and teaching of such a course by Koob for the University of California, San Diego, to juniors and seniors (circa 1996), based on Baumeister et al. [41]. Koob developed the concept of the three-stage cycle to encompass everything from compulsive gambling and compulsive sex to drug addiction. Subsequently, discussions with Michel Le Moal led to their paper [40] where they explicitly outlined the three stages in a cycle of spiraling distress. The spiral derived from a social psychology framework [41] that was integrated in the clinical phenomenology of impulse control disorders and preclinical data from animal models of addiction. Goldstein and Volkow [42] elaborated a similar three-stage cycle based largely on imaging data in humans. The neurocircuitry frameworks that are outlined above represent a confluence of these frameworks that derived from phenomena that are outlined in descriptions of addiction (both drug and nondrug addictions) and were based on the social psychology of self-regulation [41,43] and integrated neuroimaging data [42]. The conceptualization by Koob and Le Moal profited from burgeoning knowledge of preclinical and clinical neurocircuitry plasticity that occurs in addiction and animal models of addiction (Figs. 6 and 7).

3. Theories of the binge/intoxication stage

3.1 Neural substrates of alcohol self-administration: neurobiology of high alcohol drinking behavior in rodents

McBride WJ, Li TK. Animal models of alcoholism: neurobiology of high alcohol-drinking behavior in rodents. Critical Reviews in Neurobiology 1998;12:339–69 [44].

Based on an extensive series of studies with animals that were selectively bred for high alcohol consumption, a number of key brain regions and neurochemical systems were identified that contribute to circuitry that is important for alcohol reward ([44]; Fig. 8). A key focus of this circuitry framework is the ventral tegmental area and dopamine projections to the nucleus accumbens and olfactory tubercle, prefrontal cortex, and ventral pallidum in mediating alcohol drinking. Alcohol activates ventral tegmental area neurons and causes dopamine release in the nucleus accumbens (see Volume 3, *Alcohol*). Further evidence that supports a role for the ventral tegmental area as one site that mediates the

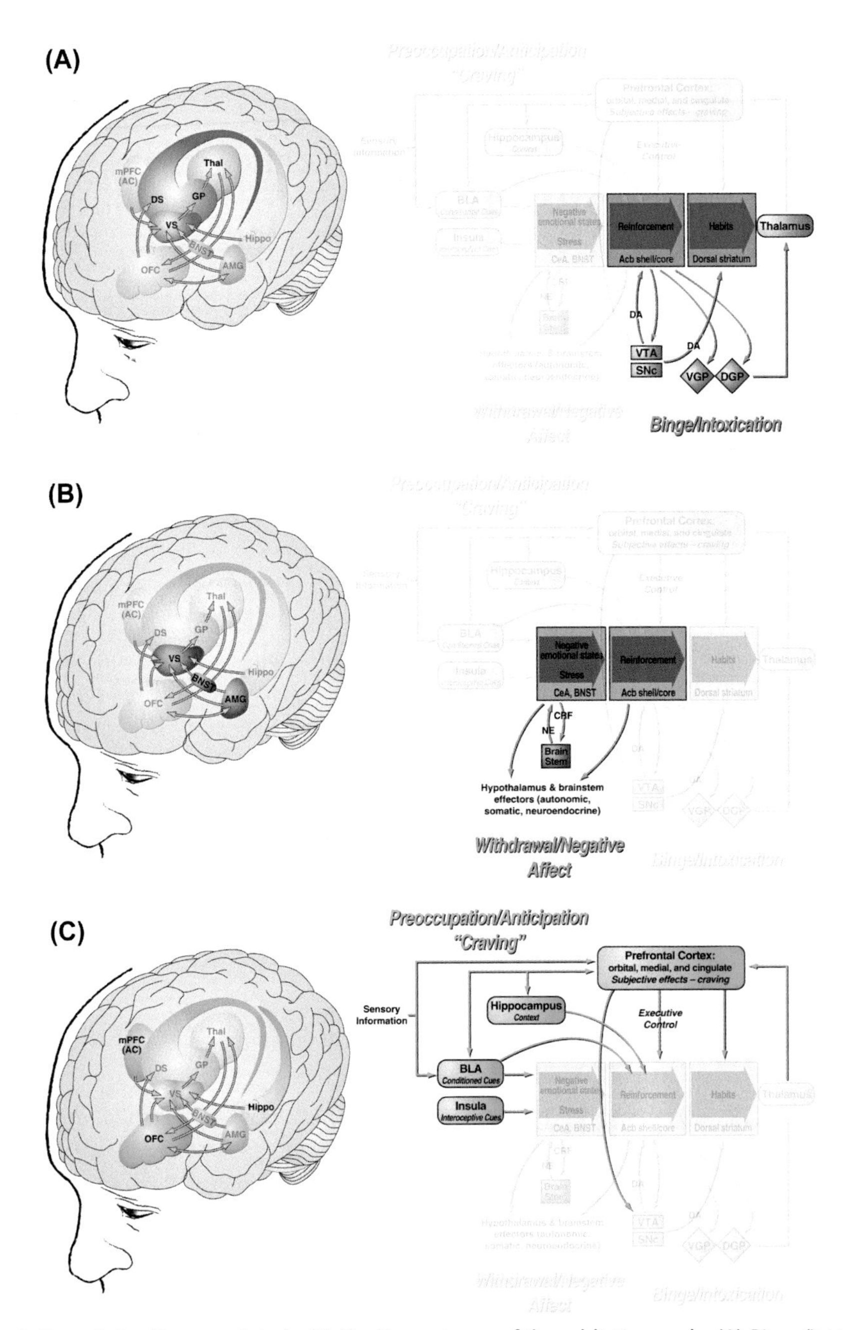

Figure 6 Neural circuitry associated with the three stages of the addiction cycle. (A) *Binge/intoxication* stage. Reinforcing effects of drugs may engage associative mechanisms and reward neurotransmitters in the nucleus accumbens shell and core and then engage stimulus-response habits that depend on the dorsal striatum. Two major neurotransmitters that mediate the rewarding effects of drugs of abuse

reinforcing actions of alcohol came from microinjection and intracranial self-administration studies. Alcohol-preferring P rats, but not nonpreferring NP rats, will self-administer 50—200 mg% alcohol directly into the ventral tegmental area [45,46], and this self-administration depends on dopamine activation [47,48]. Acetaldehyde also was self-administered into the ventral tegmental area [49], and salsolinol is self-administered into the nucleus accumbens [50]. These findings suggested that alcohol and its metabolites may interact with the mesolimbic dopamine system in alcohol-preferring rats to produce some of its reinforcing effects.

The blockade of $GABA_A$ receptors in the ventral tegmental area also blocked alcohol intake in P rats at doses that did not alter saccharin consumption [51] and was interpreted as GABA receptor antagonists that mimic the effects of alcohol by activating dopamine firing and thus substituting for the effects of alcohol [44]. Similar effects in decreasing responding for alcohol were observed with microinjections of a benzodiazepine receptor inverse agonist in the ventral tegmental area [52,53]. Others hypothesized that the decrease in alcohol intake in P rats involved the activation of muscarinic acetylcholine receptors in the ventral tegmental area [54].

Other key sites for alcohol reinforcement included the nucleus accumbens, pedunculopontine nucleus, and midbrain raphe ([44]; Fig. 8). Microinjections of both GABA receptor agonists and antagonists in the nucleus accumbens blocked alcohol self-administration [55]. Local injections of a benzodiazepine receptor inverse agonist

are dopamine and opioid peptides. (B) *Withdrawal/negative affect* stage. The negative emotional state of withdrawal may engage activation of the extended amygdala. The extended amygdala is composed of several basal forebrain structures, including the bed nucleus of the stria terminalis, central nucleus of the amygdala, and possibly a transition are in the medial portion (or shell) of the nucleus accumbens. Major neurotransmitters in the extended amygdala that are hypothesized to play a role in negative reinforcement are corticotropin-releasing factor, norepinephrine, and dynorphin. Major projections of the extended amygdala are to the hypothalamus and brainstem. (C) *Preoccupation/anticipation* (craving) stage. This stage involves the processing of conditioned reinforcement in the basolateral amygdala and the processing of contextual information by the hippocampus. Executive control depends on the prefrontal cortex and includes the representation of contingencies, the representation of outcomes, and their value and subjective states (i.e., craving and, presumably, feelings) associated with drugs. The subjective effects, termed drug craving, in humans involve activation of the orbital and anterior cingulate cortex and temporal lobe, including the amygdala, in functional imaging studies. A major neurotransmitter that is involved in the craving stage is glutamate that is localized in pathways from frontal regions and the basolateral amygdala that projects to the ventral striatum. *Green/blue arrows*, glutamatergic projections; *orange arrows*, dopaminergic projections; *pink arrows*, GABAergic projections; Acb, nucleus accumbens; BLA, basolateral amygdala; VTA, ventral tegmental area; SNc, substantia nigra pars compacta; VGP, ventral globus pallidus; DGP, dorsal globus pallidus; BNST, bed nucleus of the stria terminalis; CeA, central nucleus of the amygdala; NE, norepinephrine; CRF, corticotropin-releasing factor; PIT, Pavlovian instrumental transfer. *(Modified with permission from Koob GF, Everitt BJ, Robbins TW. Reward, motivation, and addiction. In: Squire LG, Berg D, Bloom FE, Du Lac S, Ghosh A, Spitzer N, editors. Fundamental neuroscience. 3rd ed. Amsterdam: Academic Press; 2008. p. 987—1016 [542].)*

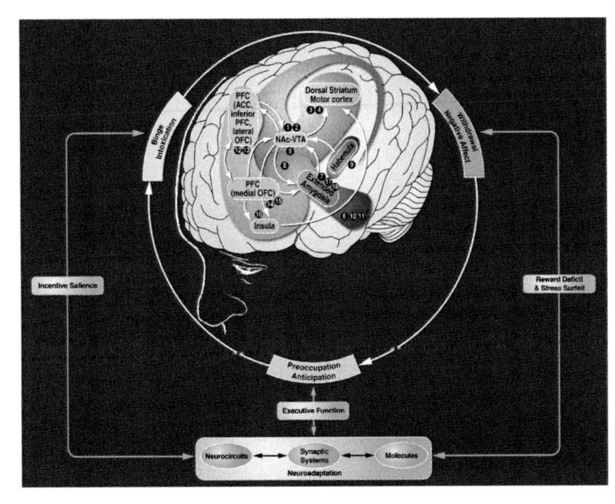

	Neurotransmitter
Binge/intoxication	
Ventral tegmental area (**circuit 1**)	Glutamate
Ventral tegmental area (**circuit 2**)	GABA
Dorsal striatum (**circuit 3**)	Dopamine
Dorsal striatum (**circuit 4**)	Glutamate
Withdrawal/negative affect	
Ventral tegmental area (**circuit 5**)	CRF
Central nucleus of the amgydala (**circuit 6**)	CRF
Bed nucleus of the stria terminalis (**circuit 7**)	Norepinephrine
Nucleus accumbens shell (**circuit 8**)	Dynorphin
Habenula (**circuit 9**)	Acetylcholine
Central nucleus of the amygdala (**circuit 10**)	Neuropeptide Y
Central nucleus of the amygdala (**circuit 11**)	Endocannabinoids
Preoccupation/anticipation	
Prefrontal cortex (**circuit 12**)	Glutamate
Prefrontal cortex (**circuit 13**)	GABA
Hippocampus (**circuit 14**)	Glutamate
Basolateral amygdala (**circuit 15**)	Glutamate
Bed nucleus of the stria terminalis (**circuit 16**)	CRF
Bed nucleus of the stria terminalis (**circuit 17**)	Norepinephrine
Insula (**circuit 18**)	CRF

Figure 7 Model proposing a network of interacting circuits, disruptions of which contribute to the complex set of compulsive-like behaviors that underlie drug addiction. The overall neurocircuitry domains correspond to the three functional domains: *binge/intoxication* (reward and incentive salience: basal ganglia [*blue*]), *withdrawal/negative affect* (negative emotional states and stress: extended amygdala and habenula [*red*]), and *preoccupation/anticipation* (craving, impulsivity, and executive function: prefrontal cortex, insula, and allocortex [*green*]). *Arrows* depict major circuit connections between the three domains, and numbers refer to neurochemical and neurocircuit-specific pathways that are known to support brain changes that contribute to the allostatic state of addiction. When these circuits are balanced, this results in proper inhibitory control and decision making and the normal functioning of reward, motivation, stress, and memory circuits. These circuits also interact with circuits that are involved in mood regulation, including stress reactivity (which involves the amygdala, hypothalamus, and habenula) and interoception (which involves the insula and anterior cingulate cortex and contributes to the awareness of negative emotional states). Drugs of abuse usurp executive function circuits, motivational circuits, and stress circuits via multiple neurotransmitter-specific neuroplasticity circuits. Key neurotransmitters that are implicated in these neuroadaptations include dopamine, enkephalins, glutamate, γ-aminobutyric acid, norepinephrine, CRF, dynorphin, neuropeptide Y, and endocannabinoids. *(Modified with permission from Koob GF, Volkow ND. Neurocircuitry of addiction. Neuropsychopharmacology Reviews 2010;35:217–38 [erratum: 35:1051] [38].)*

decreased alcohol responding in P rats [52,53]. The suppression of serotonergic function in raphe nuclei increased alcohol intake in Wistar rats [56]. The authors argued for a role for brain sites within the extended amygdala (central nucleus of the amygdala, bed nucleus of the stria terminalis, and a transition zone in the shell of the nucleus accumbens) in mediating the reinforcing effects of alcohol and implicated several neurotransmitter systems, including GABA, serotonin, CRF, and neuropeptide Y (NPY; see Volume 3,

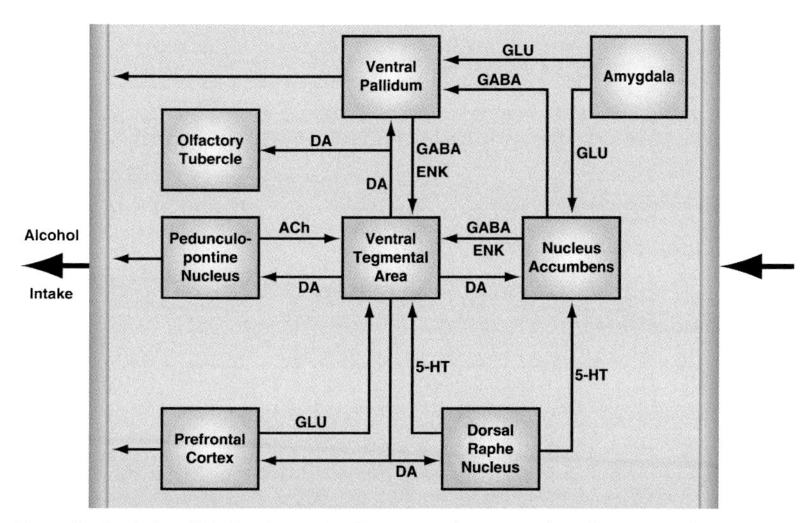

Figure 8 ***Hypothetical simplified schematic diagram of neuronal pathways and central nervous system regions that mediate alcohol drinking [44].*** The ventral tegmental area and its dopamine (DA) projections to the nucleus accumbens, ventral pallidum, prefrontal cortex, and olfactory tubercle play key roles in regulating alcohol intake. The dorsal raphe nucleus serotonin (5-HT) system serves to regulate the activity of the ventral tegmental area—nucleus accumbens dopamine pathway. Furthermore, the ventral tegmental area is regulated by γ-aminobutyric acid (GABA) and enkephalinergic (ENK) projections from the ventral pallidum and nucleus accumbens, a cholinergic (ACh) input from the pedunculopontine nucleus, and a glutamatergic (GLU) pathway from the prefrontal cortex. The ventral pallidum receives a major GABA input from the nucleus accumbens. Both of these limbic regions are regulated by glutamate projections from the amygdala. In addition to these interactions, inputs from other central nervous system regions influence the activity of these limbic structures in regulating alcohol intake. *(Taken with permission from McBride WJ, Li TK. Animal models of alcoholism: neurobiology of high alcohol-drinking behavior in rodents. Critical Reviews in Neurobiology 1998; 12:339–69 [44].)*

Alcohol). Finally, extensive neurochemical studies with P rats, Sardinian alcohol-preferring (sP) rats, and high alcohol drinking (HAD) rats (see *Animal Models of Addiction* chapter 2, Volume 1) showed inherent differences in the function of the mesolimbic dopamine system, endogenous opioid peptide systems, GABAergic systems, and serotonergic systems that were consistent with the neuropharmacological studies of alcohol reward (see Volume 3, *Alcohol*). Thus, there is a convergence of data from neuropharmacological studies and genetic animal models for a role for the circuit elements that are outlined in Fig. 8 in the reinforcing effects of alcohol [44].

Here, significant evidence was marshaled to support the hypothesis that a neurocircuit with focal points in the ventral tegmental area, nucleus accumbens, and extended amygdala mediates the acute reinforcing effects of alcohol. Neuropharmacological data from alcohol-preferring rats, combined with inherent neurochemical changes in alcohol-preferring rats, have identified four key neurochemical systems—dopamine, serotonin, GABA, and opioid peptides—in the acute reinforcing actions of alcohol. Changes in these systems at specific points in the neurocircuit may convey vulnerability to the excessive alcohol consumption that is associated with alcohol use disorder.

3.2 Differential role of the nucleus accumbens core and shell in addiction

Di Chiara G. Nucleus accumbens shell and core dopamine: differential role in behavior and addiction. Behavioural Brain Research 2002;137:75–114 [57].

Ito R, Robbins TW, Everitt BJ. Differential control over cocaine-seeking behavior by nucleus accumbens core and shell. Nature Neuroscience 2004;7:389–97 [58].

The nucleus accumbens has long been considered a key element of the locomotor and reinforcing actions of psychostimulants and other drugs of abuse [29,59] and the neuroadaptational changes that are associated with the development of addiction and reinstatement of drug-seeking behavior after extinction. However, the development of the concept of the extended amygdala, in which the shell or medial part of the nucleus accumbens forms a transition zone within the extended amygdala [34] and the strong resemblance of the core of the nucleus accumbens to the striatum and its intersection with ventral striatal-ventral pallidal loops [60,61], led to intense interest in the differential roles of the shell and core of the nucleus accumbens in addiction [57,58].

Di Chiara [57] argued that incentive-sensitization and allostatic-counteradaptive theories do not account for two basic properties of motivational disturbances that are associated with drug addiction: the narrowing of behavior toward drug incentives versus non-drug incentives and the irreversibility of such a focus on drug reinforcement. Under this conceptualization, the shell of the nucleus accumbens was hypothesized to mediate the strengthening of stimulus-drug associations. This hypothesis was based on the observation that most, if not all, drugs of abuse [62–65] share with non-drug rewards (e.g., Fonzies and chocolate; [66–70]) the ability to stimulate dopamine transmission in the shell of the nucleus accumbens (Table 1). Much less activation was observed in the core of the nucleus accumbens with both drug and non-drug rewards. Non-drug rewards also increase dopamine release in the prefrontal cortex. The activation of dopamine release shows habituation with repeated administration of non-drug (food) rewards in the nucleus accumbens shell [67–69] but not with repeated administration of non-drug (food) rewards in the nucleus accumbens core or prefrontal cortex. Indeed, the activation appears to show sensitization in these two structures [66–69]. Based on these results, Di Chiara hypothesized that the properties of dopamine responsiveness in the nucleus accumbens shell suggested a role for dopamine in associative stimulus-reward learning. The release of dopamine in the nucleus accumbens shell by unpredicted primary appetitive stimuli may serve to associate the discriminative properties of the rewarding stimulus with its

Table 1 Drug effects, non-drug effects, and lesion effects in the nucleus accumbens.

	Nucleus accumbens	
	Shell	**Core**
Acute drug-induced activation of dopamine		
Morphine	↑↑	↑
Heroin	↑↑↑	↑
Cocaine	↑↑↑	↑
Amphetamine	↑↑↑	↑↑
Δ^9-Tetrahydrocannabinol	↑↑	↑
Nicotine	↑↑	—
Non-drug-induced activation of dopamine		
Acute Fonzies	↑	↑
Chronic Fonzies	—	↑↑
Cell body-specific excitotoxic lesion effects		
Locomotor activity	↓	↑
Amphetamine locomotor activity	↓	↑
Amphetamine potentiation of conditioned response	↓	—
Discriminated approach to Pavlovian conditioned response	—	↓
Second-order schedule for drug reward	—	↓

biological outcome (e.g., the taste of food). In contrast, Di Chiara argued that the nucleus accumbens core and prefrontal cortex play a role in expression of the motivation to seek a reward and in converting motivation to action [71].

Di Chiara further argued that drugs of abuse share the effects on a non-drug reward in activating the shell of the nucleus accumbens, but this activation of the shell of the nucleus accumbens does not undergo habituation as it does with non-drug (food) rewards, imparting much more stimulus-reward learning and contributing to compulsive components of drug-seeking behavior [71]. Repeated drug exposure was also shown to induce sensitization of the drug-induced stimulation of dopamine neurotransmission in the core of the nucleus accumbens that was hypothesized to mediate instrumental performance that is involved in drug seeking [57,62].

Based on these observations, Di Chiara argued for a differential role for the dopamine system in addiction, depending on the stage of the addiction cycle [57,71]. In the stage of controlled use, the reinforcing properties of the drug facilitate further exposure to the drug via Pavlovian incentive learning, related to the release of dopamine in the shell. With repeated drug exposure, the subject enters the stage of drug abuse where repeated associations of drug reward and drug-related stimuli occur via the presence of the non-habituating stimulation of dopamine neurotransmission in the shell of the nucleus accumbens. The sensitization of dopamine neurotransmission in the core of the nucleus accumbens is hypothesized to begin in this stage. Subsequently, the tolerance and dependence stage develops, which is characterized by a negative emotional state, and the actions of the drug are amplified by potentiation of the dopamine-releasing properties of the drug (presumably superimposed on a lower baseline). In the post-addiction stage, abstinence and sensitization disappear, but Pavlovian associations remain powerful incentives for the reinstatement of drug self-administration (relapse).

A parallel series of studies on the differential role of the shell and core of the nucleus accumbens in drug seeking has involved selective lesions of the core and shell and sophisticated behavioral procedures for separating motivational components of drug reward, notably drug seeking and drug taking [58]. The self-administration of cocaine was initiated on a continuous reinforcement schedule in daily 2 h sessions. The animals were then trained on a second-order schedule of reinforcement, in which the response requirements for cocaine and a conditioned reinforcer progressively increased. Under this schedule, the rats were required to make *y* responses to obtain a single presentation of a 2 s light conditioned stimulus (conditioned reinforcer), and the completion of *x* of these responses resulted in the delivery of cocaine, illumination of the light conditioned stimulus for 20 s, retraction of both levers, and turning off the house light during a 20 s time-out period. Ultimately, the *y* was increased from 1 to 10, and the *x* was increased from 5 to 10, yielding a schedule of fixed-ratio 10 (FR10:S) in daily 2 h sessions. Lesions of the core of the nucleus accumbens profoundly impaired the acquisition of drug-seeking behavior that was maintained by drug associated conditioned reinforcers [58].

In contrast, selective excitotoxic lesions of the shell of the nucleus accumbens failed to alter the acquisition of drug-seeking behavior that was maintained by drug-associated conditioned reinforcers. Similar effects have been observed of selective nucleus accumbens core lesions on Pavlovian approach behavior [72], Pavlovian-to-instrumental transfer [73], and conditioned reinforcement [72]. Taken together with data that implicate the basolateral amygdala in similar effects on cocaine self-administration that is maintained on a second-order schedule [74], the nucleus accumbens core was hypothesized to be a component of reward-related information that derives from conditioned reinforcers [58] or essential for the influence of associative stimulus-reward information on goal-directed action.

Nucleus accumbens shell lesions, however, produced hypoactivity, attenuated the psychostimulant effects of amphetamine [72], and decreased the nucleus accumbens-administered amphetamine potentiation of conditioned reinforcement [72]. Similar effects on the amphetamine-induced potentiation of conditioned reinforcement have been observed with lesions of the central nucleus of the amygdala [75] and ventral subiculum [76]. These results suggested that the shell of the nucleus accumbens may potentiate ongoing instrumental responding in the presence of motivationally significant stimuli [72] or the psychostimulant actions of stimulant drugs.

Reconciling the microdialysis position [57] and the excitotoxic lesion position [58] may require the reassessment of the ways in which one defines the global term incentive motivation. For Di Chiara, incentive motivation is inferred by the attribution of motivational effects to changes in dopamine function in the shell of the nucleus accumbens that change with non-drug reward but show resistance to change with drug rewards. In contrast, Ito et al. [58] operationally defined the impact of conditioned reinforcement on cocaine-seeking behavior on a second-order schedule as being disrupted by lesions of the nucleus accumbens core but not shell. In contrast, using their models, the nucleus accumbens shell is more involved in the response-invigorating effects of psychomotor stimulant drugs. Overall, the results of lesion studies and microdialysis studies could be reconciled with a reversal of roles for the core and shell of the nucleus accumbens as proposed by Di Chiara [57]. Indeed, the observation that the dopamine response in the core of the nucleus accumbens sensitizes with repeated exposure to drugs of abuse could provide a neurochemical basis for the stimulus-reward association facilitation that was previously attributed to the shell of the nucleus accumbens.

In summary, based on microdialysis data, Di Chiara proposed that the shell of the nucleus accumbens is involved in the incentive motivational properties of drugs of abuse via the enhancement of stimulus-reward associations, whereas the core of the nucleus accumbens is more involved in the performance components of compulsive drug taking. Based on lesion data and a sophisticated behavioral paradigm for exploring drug seeking versus drug taking. Ito et al. [58] proposed that the shell of the nucleus accumbens is more likely to be involved in the activational (psychostimulant) component of drug effects and that the core of the nucleus accumbens is critical for imparting conditioned reinforcing properties to previously neutral stimuli. The reconciliation of these two positions may be found in the differential roles of the inputs to the shell and core subregions of the nucleus accumbens in drug-seeking and drug-taking behavior. Much data support the hypothesis that the core of the nucleus accumbens has connections more with cortical-striatal-pallidal-thalamic loops and translation to habits via the dorsal striatum and the shell of the nucleus more connected to the extended amygdala ([61,77]; see Section 3.6. Drug addiction: maladaptive incentive habits).

3.3 Narrowing of neuronal activity by changes in signal and background firing of nucleus accumbens neurons during cocaine self-administration

Peoples LL, Cavanaugh D. Differential changes in signal and background firing of accumbal neurons during cocaine self-administration. Journal of Neurophysiology 2003;90:993–1010 [78].

Peoples LL, Lynch KG, Lesnock J, Gangadhar N. Accumbal neural responses during the initiation and maintenance of intravenous cocaine self-administration. Journal of Neurophysiology 2004; 91: 314–23 [79].

Extracellular recordings from the nucleus accumbens during cocaine self-administration in awake, freely moving animals revealed a population of nucleus accumbens neurons that exhibit phasic excitatory responses that are time-locked to drug-related events, specifically responding for the drug and cues that predict drug delivery [78–84]. These phasic changes were linked to drug reward-related events. During the course of a self-administration session, phasic nucleus accumbens neurons typically decrease their firing rate, as do the majority of nucleus accumbens neurons [85,86]. However, phasic firing that is associated with drug seeking appears to be less sensitive to the inhibitory effects of cocaine on firing. During intravenous cocaine self-administration, neurons that exhibited an excitatory phasic response to cocaine did not diminish their firing rate during the self-administration session but presented large decreases in background firing compared with pre-drug baselines [78]. Prior to the recording session, rats were trained to self-administer cocaine daily in 6 h or 80 injections of cocaine for 12—17 days. Recording sessions began with a 20 min, non-drug baseline recording session and then a 6 h session. Lever-press firing rates were defined as a significant increase or decrease in average firing rate within ± 3 s of the lever press, relative to firing during the -12 s to -9 s prepress period. The background period was defined as -12 s to -4 s prepress for all neurons. Changes in signal and background firing were evaluated relative to two periods: the 30 s period that preceded the onset of the self-administration session and the 30 s period before the first press. The results showed that in neurons that exhibited an excitatory response (phasic increase in firing rate around the time of the lever press relative to background, termed "lever press neurons"), there was an increase in response rate to the cocaine infusion (relative to the 30 s pre-session), but a time-related decline in firing was observed in the background periods (relative to the 30 s pre-session; [78]; Fig. 9).

One interpretation of these results by the authors is focused on dopamine release in the nucleus accumbens by cocaine, amplifying mechanisms that normally contribute to reward-related learning, and this abnormal strengthening in turn contributes to compulsive drug seeking. However, there was no evidence of a change in the absolute magnitude of excitatory reward-related nucleus accumbens neuron firing; rather, there was a differential inhibition of signal and background firing, such that there was a net enhancement of drug reward-related signals relative to background firing. Note that with extended access to cocaine, the phasic release of dopamine as measured by fast scan cyclic voltammetry, declined in the ventral medial striatum, whereas the phasic release of dopamine in the dorsal striatum emerged progressively [87].

Dopamine inputs to the nucleus accumbens have been proposed to modulate feed-forward inhibition and as a result contribute to the selective activation of neural ensembles that are relevant to a particular behavioral situation [88]. The differential inhibition of drug reward-related firing and background firing may reflect the differential modulation of ensemble activity in the form of filtering, in which only certain neurons and ensembles are activated. This filtering, in turn, could contribute to learning by narrowing the ensembles of neurons in the nucleus accumbens to those that mediate the strengthening of particular associations.

An alternative interpretation is that during drug self-administration sessions, dopamine-mediated drug effects may weaken synaptic connections and neural responses that are involved in the

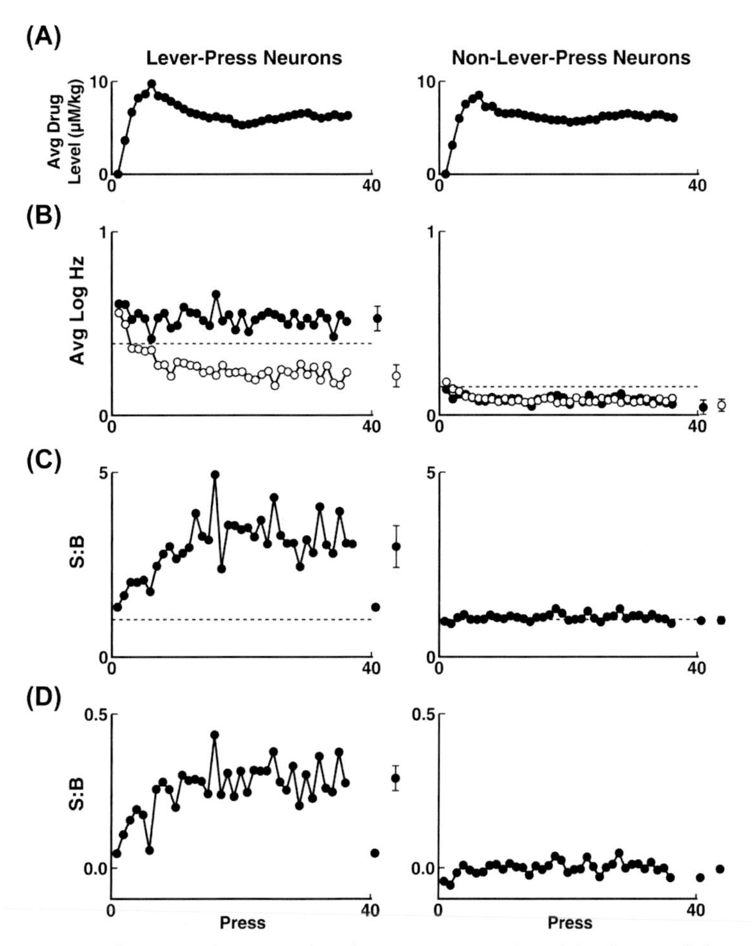

Figure 9 ***Narrowing of neuronal activity by changes in signal and background firing of nucleus accumbens neurons during cocaine self-administration [78].*** Changes in signal and background for tonically inhibited neurons in the nucleus accumbens during extracellular recordings of intravenous cocaine self-administration sessions. Rats were trained to intravenously self-administer cocaine in 6 h (or 80 infusion) sessions. A recording session started with a 20 min non-drug baseline recording period, followed by the typical daily self-administration session and ended by a 40 min non-drug recovery period. Left and right: tonically inhibited lever-press neurons (n = 14) and non-lever-press neurons (n = 31), respectively. (A) Average drug level at the time of each press is plotted as a function of press number. (B) Average log Hz during the signal (*closed circles*) and background (*open circles*) periods is plotted as a function of lever-press number. *Dashed line*, average firing rate during the 30 s pre-S^D. To the right of the plot, standard error bars are shown for the median average log Hz during presses 11–36 for the signal (*closed circles*, left) and background (*open circles*, right) periods. (C) Average signal-to-background ratio (S:B) is plotted as a function of lever-press number. *Dotted line*, a signal to background ratio of 1. To the right of the plot, standard error bars are shown for the first press (*light gray bars*, left) and the press at which the median ratio during presses 11–36 was observed (*black bars*, right). (D) Average difference between signal and background (S-B) plotted as a function of lever-press number. To the right of the plot, standard error bars are shown for the first press (*light gray bars*, left) and the press at which the median S-B difference during presses 11–36 was observed (*black bars*, right). *(Taken with permission from Peoples LL, Cavanaugh D. Differential changes in signal and background firing of accumbal neurons during cocaine self-administration. Journal of Neurophysiology 2003;90:993–1010 [78].)*

transmission of signals that are unrelated to reward and a highly selective sparing of neural responses that are related to drug reward. This, in turn, could lead to a progressive decline in the throughput of non-drug reward-related signals in circuits that involve the nucleus accumbens, such as cortico-striato-pallido-thalamo-cortical loops. Such a narrowing of information flow could also potentially contribute to progressive narrowing of the behavioral repertoire, lower reward function, dysphoria, and cognitive deficits that are associated with the development of drug addiction [89–92].

In summary, Peoples and colleagues argued that narrowing the behavioral repertoire to drug-related rewards rather than natural rewards during the development of addiction is dependent on phasic response neurons in the nucleus accumbens that differentially respond to repeated cocaine compared with nonphasic neurons, with a stable response to cocaine that is superimposed on a time-related decline in background firing.

3.4 Synaptic plasticity in the mesolimbic dopamine system and drugs of abuse

Hyman SE, Malenka RC. Addiction and the brain: the neurobiology of compulsion and its persistence. Nature Reviews Neuroscience 2001;2:695–703 [93].

Thomas MJ, Malenka RC. Synaptic plasticity in the mesolimbic dopamine system. Transactions of the Royal Society of London B Biological Sciences 2003;358:815–9 [94].

The basic underlying thesis of this conceptual framework is that the compulsive characteristics of drug abuse and its persistence even after abstinence are based on the pathological disruption of molecular and physiological mechanisms that are involved in memory [94]. Several premises guide this conceptual framework. First, the mesolimbic dopamine system that originates in the ventral tegmental area and terminates in the nucleus accumbens is viewed as the major, though not exclusive, substrate for reward and the reinforcement of both drugs of abuse and natural rewards. Second, repeated exposure to drugs is argued to increase their rewarding and locomotor-activating properties [95]. This "sensitization" is hypothesized to be mediated by changes in the mesolimbic dopamine system, with the induction of sensitization that depends on functional changes in the ventral tegmental area [96], whereas the long-term maintenance of sensitization involved adaptations in the nucleus accumbens [97,98]. Third, synaptic plasticity was demonstrated in the mesolimbic dopamine system. Long-term potentiation (LTP) and long-term depression (LTD) can be elicited at excitatory synapses on medium spiny neurons in the nucleus accumbens [99], and excitatory synapses on dopamine neurons in the ventral tegmental area also undergo both LTP and LTD [100].

The similarities between LTP and LTD that are observed in the nucleus accumbens and those that are observed in the CA1 region of the hippocampus led to the hypothesis that associative synapse-specific plasticity can occur in the mesolimbic dopamine system and link this "reward" system to a role in the development of addiction ([93]; Fig. 10). The high-frequency titanic stimulation of presynaptic fibers induced LTP, but low-

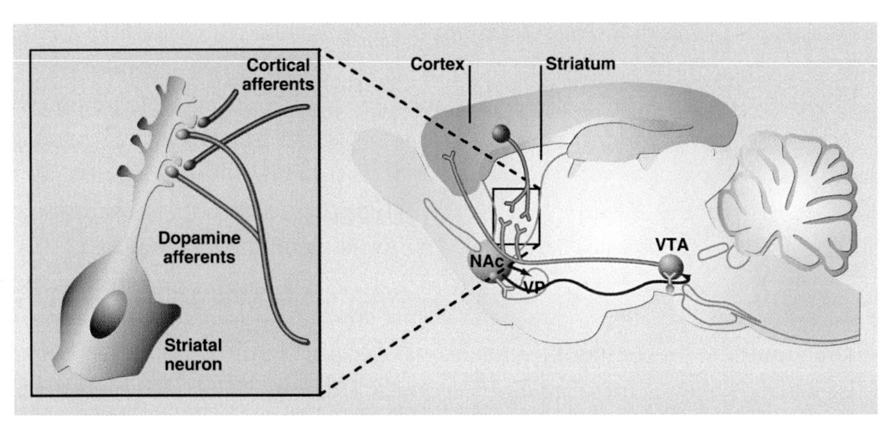

Figure 10 ***Dopamine-glutamate interactions in the striatum that are hypothesized to be involved in the reorganization of neurocircuitry of drug dependence [93].*** Approximately 95% of neurons in the dorsal striatum and nucleus accumbens (NAc) are medium-sized spiny projection neurons, which use γ-aminobutyric acid (GABA) as their main neurotransmitter. These neurons receive glutamatergic projections from the cerebral cortex, which form well-defined synapses on the heads of dendritic spines. Dopaminergic axons from the midbrain pass by the necks of spines, where they release neurotransmitter; however, dopamine receptors are widely distributed on the cell membrane, including the soma. Changes in synaptic strength may result from a change in neurotransmitter release, neurotransmitter receptors, or receptor-mediated signaling. Alternatively, changes in the intrinsic excitability of neurons might follow changes in the properties or numbers of voltage-dependent ion channels. A third possibility is that morphological changes, such as the generation of new synaptic connections or pruning away of existing ones, might be initiated by various forms of synaptic plasticity. VP, ventral pallidum; VTA, ventral tegmental area; NAc, nucleus accumbens. *(Taken with permission from Hyman SE, Malenka RC. Addiction and the brain: the neurobiology of compulsion and its persistence. Nature Reviews Neuroscience 2001;2:695–703 [93].)*

frequency stimulation during modest depolarization of the postsynaptic cell induced LTD. Both LTP and LTD in the nucleus accumbens require the activation of *N*-methyl-D-aspartate (NMDA) glutamate receptors, and the excitatory input appears to derive from prelimbic cortical afferents [99,101,102]. Long-term potentiation in the ventral tegmental area is also NMDA receptor-dependent [100]. Long-term depression is generated in the ventral tegmental area by the activation of voltage-gated calcium channels and does not require the activation of dopamine receptors [100,103]. However, LTD is blocked by dopamine or amphetamine that acts through dopamine D_2 receptors [103].

The demonstration of synaptic plasticity at excitatory synapses in the source and terminal areas of the mesolimbic dopamine system was argued to be evidence of a role for synaptic plasticity in addiction. Supporting this hypothesis, a single *in vivo* administration of cocaine markedly potentiated synaptic strength in the ventral tegmental area in midbrain slices using whole-cell recording techniques, which was attributable to the upregulation of α-amino-3-hydroxy-5-methyl-4-isoxale propionic acid (AMPA) glutamate receptors. Similar to locomotor sensitization to cocaine, this potentiation lasted up to 5 days post-cocaine [104]. This potentiation did not occur in ventral tegmental area GABAergic neurons or hippocampal CA1 neurons. The resemblance to sensitization was further argued by the observation that, like sensitization, cocaine-induced synaptic potentiation was blocked by the co-administration of cocaine with an NMDA receptor antagonist. Similar synaptic potentiation was observed with many different drugs of abuse, including amphetamine, morphine, alcohol, and nicotine [105]. Acute stressor exposure also caused a robust increase in excitatory synaptic strength in the ventral tegmental area [105].

In contrast, pre-exposure to cocaine *in vivo* decreased synaptic strength in neurons from prefrontal cortical afferents in slices of the nucleus accumbens shell but not the core. The amplitude of excitatory postsynaptic potentials and LTD were decreased at these excitatory synapses from prelimbic cortical afferents [106]. The authors argued that the development of an LTD-like process in the nucleus accumbens contributes to the reorganization of neural circuitry that underlies behavioral sensitization to cocaine and thus may be important for the development of addiction. These results suggested to the authors that increasing synaptic drive onto mesolimbic dopamine cells enhances the motivational significance of the drugs themselves, as well as stimuli that are closely associated with drug seeking. They further argued that the results that showed LTP and LTD and changes in synaptic strength with drug pre-exposure demonstrated that persistent drug-induced behavioral changes, such as sensitization, likely occur because of the ability of drugs to elicit long-lasting changes in synaptic weights in crucial brain circuits, notably the mesolimbic dopamine circuit [93,94].

Here, the development of drug addiction is conceptualized as being reflected by plasticity in both the ventral tegmental area and nucleus accumbens, demonstrated by drug-induced LTP and LTD with increases in synaptic strength of the mesolimbic dopamine system and decreases in synaptic strength of prefrontal-nucleus accumbens synapses that parallel the development of locomotor sensitization. These seminal studies laid the foundation for much of the work that followed that showed that dynamic changes occur in the nucleus accumbens glutamatergic system that persist long after abstinence from engagement with drugs of abuse. These range from the expression of AMPA receptors (see Section 5.11. Neurobiology of cue-induced incubation ["craving"]: synaptic mechanism hypothesis) to changes in the glutamate transporter molecular mechanisms (see Section 5.8. Glutamate systems in cocaine addiction).

3.5 Choice and the delayed reward hypothesis

Ahmed SH. Individual decision-making in the causal pathway to addiction: contributions and limitations of rodent models. Pharmacology Biochemistry and Behavior 2018;164:22–31 [107].

The hypothesis here is that two major types of decisions are made by an individual when seeking and taking a drug of abuse. There is the initial decision to use a drug for the first time, and this decision is influenced by prior knowledge about the drug and the effects that are expected and by prior self-knowledge of one's own vulnerability, if any. After an individual has used a drug for the first time, there is the decision to repeat drug use. This decision is influenced by the way in which the drug changes the brain after the first administration both acutely and also by neuroadaptive responses in a way that could bias subsequent decision-making in favor of repeated drug use. Animal models of drug seeking, notably studies of psychostimulant self-administration, have provided unique insights into these decisions. In rodents, prior psychostimulant use can increase impulsive, risky, and/or potentially harmful decision-making, yet when rodents have the choice between a drug and a competing, nondrug option, such impulsivity does not translate into more drug use. The delayed drug reward hypothesis was formulated to explain this apparent discrepancy.

The argument is that the extrapolation of rodent research to humans should consider inherent species-specific differences in individual decision-making. In humans, the intravenous route of administration is associated with a relative increase in the risk of developing addiction within a drug class [108]. The explanation is that the intravenous route allows the drug to reach and act rapidly on brain reward circuits, without undergoing metabolism or clearance. However, the argument in the delayed response hypothesis is that the delay to the onset and delay to the peak of drug reward are not instantaneous; indeed, such delays appear to be sufficiently long to produce a large time-discounting of the value of the intravenous drug reward at the time of choice. A "high sensitivity to delay" in rodents is hypothesized because they have, as a species, relatively high sensitivity to delay coupled with a limited ability to foresee the future [109], attributed to the fact that rats live in a different framework of time scale as a function of their relatively short lifespan [110]. Thus, rodents typically prefer a more immediate nondrug reward when available as a choice, as has been amply demonstrated [111]. Here, although an intravenous drug infusion induces high rewarding stimulation via a direct action on brain reward circuits, its relative value at the time of choice would be largely offset by its associated long delay (i.e., delay-discounting; [112]). The argument is that this hypothesis not only helps reconcile drug choice data in rodents with many neurobiologically inspired theories of drug addiction—it also sheds new light on some species-specific differences in drug choices.

This is an intriguing hypothesis, emphasizing the fact that rats are not humans, and such considerations may contribute to current concerns about a translational disconnect. It also supports the hypothesis that drug addiction forms a spectrum disorder, and we would argue that as a subject moves through the addiction cycle, the sources of reinforcement change, and the positive hedonic effects of the drug give way to the negative-reinforcing properties of drug withdrawal. As a result, and as published by Ahmed and colleagues, increasing dependence on opioids can move choice from nondrug to drug [113]. One other thought is that the delayed response hypothesis may not only be mediated by the delay in intravenous drug that reaches the brain but also by the rapid onset of an opponent process that follows cocaine self-administration [114].

3.6 Drug addiction: maladaptive incentive habits

Belin D, Belin-Rauscent A, Murray JE, Everitt BJ. Addiction: failure of control over maladaptive incentive habits. Current Opinion in Neurobiology 2013;23:564–72 [115].

These authors hypothesized that incentive habits result from a pathological coupling of drug-influenced motivational states and a rigid stimulus response habit system; as a result, drug-associated stimuli, through automatic processes, elicit and maintain drug seeking [115]. Thus, the authors argue that drug addiction does not only involve taking drugs; individuals with addiction spend much of their time foraging for the drug over long periods of time [116]. The compulsive feature of addiction is expressed during these long periods of drug-seeking behavior, often supported and eventually controlled by drug-associated conditioned stimuli (CSs) that act as conditioned reinforcers (Fig. 11). Goal-directed actions (A-O) and habits (S-R) are operationally defined as sensitive or resistant to reinforcer devaluation or contingency degradation that is tested under extinction conditions, respectively [117].

To address this issue in the studies by Belin and colleagues, a two-linked chain schedule of reinforcement was used, in which rats responded on a first lever (the seeking lever) under a random interval (RI) schedule of reinforcement. The first seeking lever press that occurred after the RI elapsed resulted in retraction of the seeking lever and insertion of the taking lever, a response at which (under a fixed-ratio 1 [FR1] schedule) yielded a drug infusion. Early performance (10–15 days) of cocaine-seeking behavior was sensitive to extinction of the taking lever, a manipulation that resulted in devaluation of the outcome of the seeking response—a hallmark of goal-directed behavior. It then became insensitive to such manipulation after extended training (40 daily sessions).

The authors further argued that drug-seeking and -taking habits in individuals with addiction to drugs are not dissociated from motivational processes, in which they are also reinforced by Pavlovian-conditioned, drug-associated CSs in the environment that act as conditioned reinforcers, supporting protracted sequences of behavior, often in the absence of the outcome. Here, cue-controlled cocaine seeking is operationalized in second-order schedules of reinforcement under which drug-seeking responses by rats during prolonged drug-free periods (classically FI15 min schedules) are reinforced by contingent presentations of drug-associated CSs (upon every 10th lever press). Drug-seeking performance depends as much on the presentation of these CSs as on the drug itself, thereby suggesting that the former indeed facilitates the development of aberrant "incentive habits."

The authors argue that goal-directed actions and habits depend on different neural networks ([115,118]; Fig. 11). A distributed corticostriatal network, including the prelimbic cortex [119], basolateral amygdala [119,120], core of the nucleus accumbens, and particularly the dorsomedial striatum [118,121], is hypothesized to be critical for goal-directed behavior toward natural rewards. In contrast, the dorsolateral striatum [122] and its dopaminergic innervation from the substantia nigra pars compacta [123] are hypothesized to mediate the control of stimulus-response habits over behavior, and the dorsolateral striatum is functionally connected with the central nucleus of the amygdala [124] and infralimbic cortex [125]. It can be hypothesized, therefore, that there may be an apparent shift from a prelimbic cortex—basolateral amygdala—nucleus accumbens core—dorsomedial striatum network to an infralimbic cortex—central nucleus of the amygdala—substantia nigra pars compacta—dorsolateral striatum network in the transition from goal-directed actions and habits in controlling drug-seeking behavior (Figs. 11 and 12). Thus, incentive habits may depend on an interaction between the basolateral amygdala and nucleus accumbens core, together with the progressive development of ventrolateral-to-dorsolateral striatum functional coupling through the recruitment of striato-nigro-striatal dopamine-dependent loop circuitry. Additionally, both ventral striatal and central nucleus pathways from the amygdala may be required for the recruitment of dorsolateral striatum-dependent control over incentive habits.

3.7 Dopamine prediction errors in reward learning and addiction: from theory to neural circuitry

Keiflin R, Janak PH. Dopamine prediction errors in reward learning and addiction: from theory to neural circuitry. Neuron 2015;88:247—63 [126].

Dopamine function is enhanced by drugs of abuse that act directly or indirectly on midbrain dopamine neurons to transiently increase extracellular dopamine concentrations. One prominent hypothesis of dopamine function is the signaling of reward prediction errors (RPEs) by dopamine neurons [126]. Early on, Schultz and colleagues reported that dopamine neurons in the ventral tegmental area and substantia nigra pars compacta respond to natural rewards, such as palatable food. They reported that the sign (positive or negative) and magnitude of the dopamine neuron response to reward are modulated by the degree to which the reward is expected [23,24]. Early in learning, when cue-reward associations are weak, dopamine neurons were shown to respond strongly to the presentation of reward and less strongly to reward-predictive cues. As learning progressed, neural responses to the cue became more pronounced, and reward responses diminished [23]. The gradual transfer of neural activation from reward delivery to cue onset during associative learning is the hallmark of the dopamine-RPE theory. Thus, dopamine neurons encode the discrepancy between reward predictions and information about the actual reward received (i.e., the RPE) and carry this signal to downstream brain regions that are involved in reward learning. The authors argued that the dopamine-RPE theory has important implications for addiction for one prominent aspect of dopamine function, namely the signaling of RPEs by dopamine neurons.

The neurocircuitry that is hypothesized to mediate dopamine-RPEs is hypothesized to be embedded in a circuit that allows extraction of the error signal from neural elements that encode the expected reward and actual reward. Such an RPE-generating circuit includes the connectivity of midbrain dopamine neurons from the ventral tegmental area and substantia nigra, as well as inputs from the prefrontal cortex and brainstem with a complex local circuitry in the basal ganglia that involves GABAergic and glutamatergic neurons (Figs. 13 and 14).

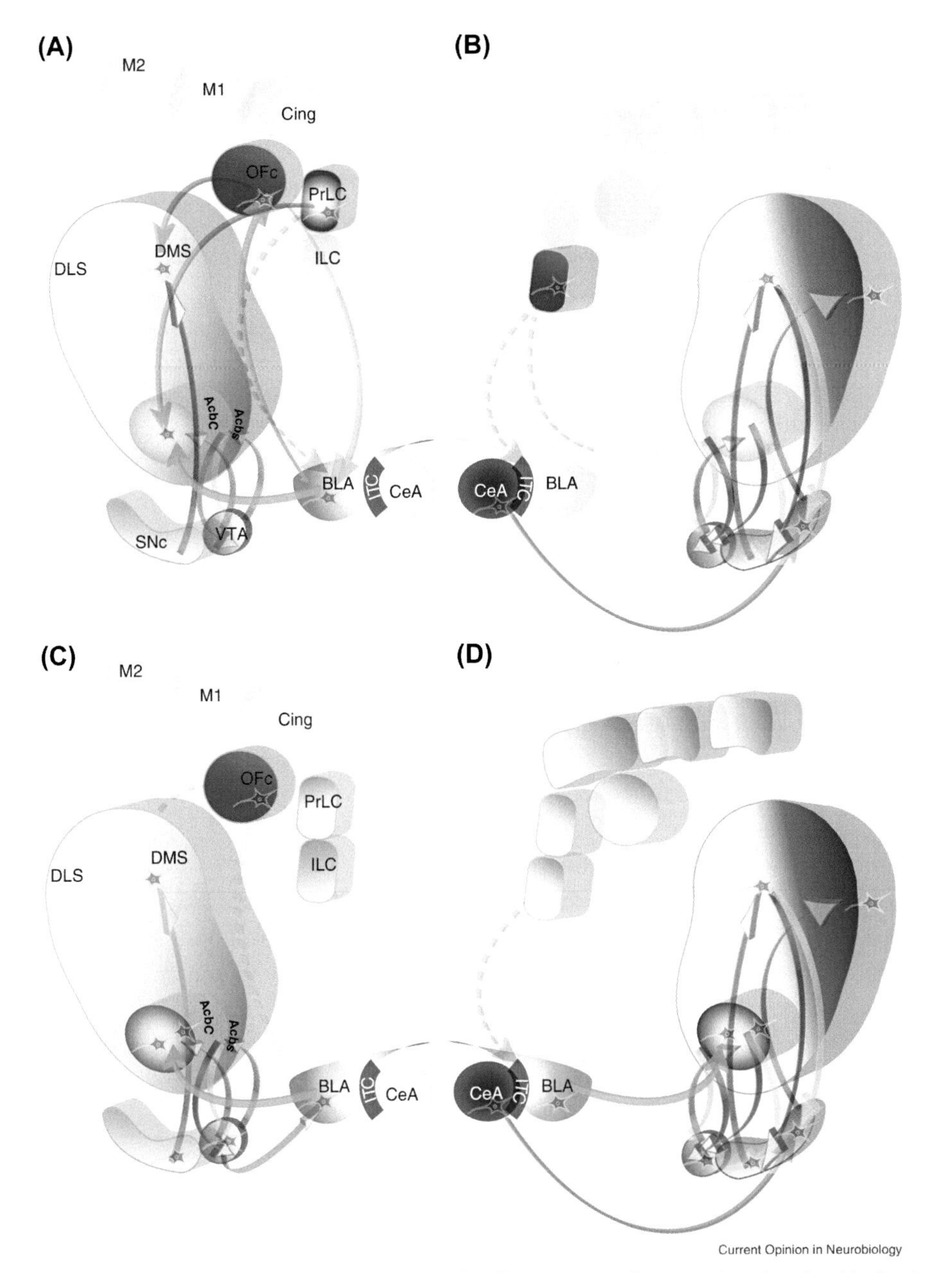

Figure 11 From actions to incentive habits: models of corticostriatal interactions involved in food and drug seeking processes [115]: (A) Corticostriatal interactions involved in goal-directed actions for natural rewards. Goal-directed actions reinforced by natural rewards depend on corticostriatal networks, including the orbitofrontal cortex and dorsomedial striatum (which receives projections from dopaminergic neurons in the medial substantia nigra pars compacta [SNc]), as well as the basolateral amygdala (BLA) and core of the nucleus accumbens (AcbC), both of which may be under the influence of

By applying the dopamine-RPE theory to addiction, the authors argued that the independent, generally dramatic surge of dopamine that is induced by drugs of abuse could contribute to drug taking and perhaps addiction. If drugs of abuse mimic a dopamine-RPE every time the drug is consumed [127], then the repetition of these dopamine signals over repeated drug use would continue to reinforce drug-related cues and actions to pathological levels, thus biasing future decision-making toward drug seeking and taking. This would be in contrast to natural rewards that produce error-correcting dopamine-RPE signals only until the predictions match the actual events. The authors further argued that the presence of a persistent dopamine-RPE could explain several aspects of addiction, such as the progressive allocation of time and effort to seek and take drugs of abuse, often at the expense of natural rewards and despite negative consequences.

A behavioral phenomenon, termed behavioral sensitization (i.e., a progressive increase in a behavioral response to a drug of abuse), usually in the domain of the psychomotor effects of psychostimulant drugs, has often been interpreted as the sensitization of motivational or incentive properties of rewarding stimuli [128]. Others argue that this

the prelimbic cortex (PrLC; [119]). Functional connectivity between the BLA and AcbC and DMS may be one mechanism that provides the link between motivation, conditioned incentives, and goal-directed action. (B) Corticostriatal interactions involved in habitual responding for natural rewards. Stimulus-response (S-R) control over behavior depends instead on the dorsolateral striatum (DLS; [121]), which receives projections from the primary motor cortex (M1) together with its functional connection with the central nucleus of the amygdala (CeN) via dopaminergic projections from the SNc and excitatory influence from the infralimbic cortex (ILC; [125]). Thus, as instrumental behavior comes progressively to emerge as a stimulus-response habit, we speculate that ILC influences are strengthened at the expense of the PrLC input. (C) Corticostriatal interactions subserving goal-directed drug seeking behavior. Drug seeking responses, early on in training, are goal-directed [481–483] and depend on a similar network to that described for natural rewards (A) that includes the OFC, AcbC, and its functional interactions with the BLA [484,485], and dopamine transmission in the DMS [481], especially its posterior part in the case of stimulants [486]. (D) Corticostriatal interactions involved in incentive habits. After protracted exposure to drugs, neurobiological adaptations that were initially restricted to the ventral striatum and ventromedial prefrontal cortex spread to the more dorsolateral areas of the corticostriatal circuitry [487,488] as illustrated by the *blue* colored-structures. Well-established cue-controlled drug seeking habits, or incentive habits, thus depend on recruitment of the DLS [87,483,485,487–491], potentially through the dopamine-dependent striato-nigro-striatal ascending spiraling circuitry that functionally links the ventral striatum and DLS [490]. This aberrant functional coupling between the AcbC and DLS relays to the DLS Pavlovian associative influences from the BLA that projects to the AcbC, thereby resulting in incentive habits. However, recent evidence also suggests that the CeN–SNc–DLS pathway, potentially driving general arousal and mediating the influence of drug-recruited stress processes on instrumental performance, is also necessary for the expression of incentive habits [492]. The role or drug-induced alteration of the function of each corticostriatal structure is represented as a scale of blue intensity. *Dotted arrows* represent putative functional relationships involved in actions and habits for natural rewards or drugs. ITC, intercalated cells. *(Taken with permission from Belin D, Belin-Rauscent A, Murray JE, Everitt BJ. Addiction: failure of control over maladaptive incentive habits. Current Opinion in Neurobiology 2013;23: 564–72 [115].)*

Addiction: loss of control over actions

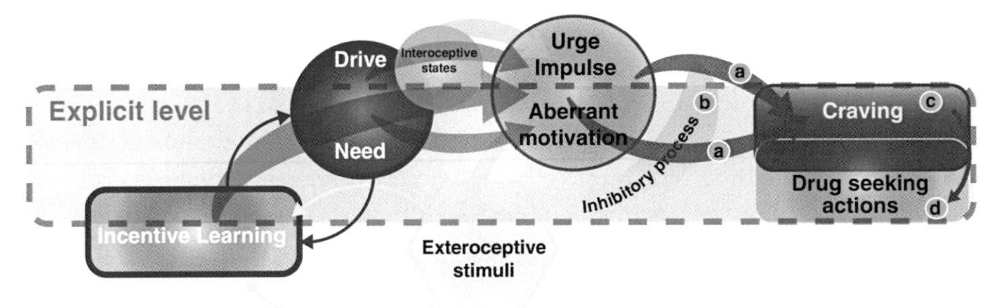

Addiction: loss of control over incentive habits

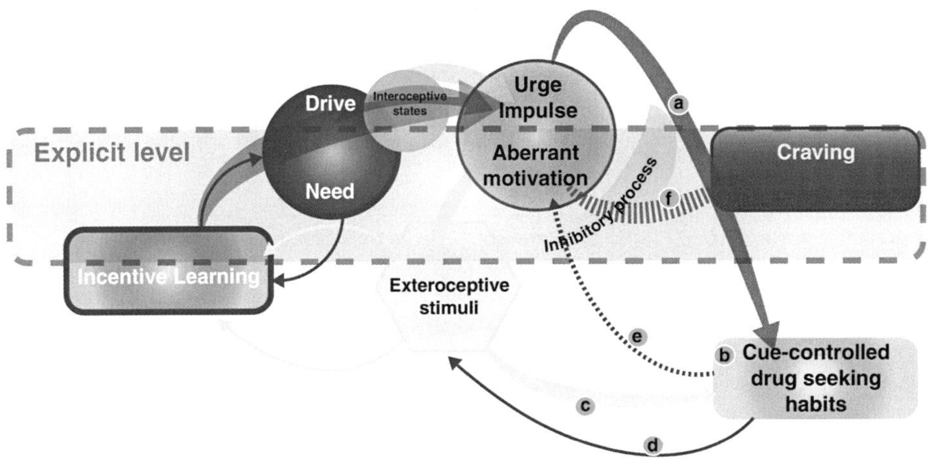

Figure 12 ***Addictions: loss of control over actions or habits? [115].*** Drugs hijack incentive learning processes, thereby resulting in the aberrant incentive control of explicit or implicit impulses, often termed aberrant motivation. Incentive learning also interacts with drug-induced interoceptive states that contribute in both the explicit and implicit domains to the genesis of drug-related urges and associated motivation. If addiction is considered to be a loss of control over actions (top panel), then alterations of the interoceptive state are interpreted as an impulsive urge-to-use, with intense motivation that leads to drug craving that is consciously accessible (A). At this level of analysis, executive inhibitory processes that were formerly able to attenuate the impact of urge-to-use on the experience of craving (B) are impaired. The convergence of impulses and motivation that are driven by alterations of interoceptive states as a result of incentive learning into an explicit state of craving, considered a compulsive, incentive motivational response (C), may lead the addicted individual to engage in drug-seeking behavior (D) that is under the control of goal-directed schemata. If addiction is considered to reflect the loss of control over incentive habits (bottom panel), then the impulsive urge-to-use and aberrant motivation directly initiate drug-seeking behavior (A) implicitly, bypassing a likely compromised executive inhibitory control system that is thus "blind" to this impulse-action coupling (B). Habitual drug seeking responses are not only triggered by these non-declarative processes that are elicited by exposure to drug-associated exteroceptive stimuli and interoceptive states; they are also reinforced, while engaged, by these stimuli (C), thus strengthening their conditioned reinforcing properties (D). If drugs are not accessible, then the drug-seeking behavior may be assigned post hoc an explanation of greater motivation (E) that cannot be filtered by the impairment in inhibitory control processes and is thus interpreted consciously as craving (F). *(Taken with permission from Belin D, Belin-Rauscent A, Murray JE, Everitt BJ. Addiction: failure of control over maladaptive incentive habits. Current Opinion in Neurobiology 2013;23:564–72 [115].)*

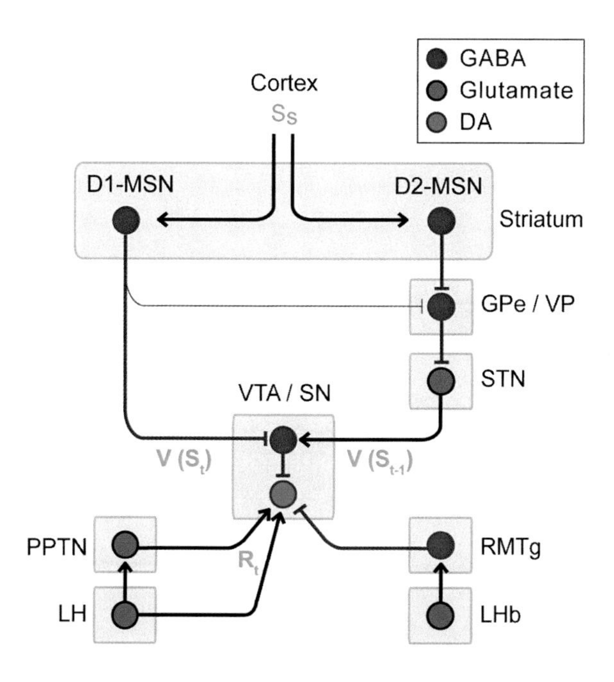

Figure 13 ***Simplified neural circuit diagram for computation of the reward prediction error (RPE) by dopamine (DA) neurons [126].*** The coordinated activation of this circuit by rewards and their predictors could result in DA neuron firing in accordance with the RPE model. GPe, external globus pallidus; LHb, lateral habenula; LH, lateral hypothalamus; MSN, medium spiny neurons; PPTN, pedunculopontine tegmental nucleus; RMTg, rostromedial tegmental nucleus; SN, substantia nigra; STN, subthalamic nucleus; VP, ventral pallidum; VTA, ventral tegmental area. *(Taken with permission from Keiflin R, Janak PH. Dopamine prediction errors in reward learning and addiction: from theory to neural circuitry. Neuron 2015;88:247–63 [126].)*

progressive increase in behavior could also be interpreted as the sensitization of reinforcing properties (i.e., ability to act as a teaching signal) of rewarding drugs and their associated cues [127,129]. Here, drugs of abuse not only mimic RPEs via their acute effect on dopamine transmission; they also potentiate future drug- and cue-induced RPEs via the long-lasting sensitization of mesolimbic dopamine circuits that produces an accelerated progression of the value of drug-related cues [126].

The role of dopamine-RPEs in learning must include neuroanatomical regional changes in the progression of dopamine transients and perhaps dopamine-RPEs over the course of training. Here, reward-evoked dopamine-RPEs that are initially restricted to the nucleus accumbens contribute to greater value of the cue that consistently precedes reward delivery. However, as the value of the cue increases, its presentation activates direct pathway medium spiny neurons in the nucleus accumbens that act to disinhibit

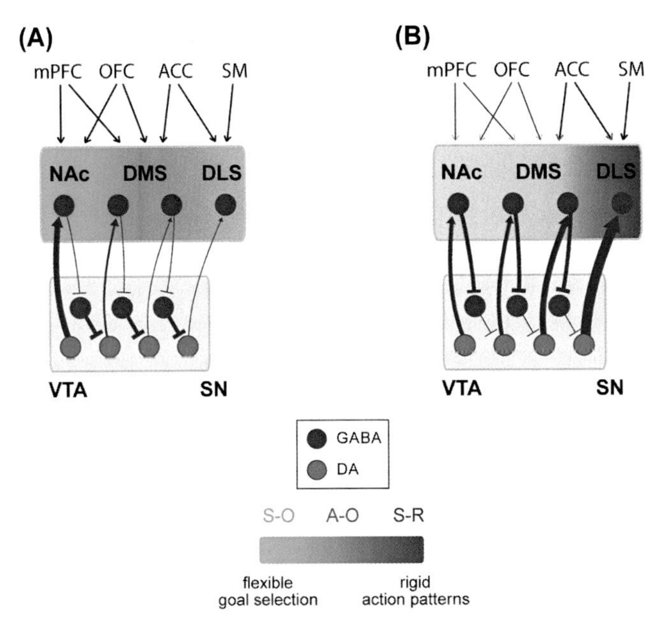

Figure 14 ***Proposed mechanism for the accelerated propagation of RPE-dopamine (DA) signals in different striatal domains following drug exposure and its consequence for learning [126].*** (A) In drug-naive animals, the activity of midbrain DA neurons is tightly regulated by local inhibitory GABA neurons. Unexpected reward activates dopamine neurons in the VTA; the resulting DA release in the NAc promotes Pavlovian (S-O) learning. (B) Repeated exposure to cocaine increases the excitability of DA neurons by potentiating striatal inhibitory inputs on midbrain local GABA neurons. Striatal feedback on DA neurons progressively recruits more lateral DA neurons from the VTA to SN to encode the RPE. The resulting emergence of DA signals in the sensorimotor DLS reinforces S-R associations that contribute to rigid and possibly compulsive drug-seeking behavior. ACC, anterior cingulated cortex; DLS, dorsolateral striatum; DMS, dorsomedial striatum; NAc, nucleus accumbens; OFC, orbitofrontal cortex; SM, sensorimotor cortex; SN, substantia nigra; VTA, ventral tegmental area; S-O, stimulus-outcome; A-O, action-outcome; S-R, stimulus-response. *(Taken with permission from Keiflin R, Janak PH. Dopamine prediction errors in reward learning and addiction: from theory to neural circuitry. Neuron 2015;88:247–63 [126].)*

ventral but also more lateral dopamine neurons in the midbrain, activating cue-evoked dopamine-RPEs not only in the nucleus accumbens but also in the dorsolateral striatum. Thus, the architecture of the dopamine-RPE circuit involves the regional propagation of dopamine-RPEs from the more ventromedial to more dorsal and lateral striatal domains and may contribute to the shift in goal-directed responding to habitual responding that is outlined by other theoretical frameworks that have been proposed by other authors (e.g., see Everitt and Belin theories in this chapter).

The dopamine-RPE theory has received widespread acceptance in the field from the perspective of why drugs of abuse are so rewarding initially and the ways in which the environment comes to control drug-seeking behavior independent of the drug (originally termed the facilitation of conditioned reinforcement [130], now often termed the facilitation of incentive salience [128]). Keiflin and Janak

[126] provide both theoretical and neurocircuitry explanations for the mechanisms that are involved and challenges in the field to understand "how these acute and long-term pharmacological effects of drugs alter the ongoing and future probability of drug-cue responsiveness and of actions made to obtain drug." One possible flaw in the persistent dopamine-RPE hypothesis, however, is that quite dramatic tolerance develops to the actual acute release of dopamine, at least in the context of extended access to psychostimulants [87,131]. Although persistent dopamine-RPEs may exist with limited access to drugs, once compulsive-like habitual responding develops, it may not be a key contributor. Again, this observation provides additional support for the cascade of neuroadaptations that occur in addiction and emphasize that the role of dopamine-RPEs may be key for early stages of addiction and more important for certain drugs, such as psychostimulants, that do not enter the cycle from a negative reinforcement perspective [132,133].

3.8 Multistep general theory of the transition to addiction

Piazza PV, Deroche-Gamonet V. A multistep general theory of transition to addiction. Psychopharmacology 2013;229:387–413 [134].

These authors proposed a general theory of the transition to addiction. They synthesized knowledge that was generated in the field of addiction with what was described as a unitary explanatory framework. Their theory of the transition to addiction was hypothesized to consist of three principles: (*i*) the transition to addiction depends on an interaction between individual vulnerability and drug exposure, (*ii*) the transition to addiction involves at least three steps (i.e., recreational/sporadic drug use, intensified/sustained/escalated drug use, and loss of control/full addiction), and (*iii*) addiction is a true psychiatric disease ([134]; Fig. 15).

Piazza and Deroche-Gamonet argued the importance of defining the criteria that need to be met to test the validity of a novel theory. A scientific theory is a well-substantiated explanation of a phenomenon, based on knowledge that has been repeatedly confirmed through observation and experimentation. Its strength is related to the diversity of the phenomena that it can explain and the accuracy in predicting outcomes. A scientific theory should allow for falsifiable predictions. Finally, a new theory should better explain experimental observations than previous theories and result in further testable predictions that can be confirmed. We will review each principle that is proposed by Piazza and Deroche-Gamonet based on each of these criteria.

Piazza and Deroche-Gamonet hypothesized that the transition to addiction results from an interaction between individual vulnerability and the degree or amount of exposure. Extensive research has repeatedly demonstrated that behavioral phenotypes, including drug-related behaviors, are modulated by the interaction between individual vulnerability (whether or not genetic) and the environment, including drug availability/exposure (for review, see Refs. [135,136]).

They also hypothesized, "The transition to addiction is composed of at least three steps: recreational sporadic (ReS), intensified, sustained, escalated (ISuE), and loss of control (LoC)." Again, this formulation parallels the field where numerous authors have

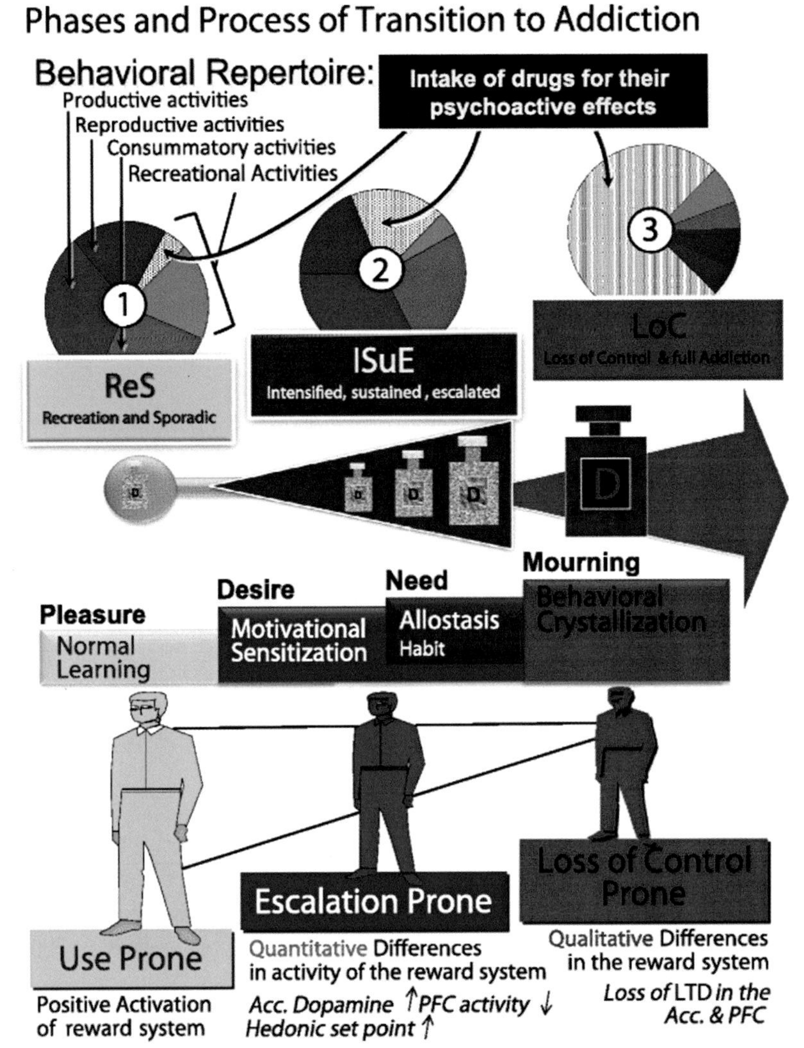

Figure 15 ***Summary of the phases and processes of the transition to addiction [134].*** The transition to addiction is a progression of three consecutive phases: (1) Recreational, sporadic (ReS) drug use, in which drug intake is moderate and sporadic and still one among many recreational activities of the individual. (2) Intensified, sustained, escalated (ISuE) drug use, in which drug intake intensifies and is now sustained and frequent and becomes the principal recreational activity of the individual; although some decreased societal and personal functioning starts appearing, behavior is still largely organized. (3) Loss of control (LoC) of drug use and full addiction that results in disorganization of the addict's behavior; drug-devoted activities are now the principal occupations of the individual. The three phases are consecutive but independent: entering one phase is necessary but not sufficient to progress toward the next phase, because specific individual vulnerabilities are needed. The first phase (ReS) occurs in most individuals (use prone); drugs overactivate the same substrates of natural rewards and therefore are perceived as extremely salient and likable stimuli. The second phase (ISuE) occurs in a vulnerable subset of individuals (escalation prone) because of quantitative differences in

theorized that the transition to addiction involves the same three steps: use (i.e., the initial first step of reinforcement that is recreational and social and involves learning the reinforcing effects of the drug), abuse (i.e., the second step of an increase in seeking, the consolidation of learning, and the escalation of drug intake), and dependence (i.e., the third step with maintenance of escalated intake with loss of control, full addiction, or dependence).

In their third principle, Piazza and Deroche-Gamonet argued that the transition to addiction is a true psychiatric disease. The overwhelming majority of neuroscientists would agree that addiction is a true psychiatric disorder. The debate between the iatrogenic and psychiatric views of addiction as outlined by Piazza and Deroche-Gamonet has added value, however, in elegantly applying multiple arguments with which to persuade those few academic holdouts and the general public.

A key part of the validation of their theory focused on the argument that their animal model of the loss of control exhibits the same percentage of vulnerable animals (~ 15–20%) as the percentage of vulnerable individuals in the human population, and this observation is used as a powerful argument for face validity.

The authors indeed embrace hedonic allostasis but discard "withdrawal," arguing that tolerance and withdrawal are not necessary or sufficient conditions for a diagnosis of the disease. They also repeatedly invoke negative reinforcement in the form of self-medication as a mechanism for enjoying the pleasures of drug taking in the human population, but in a relatively narrow sense.

Although not exactly a neurobiological theory, the theory of Piazza and Deroche-Gamonet was included here because it has had a significant influence on the field and generated a framework for defining individual differences in neurobiological vulnerability. It is included here in the binge/intoxication stage section as a transition to the other stages of the addiction cycle. One criticism is that their analysis of the literature is accurate, but their view that only animal models that reflect the Diagnostic and Statistical Manual of Mental Disorders, fourth or fifth edition, are relevant for the transition to addiction is limited to the face validity of symptoms of the disorder and not the underlying mechanisms. The key to solve the challenge of drug addiction is not whether we

---

the activity of the brain reward-related system, which increases the motivational effects of the drug; for example, a hyperactive (sensitized) dopaminergic system and an impaired prefrontal cortex. The ISuE phase is then stabilized by additional drug-induced adaptations, inducing an allostatic state that makes drugs not only strongly wanted but also needed in order for the individual to function normally and, in certain cases, by habit formation. The last phase (LoC) leads to full addiction and is due to a second vulnerable phenotype that we term loss of control prone. This phenotype is characterized by a persistent loss of long-term depression of synaptic transmission (LTD) in reward-related brain areas, which can induce a crystallization of behavior around drug-taking, resulting in losing control of drug intake. In the addicted state, the presence of the drug is not only needed to function normally, as at the end of the ISuE phase, but its absence is experienced as an irreplaceable loss and strongly pathologically mourned. When the individual goes from liking drugs to pathologically mourning them when they are not available, the process of transition to addiction is complete. *(Taken with permission from Piazza PV, Deroche-Gamonet V. A multistep general theory of transition to addiction. Psychopharmacology 2013;229:387–413 [134].)*

can produce addiction in every single individual, whether there are more than two steps in the transition to addiction, or whether addiction is a psychiatric disorder. All of these questions have already been answered. Instead, one needs to investigate neuroadaptations that are associated with different aspects of the transition to addiction, including incentive-salience, tolerance, motivational withdrawal, escalation, cognitive impairment, and loss of control, not only over drug intake but also the loss of control over emotion, stress, and pain. One could argue that their three-stage model provides such a framework but so do all the models that are outlined in this chapter and Chapter 2, Animal Models of Addiction. One also needs to determine the neuronal networks and plasticity (or lack thereof) that mediate the vulnerability to seek and take drugs at every single step of the addiction process as well as relapse after abstinence. Finally, one needs to develop novel therapeutic approaches that can reduce compulsive drug seeking and taking in individuals with addiction and return the brain motivational systems to homeostasis and to use various animal and human models for every stage of the addiction process to identify resistance to the transition to addiction and provide an evidence-based approach to prevention [132]. Additionally, substantial arguments can be made that the 15–20% of rats that are vulnerable may be mathematically constrained to produce a very restricted range of values [132], based on 40-year-old human data, based on one drug, and based on a corruption of face validity (i.e., one is taking face validity too seriously). It is of value to select vulnerable subgroups for study, but one subgroup addiction does not make. Finally, although Piazza and Deroche-Gamonet elegantly and correctly outlined the allostatic view of addiction of Koob and Le Moal, they left out half of the story. We have argued, with some substantial evidence, that as dependence and withdrawal develop, brain anti-reward systems, such as CRF, norepinephrine, and dynorphin, are recruited in the extended amygdala to produce a negative emotional state from the side of stress, malaise, and pain that we believe also accounts for a significant amount of motivational withdrawal or what Piazza and Deroche-Gamonet call "mourning."

4. Theories of the withdrawal/negative affect stage

4.1 Stress, dysregulation of drug reward pathways, and drug addiction

Piazza PV, Le Moal ML. Pathophysiological basis of vulnerability to drug abuse: role of an interaction between stress, glucocorticoids, and dopaminergic neurons. Annual Review of Pharmacology and Toxicology 1996;36:359–78 [137].

Shaham Y, Shalev U, Lu L, De Wit H, Stewart J. The reinstatement model of drug relapse: history, methodology and major findings. Psychopharmacology 2003;168:3–20 [138].

Aston-Jones G, Harris GC. Brain substrates for increased drug seeking during protracted withdrawal. Neuropharmacology 2004;47(Suppl. 1):167–79 [139].

Kreek MJ, Koob GF. Drug dependence: stress and dysregulation of brain reward pathways. Drug and Alcohol Dependence 1998;51:23–47 [140].

Koob GF, Kreek MJ. Stress, dysregulation of drug reward pathways, and the transition to drug dependence. American Journal of Psychiatry 2007;164:1149–59 [141].

Drugs of abuse acutely activate the hypothalamic-pituitary-adrenal (HPA) response to stress and ultimately engage brain stress systems as dependence develops. These basic observations led to the hypothesis that the brain and pituitary stress systems play a role in the initial vulnerability to drugs of abuse [137,140,141], the development of dependence [140,141], and the vulnerability to stress-induced relapse [138]. Stressors facilitated the acquisition of cocaine and amphetamine self-administration, and food restriction increased the self-administration of most drugs of abuse ([142,143]; Table 2). Rats that were bred for higher basal exploration of a novel environment and a high initial corticosterone response were much more likely to self-administer psychostimulant drugs [137]. There is a positive correlation between the locomotor response to novelty, behavioral reactivity to stress, and the amount of amphetamine that is self-administered by individual rats [144,145]. Rats that received repeated injections of corticosterone acquired cocaine self-administration at a lower dose compared with rats that received vehicle [146]. Corticosterone administration caused rats that would not self-administer cocaine at low doses to then self-administer cocaine [145]. Glucocorticoids have been shown to facilitate the locomotor activation that is associated with drugs of abuse [147,148]. The acute blockade of corticosterone secretion decreased the psychomotor response to cocaine [149]. The greater propensity to self-administer drugs of abuse that was produced by stressors has been linked to greater activation of the mesolimbic dopamine system that is mediated by glucocorticoid release [143]. Glucocorticoids facilitate dopamine-dependent behaviors and act by modulating dopamine transmission in the ventral striatum. The suppression of glucocorticoids by adrenalectomy reduced extracellular concentrations of dopamine in the ventral striatum, particularly the shell [150], under basal conditions and after psychostimulant administration. These effects were reversed by corticosterone replacement and selectively mediated by glucocorticoid receptors [150].

These initial responses of the HPA system to drugs change dramatically when the animals are experimenter-administered psychostimulant drugs or self-administer binge-like amounts of psychostimulant drugs [151–153]. There are large increases in adrenocorticotropic hormone and corticosterone in rats during an acute binge [152,153]. However, this increase shows “tolerance” in the chronic binge stage, followed by reactivation during acute withdrawal and dysregulation that persists into protracted abstinence [151,152].

The escalation of drug intake either with extended access or dependence induction activates the brain stress system and norepinephrine stress system outside the hypothalamus in the extended amygdala [139,154,155]. Acute withdrawal from cocaine [156], alcohol [157,158], nicotine [159], cannabinoids [160], and opioids [161] increases CRF release, measured by *in vivo* microdialysis, in the central nucleus of the amygdala and in some cases in the bed nucleus of the stria terminalis. Corticotropin-releasing factor receptor antagonists block the increase in anxiety-like or aversive responses that are associated with alcohol withdrawal [162,163], precipitated opioid withdrawal [164,165], and cocaine withdrawal [166,167]. These increases in CRF may have motivational significance, in which a competitive CRF receptor antagonist blocks excessive drinking in dependent rats but not baseline nondependence drinking [168]. A competitive CRF_1 receptor antagonist blocked the development of place aversion to precipitated opioid withdrawal [165].

Table 2 Stressors that increase drug self-administration.

	Approaches to the study of drug self-administration				
Stressor	**Acquisition**	**Dose-response**	**Progressive-ratio**	**Reinstatement**	**Reference**
Food restriction	↑ Psychostimulants ↑ Opiates ↑ Alcohol	↑ Psychostimulants ↑ Opiates ↑ Alcohol			[142,511]
Tail pinch	↑ Amphetamine				[512]
Footshock	↑ Cocaine		↑ Heroin	↑ Heroin ↑ Cocaine	[176,179,511,513]
Restraint	↑ Morphine				[514]
Social aggression	↑ Cocaine	↑ Cocaine			[515,516]
Social competition	↑ Amphetamine				[517,518]
Social isolation	↑ Opiates ↑ Alcohol	↑ Cocaine[a] ↑ Heroin			[519–525]
Witnessing stress	↑ Cocaine				[526]
Prenatal stress	↑ Amphetamine				[527]

↑, Depending on the type of self-administration used: facilitation of acquisition; upward shift of the dose-response curve; higher break point; induction of responding on the device previously associated with the infusion of the drug.
[a]Note that for social isolation, slightly higher [528], equal [524], and lower [529] sensitivities to the reinforcing effects of cocaine also have been reported.
Taken with permission from Piazza PV, Le Moal M. The role of stress in drug self-administration. Trends in Pharmacological Sciences 1998;19:67–74 [143].

Norepinephrine projections to the bed nucleus of the stria terminalis, which are known to activate CRF systems, have also been shown to be critically involved in the motivational aspects of opioid withdrawal [139,169,170]. Microinjection of β-adrenergic receptor antagonists in the bed nucleus of the stria terminalis blocked opioid withdrawal-induced conditioned place aversion [170]. The origin of the norepinephrine projections for this effect appear to be from the ventral noradrenergic bundle from noradrenergic cell groups of the caudal medulla. Lesions of this ventral system but not the dorsal noradrenergic bundle from the locus coeruleus blocked opioid withdrawal-induced conditioned place aversion [170]. These results suggest that the brain stress systems that involve CRF and norepinephrine may be activated during the development of dependence and contribute to the motivation for excessive drug seeking that is associated with dependence ([171]; Fig. 16). Other anti-stress neural systems, such as NPY, may be dysregulated as a "brake" on the stress system, which would further exacerbate dysregulation of the brain stress systems [172].

Some of these dysregulations of the CRF and norepinephrine systems have been shown to persist during protracted abstinence and can be reinstated by the reinstatement of an acute stressor ([139,139a]). Rats that were made dependent on alcohol and tested 2–6 weeks post-acute withdrawal exhibited an increase in anxiety-like response in the elevated plus maze that was blocked by the intracerebroventricular administration of a competitive CRF receptor antagonist [173]. Parallel results were reported by studies of chronic morphine and conditioned place preference, in which chronic morphine treatment enhanced morphine-induced conditioned place preference for weeks after acute withdrawal, and this increase in place preference was accompanied by greater anxiety-like responses and increases in Fos staining in the medial prefrontal cortex (cingulate cortex), basolateral amygdala, and ventral part of the bed nucleus of the stria terminalis [174]. Aston-Jones and Harris suggested that the conditioned release of norepinephrine in the bed nucleus of the stria terminalis in response to stressors may elevate anxiety, which then augments the reward value of drugs through negative reinforcement mechanisms [139,171].

The same brain stress systems have been strongly implicated in the stress-induced reinstatement of drug-seeking behavior. Clear evidence showed that rats that were previously trained to self-administer cocaine and heroin and then extinguished reinstated their responding when exposed to a mild footshock immediately before the test session [175–179]. This stress-induced reinstatement was not blocked by the removal of glucocorticoids [180] but was blocked by the pretreatment of CRF receptor antagonists that were administered in the brain [180–183]. This effect appears to be mediated by CRF_1 receptors [182–184]. The brain site that is responsible for the actions of CRF receptor antagonists on cocaine reinstatement appears to be the ventral bed nucleus of the stria terminalis. Infusions of CRF into this area reinstated responding, and the local administration of CRF receptor antagonists in the ventral bed nucleus of the stria

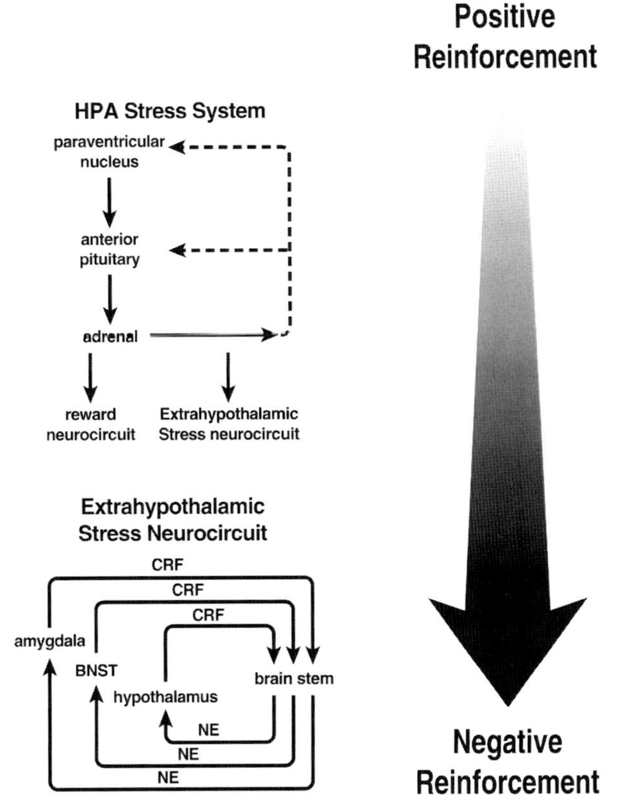

Figure 16 Hypothalamic-pituitary-adrenal and brain stress circuits hypothesized to be recruited at different stages of the addiction cycle as addiction moves from positive reinforcement to negative reinforcement (Koob and Kreek, 2005; [138,139]). The top circuit refers to the HPA axis which (a) feeds back to regulate itself, (b) activates the brain reward neurocircuit, and (c) facilitates the extrahypothalamic stress neurocircuit. The bottom circuit refers to the brain stress circuits in feed-forward loops. Here, corticotropin-releasing factor in the extended amygdala drives norepinephrine systems in the pons-medulla of the brainstem, which in turn drives CRF in the extended amygdala (see Ref. [154]). *(Adapted from Koob GF, Le Moal M. Drug addiction and allostasis. In: Schulkin J, editor. Allostasis, homeostasis, and the costs of physiological adaptation. New York: Cambridge University Press; 2004. p. 150–63 [171].)*

terminalis blocked footshock-induced reinstatement, although administration in the central nucleus of the amygdala had no effect [185]. However, the reversible activation of both the central nucleus of the amygdala and bed nucleus of the stria terminalis with tetrodotoxin blocked the footshock-induced reinstatement of heroin responding [186]. An asymmetric lesion procedure to functionally disconnect the CRF-containing pathway from the central nucleus of the amygdala to the bed nucleus of the stria terminalis significantly reduced footshock-induced reinstatement, suggesting that an important origin of CRF terminals in the bed nucleus of the stria terminalis for cocaine-induced reinstatement may be the central nucleus of the amygdala [187].

Noradrenergic receptor functional antagonists also block footshock-induced reinstatement [188–190]. The brain sites for these effects appear to be ventral noradrenergic bundle projections to the bed nucleus of the stria terminalis. Neurotoxin-specific lesions of the ventral noradrenergic bundle attenuated the footshock-induced reinstatement of heroin responding [190]. A local injection of a β-adrenergic receptor antagonist in the bed nucleus of the stria terminalis also blocked footshock-induced reinstatement in cocaine-trained rats [191]. The exact interaction between the noradrenergic ventral bundle projections and intrinsic CRF systems of the central nucleus of the amygdala and bed nucleus of the stria terminalis are unknown but may involve activation by the noradrenergic pathway of CRF to the bed nucleus of the stria terminalis pathway [192], which in turn may activate noradrenergic systems in the ventral medulla [154]. Such a feed-forward system has been hypothesized for CRF/noradrenergic interactions for anxiety- and stress-like responses and may also become activated during the development of dependence.

In summary, the HPA hormonal stress system and brain stress systems are engaged by drugs of abuse and may contribute to the initial vulnerability to take drugs and the development of dependence and vulnerability to relapse. The HPA hormonal stress system appears to play an important role in the initiation of drug seeking and maintenance of drug-taking behavior. However, brain extrahypothalamic stress systems appear to play a more important role in the motivational effects of both acute withdrawal and protracted abstinence and stress-induced reinstatement.

4.2 Neurocircuits for reward system dysregulation in addiction or the extended amygdala: a common substrate for dysregulation of reward and stress function in addiction

Koob GF, Le Moal M. Drug addiction, dysregulation of reward, and allostasis. Neuropsychopharmacology 2001;24:97–129 [90].

Although changes in reward function are components of other conceptual frameworks in the neurocircuitry of addiction, for Koob and Le Moal [90] the primary "deficit" is a neuroadaptational shift in the way in which rewards are processed. A loss of positive reinforcement and the recruitment of negative reinforcement are hypothesized to occur within a specific basal forebrain circuit termed the extended amygdala. The extended amygdala has been identified by neuroanatomical studies [34,193] as a separate entity within the basal forebrain and has been hypothesized to be a common neural circuitry for the reinforcing actions of drugs [34]. The term *extended amygdala* was originally described by Johnston [194] and represents a macrostructure that is composed of several basal forebrain structures: lateral and medial bed nucleus of the stria terminalis, central and medial nuclei of the amygdala, sublenticular substantia innominata, and a transition zone that forms the medial posterior part (e.g., shell) of the nucleus accumbens [77,195]. These structures have similarities in morphology, immunohistochemistry, and connectivity ([34,77]; Fig. 17).

However, further examination of this anatomical system reveals two major divisions: central division and medial division. These two divisions have important differences in structure and afferent and efferent connections [195] that may have heuristic value. The central division of the extended amygdala includes the central nucleus of the amygdala, the central sublenticular extended amygdala, the lateral bed nucleus of the stria terminalis, and a transition area in the medial and caudal portions of the nucleus accumbens. These structures are within the central division and are largely defined by their network of intrinsic connections and extensive connections to the lateral hypothalamus [196]. The medial division of the extended amygdala includes the medial bed nucleus of the stria terminalis, medial nucleus of the amygdala, and the medial sublenticular extended amygdala. These structures have been largely defined as the medial division by their network of intrinsic associative connections and extensive relations to the medial hypothalamus [195]. The lateral bed nucleus of the stria terminalis that forms a key element of the central division of the extended amygdala has high amounts of dopamine and norepinephrine terminals, CRF terminals, CRF cell bodies, NPY terminals, and galanin cell bodies and receives afferents from the prefrontal cortex, insular cortex, and amygdalopiriform area. The medial bed nucleus of the stria terminalis, in contrast, contains high amounts of vasopressin, is sexually dimorphic, and receives afferents from such structures as the infralimbic cortex, entorhinal cortex, and subiculum [197–202]. Evidence suggests that the central division may be involved in receiving cortical information and regulating the HPA axis [203], whereas the medial division may be more involved in sympathetic and physiological responses and receiving olfactory information [204–206]. Most motivational manipulations that result in modification of the reinforcing effects of drugs of abuse have been in the central nucleus of the amygdala and lateral nucleus of the bed nucleus of the stria terminalis.

Specific sites within the extended amygdala and selective neurochemical and neuropharmacological actions have been identified for both the acute positive reinforcing effects of drugs of abuse and in the negative reinforcement that is associated with drug abstinence. Microinjections of dopamine D_1 receptor antagonists directly in the shell of the nucleus accumbens, central nucleus of the amygdala [207], and bed nucleus of the stria terminalis [208] are particularly effective in blocking cocaine self-administration. *In vivo* microdialysis studies show the selective activation of dopaminergic transmission in the shell of the nucleus accumbens in response to acute administration of virtually all major drugs of abuse [63–65]. Additionally, the acute reinforcing effects of alcohol are blocked by the administration of GABAergic and opioidergic competitive antagonists in the central nucleus of the amygdala [32,209]. Lesions of cell bodies within this structure markedly suppress alcohol self-administration ([210]; Table 3). Thus, the

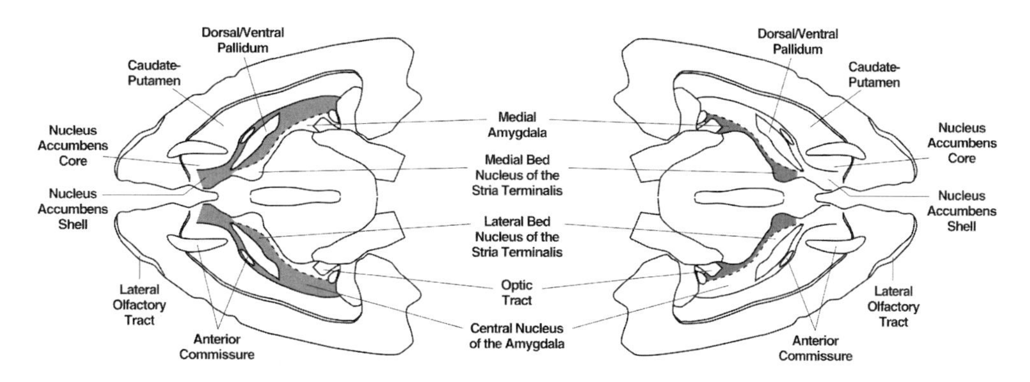

Central Division

Neuropharmacological Component
- Dopamine terminals
- Norepinephrine terminals
- CRF terminals and cell bodies
- NPY terminals
- Galanin cell bodies

Afferent/Efferent
- Receives cortical information from prefrontal cortex and insular cortex
- Sends information to lateral hypothalamus, ventral tegmental area, and pedunculopontine tegmental nucleus

Functional Attributes
- Regulates pituitary adrenal axis
- Major role in mediating reinforcing actions of drugs of abuse and ethanol

Medial Division

Neuropharmacological Component
- Vasopressin terminals
- Sex hormone receptors

Afferent/Efferent
- Receives olfactory information from infralimbic cortex, entorhinal cortex and subiculum
- Sends information to ventral striatum, ventromedial hypothalamus and mesencephalic central gray

Functional Attributes
- Sexually dimorphic – larger in males than females
- Major role in mediating male and female sexual behavior
- Regulates sympathetic and physiological responses

Figure 17 Neuroanatomical connections of the extended amygdala relevant to the positive reinforcing effects of drugs and the negative reinforcing effects of addiction. Neuroanatomical studies reveal two major divisions: the central division (A) and the medial division (B). These two divisions have important differences in structure and afferent and efferent connections. The central division of the extended amygdala includes the central nucleus of the amygdala, the central sublenticular extended amygdala, the lateral bed nucleus of the stria terminalis, and a transition zone in the medial and caudal portions of the nucleus accumbens. These structures are within the central division and are largely defined by their network of intrinsic connections and extensive connections to the lateral hypothalamus. The medial division of the extended amygdala includes the medial bed nucleus of the stria terminalis, medial nucleus of the amygdala, and medial sublenticular extended amygdala. These structures have been largely defined as the medial division by their network of intrinsic associative connections and extensive relations to the medial hypothalamus [195]. The lateral bed nucleus of the stria terminalis that forms a key element of the central division of the extended amygdala has high amounts of dopamine and norepinephrine terminals, CRF terminals, CRF cell bodies, NPY terminals, and galanin cell bodies and receives afferents from the prefrontal cortex, insular cortex, and amygdalopiriform area. The medial bed nucleus of the stria terminalis, in contrast, contains high amounts of vasopressin, is sexually dimorphic, and receives afferents from such structures as the infralimbic cortex, entorhinal cortex, and subiculum [197–202]. Evidence suggests that the central division may be involved in receiving cortical information and regulating the hypothalamic-pituitary-adrenal axis [203], whereas the medial division may be more involved in sympathetic and physiological responses and receiving olfactory information [204–206]. Most neuropharmacological effects that result from motivational manipulations that are associated with the reinforcing effects of drugs of abuse have been in the central nucleus of the amygdala and the lateral bed nucleus of the stria terminalis. *(Brain sections modified with permission from Heimer L, Alheid G. Piecing together the puzzle of basal forebrain anatomy. In: Napier TC, Kalivas PW, Hanin I, editors. The basal forebrain: anatomy to function. Advances in experimental medicine and biology, vol. 295. New York: Plenum Press; 1991. p. 1–42 [77].)*

Table 3 Behavioral effects relevant to drug of abuse action of lesions of basolateral and central nuclei of the amygdala.

Functional effect	Reference
Central nucleus of the amygdala	
↓ Amphetamine-induced potentiation of conditioned reward	[75,263]
↓ Naloxone-precipitated withdrawal-induced place aversion	[530]
↓ Morphine antinociception	[531,532]
↓ Anxiety-like effects of benzodiazepines	[533]
Basolateral nucleus of the amygdala	
↓ Cocaine-seeking behavior under a second-order schedule of reinforcement	[74]
↓ Cue-induced reinstatement of drug responding in rats	[268]
↓ Modulation of intracranial self-stimulation thresholds by drug-associated cues	[534]
↓ Cue-induced and cocaine-induced reinstatement in a discriminative stimulus task	[535]
↓ Acquisition and extinction of cocaine conditioned place preference	[536]
↓ Conditioned withdrawal from opiates	[537]
↓ Self-administration of ethanol	[210]

activation of dopamine and opioid peptides and potential contributions from other neurotransmitters, such as GABA and serotonin, and neuromodulators, such as endocannabinoids, are importantly involved in the acute reinforcing effects of drugs of abuse (Table 4, *binge/intoxication* stage).

A role for the extended amygdala in the aversive stimulus effects of drug withdrawal includes changes in opioidergic, GABAergic, and CRF neurotransmission during acute withdrawal. There is enhanced sensitivity of alcohol-dependent rats to GABA receptor agonists during acute withdrawal [211], and the CRF systems in the central nucleus of the amygdala are activated during acute cocaine, alcohol, opioid, Δ^9-tetrahydrocannabinol, and nicotine withdrawal as measured by *in vivo* microdialysis and neuropharmacological probes [157,159,164]. Similar effects have been observed with alcohol and opioids in the lateral bed nucleus of the stria terminalis ([139,158]; Table 4, *withdrawal/negative affect* stage). However, decreases in the function of dopamine, serotonin, and opioid peptides, as well as the recruitment of brain stress systems such as CRF, are hypothesized to contribute to a shift in reward set point that characterizes the transition to the addictive state.

Thus, acute withdrawal from drugs of abuse produces opponent process-like changes in reward neurotransmitters in specific elements of reward circuitry that is associated with the extended amygdala and the recruitment of brain stress systems that motivationally oppose the hedonic effects of drugs of abuse. Such changes in these brain systems that are associated with the development of motivational aspects of withdrawal are hypothesized to be a major source of neuroadaptive changes that drive and

Table 4 Neurochemical systems in the extended amygdala involved in the motivational effects of different stages of the addiction cycle.

Stage of addiction cycle	Neurochemical system	Functional effect
Binge/intoxication	↑ Dopamine	"Euphoria"
	↑ Opioid peptides	"Euphoria"
	↑ GABA	Anti-anxiety
	↑ NPY	Anti-stress
Withdrawal/negative affect	↓ Dopamine	"Dysphoria"
	↓ Serotonin	"Dysphoria"
	↓ GABA	Anxiety, panic attacks
	↓ NPY	Anti-stress
	↑ Dynorphin	"Dysphoria"
	↑ CRF	Stress
	↑ Norepinephrine	Stress
Preoccupation/anticipation		
Cue-induced craving	↑ Dopamine	"Craving"
	↑ Opioid peptides	"Craving"
	↓ Glutamate	"Craving
Residual negative affective state	↑ Dynorphin	"Dysphoria"
	↑ CRF	Stress
	↑ Norepinephrine	Stress
	↓ GABA	Anxiety, panic attacks

maintain addiction. All of these changes are hypothesized to be focused on the dysregulation of function within the neurocircuitry of the basal forebrain macrostructure of the extended amygdala.

In summary, the extended amygdala circuit for drug addiction puts a major focus on the role of the shell of the nucleus accumbens, bed nucleus of the stria terminalis, and central nucleus of the amygdala for changes in hedonic tone that are associated with the transition from positive reinforcement to negative reinforcement as addiction develops.

4.3 Addiction and stress: an allostatic view

Koob GF, Schulkin J. Addiction and stress: an allostatic view. Neuroscience & Biobehavioral Reviews 2018 [in press] [212].

In this view, the role of stress in addiction is explicitly linked to an allostatic process in the brain and pituitary stress axis that begins with activation of the HPA axis. Stress is argued to impact all three stages of the addiction cycle—*binge/intoxication*, *withdrawal/negative affect*, and *preoccupation/anticipation*—exposing the animal to an emotional allostatic load and allostatic state that forms the growing motivational pathology of addiction. In the *binge/intoxication* stage, excessive drug use can initiate neuroadaptations that feed allostasis via activation of the HPA axis to facilitate reward and then subsequently via the sensitization of extrahypothalamic CRF systems in the extended amygdala to facilitate incentive salience (see *Section 4.1. Stress, dysregulation of drug reward pathways, and drug addiction*). Allostatic changes in the HPA axis, with subsequent blunting of the HPA axis and the sensitization of extrahypothalamic CRF, are further exaggerated by repeated binge-withdrawal cycles of drug taking, such that progressively greater negative emotional states are generated that drive negative reinforcement (Fig. 18). Additionally, activation of the HPA axis and extrahypothalamic CRF system is hypothesized to negatively impact the prefrontal cortex to impair this top-down connectivity and help feed growing allostatic changes in the extended amygdala brain stress systems and residual vulnerability to stress-induced relapse.

The advance here in conceptually thinking of the role of the brain stress systems in addiction is the explicit view of addiction as an allostatic mechanism, thus providing key insights into the ways in which dysregulated neurocircuitry that is involved in basic motivational systems can transition to pathophysiology. Perhaps even more compelling is the focus on the effect of glucocorticoids in downregulating the HPA axis while sensitizing extrahypothalamic CRF systems. Evidence shows that the HPA axis may also play a key role in driving incentive salience in the binge/intoxication stage and prefrontal dysfunction in the preoccupation/anticipation stage. Such a formulation provides a mechanism for a long hypothesized but previously not mechanistically outlined key role of an allostatically dysregulated HPA axis in the development and persistence of addiction.

4.4 Negative reinforcement in drug addiction: the darkness within

Koob GF. Negative reinforcement in drug addiction: the darkness within. Current Opinion in Neurobiology 2013a;23:559–63 [133].

Koob argues that drug seeking involves a long-neglected other source of reinforcement during the course of the three-stage addiction cycle [40]. Negative reinforcement is defined as the process by which the removal of an aversive stimulus (or aversive state of withdrawal in the case of addiction) increases the probability of a response. The hypothesis further elaborates that the state that drives negative reinforcement is a negative emotional state that is composed of dysphoria, malaise, irritability, hyperkatifeia [213], and sleep disturbances [133]. The mechanism at the neurocircuitry level is hypothesized to be the loss of function in normal reward-related neurocircuitry and persistent recruitment of the brain stress systems ([133]; Fig. 19). Koob further argues that the combination of lower reward function and higher stress function forms the antireward systems or "darkness within." He further argues that this neuroplasticity explains the misery of addiction, its role in the transition to addiction, and its role in the persistence of addiction.

An important conceptual framework for the position that is outlined by Koob is that the negative emotional state that is hypothesized to drive negative reinforcement is an opponent process [214] and thus only follows in the drug taking and seeking cycle if there is excessive engagement by the drug in promoting reward or incentive salience neurocircuitry. Although opponent processes can be manifest with the single administration of a drug, they are short-lasting and dissipate rapidly. However,

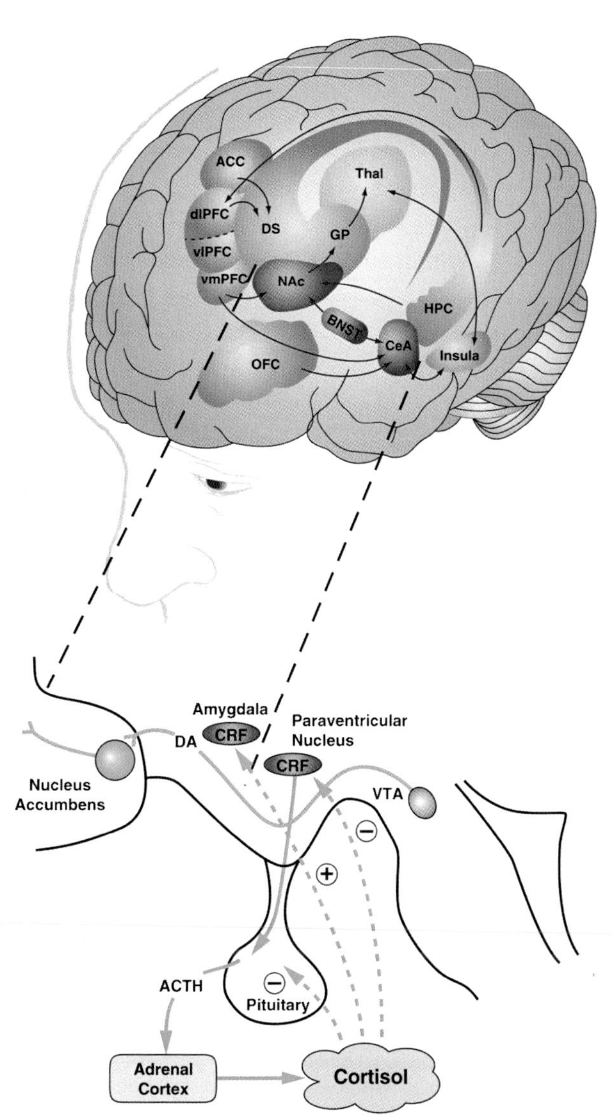

Figure 18 Conceptual framework of the way in which dysregulation of the hypothalamic-pituitary-adrenal axis and extrahypothalamic CRF systems can influence the *withdrawal/negative affect* stage of the addiction cycle to drive allostasis in addiction. Here, the activation of glucocorticoids (bottom) inhibits the paraventricular nucleus but drives CRF in the extended amygdala, triggering hyperkatifeia via extrahypothalamic stress systems. The activation of CRF in the central nucleus of the amygdala and bed nucleus of the stria terminalis produces increases in hyperkatifeia-like responses in animals during acute and protracted withdrawal. ACC, anterior cingulate cortex; dlPFC, dorsolateral prefrontal cortex; vlPFC, ventrolateral prefrontal cortex; vmPFC, ventromedial prefrontal cortex; DS, dorsal striatum; NAc, nucleus accumbens; GP, globus pallidus, Thal, thalamus; OFC, orbitofrontal cortex; BNST, bed nucleus of the stria terminalis; CeA, central nucleus of the amygdala; HPC, hippocampus; CRF, corticotropin-releasing factor, DA, dopamine, ACTH, adrenocorticotropic hormone. *(Taken with permission from Koob GF, Schulkin J. Addiction and stress: an allostatic view. Neuroscience and Biobehavioral Reviews 2018 [in press] [212].)*

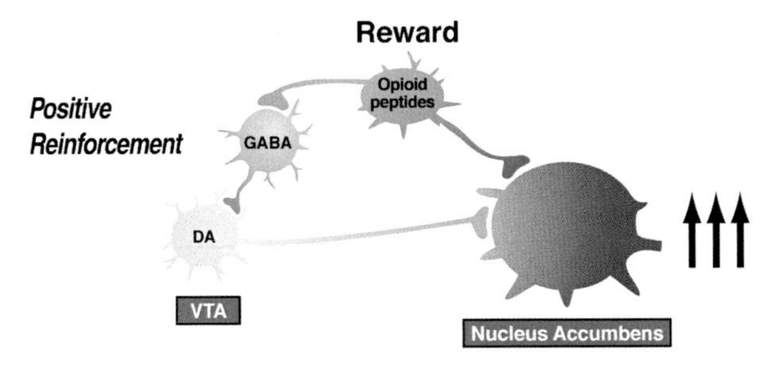

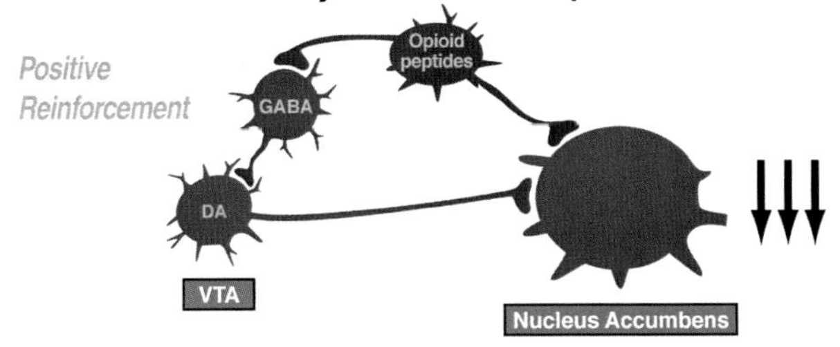

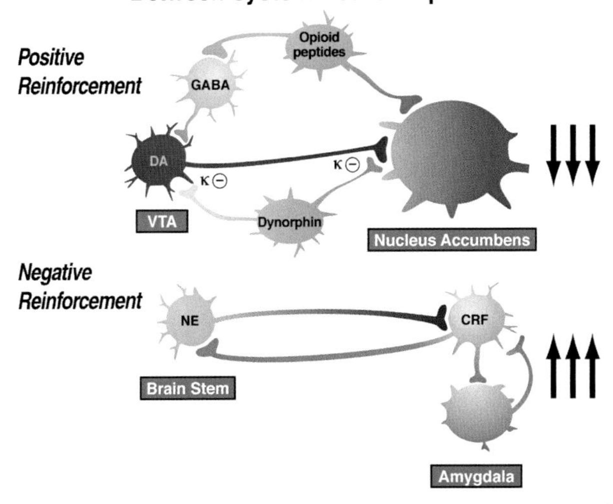

Figure 19 Diagram of the hypothetical "within-system" and "between-system" changes that lead to the "darkness within." (Top) Circuitry for drug reward with major contributions from mesolimbic dopamine and opioid peptides that converge on the nucleus accumbens. During the *binge/intoxication* stage of the addiction cycle, the reward circuitry is excessively engaged. (Middle) Such excessive activation of the reward system triggers "within-system" neurobiological adaptations during the *withdrawal/negative affect* stage, including the activation of cyclic adenosine monophosphate (cAMP) and cAMP response element binding protein (CREB), the downregulation of dopamine D_2 receptors, and a decrease in the firing of ventral tegmental area (VTA) dopaminergic neurons. (Bottom) As dependence progresses and the withdrawal/negative affect stage is repeated, two major between-system neuroadaptations occur. One is activation of dynorphin feedback that further decreases dopaminergic activity. The other is the recruitment of extrahypothalamic norepinephrine (NE)-corticotropin-releasing factor (CRF) systems in the extended amygdala. Facilitation of the brain stress system in the prefrontal cortex is hypothesized to exacerbate between-system neuroadaptations while contributing to the persistence of the dark side into the *preoccupation/anticipation* stage of the addiction cycle. *(Taken with permission from Koob GF. Negative reinforcement in drug addiction: the darkness within. Current Opinion in Neurobiology 2013a;23:559–563 [133]. Koob GF, editor. Impulse control disorders. San Diego: Cognella; 2013b [ISBN: 978-1-62131-212-3] [43].)*

extended or repeated self-administration produces progressively greater deviations from hedonic baseline [114]. Note that others have argued from the human condition perspective that negative reinforcement is a compelling driving force in tobacco addiction [215], and the opioid overdose crisis in the United States and concomitant "deaths of despair" [216] point to negative emotional states as a driving force in certain addictions.

4.5 Neurobiology of opioid dependence in creating addiction vulnerability

Evans CJ, Cahill CM. Neurobiology of opioid dependence in creating addiction vulnerability. F1000Research 2016;5. F1000 Faculty Rev-1748 [217].

Evans and Cahill extended the consequences of allostatic changes (cellular, circuit, and system adaptations) that accompany the drug-dependent state to focus on drug withdrawal, specifically opioid withdrawal, as a model of negative reinforcement vulnerability and delineate a molecular neurocircuitry to explain their hypothesis. They argue that drug withdrawal triggers a temporally defined allostatic reestablishment of neuronal systems. This reestablishment follows a classic opponent process, manifested as physiological and psychological effects that oppose those that are manifest during acute drug intoxication. The physical symptoms of withdrawal (e.g., sweating, shaking, and diarrhea) resolve within days, but the "psychological" symptoms, such as dysphoria, insomnia, and anxiety, can linger for months. Some adaptations, such as learned associations, may be present for life (Figs. 20 and 21). The cellular mechanisms and neural circuitry that contribute to the opioid drug-dependent state range from changes in the μ-signalosome, neuronal proteome, and transcriptome to modifications of neuronal morphology and an emerging role for neuroinflammation. Based on these data, the authors argue that opioid addictive behaviors result from a learned association between opioids and relief from an existing withdrawal-induced anxiogenic and/or dysphoric state. This learning would be characterized as driven by negative reinforcement [133]. The authors carry the argument further and suggest that future stressful life events can generalize to such an anxiogenic and/or dysphoric state and thus can recall the memory that opioid drugs alleviate negative affect (e.g., despair, sadness, and anxiety), thereby precipitating craving and resulting in relapse.

This theory of learned negative reinforcement is very timely, given the continuing opioid overdose crisis in the United States, and may be one of the keys to the burgeoning deaths of despair in certain subgroups in the United States. Thus, the ready availability of opioid drugs and synthetic opioids of very high potency, alcohol abuse, depression, and suicide may all be linked in the domain of the dark side of addiction. Conditioned negative reinforcement or conditioned withdrawal has been known for some time in the clinical setting. Methadone-maintained individuals who were conditioned to a previously neutral stimulus (e.g., peppermint smell) during naloxone-precipitated withdrawal presented a subsequent withdrawal response to the peppermint smell [218]. This conditioned withdrawal was hypothesized to contribute to craving in a real-world setting. These results were replicated in an intravenous self-administration model of extended access to heroin, in which stimuli that were paired with

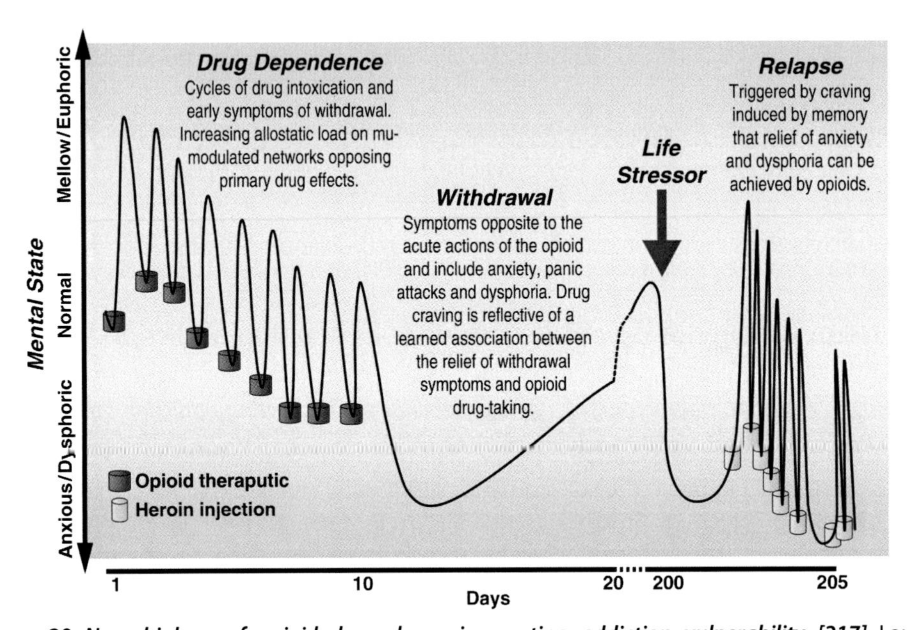

Figure 20 ***Neurobiology of opioid dependence in creating addiction vulnerability [217].*** Laura's pathway to heroin addiction, showing a hypothetical scenario of how learned association of relief of aversive states could lead to the development of addiction and the key role of opioid dependence. Let us consider a disappointed teenager (Laura) who sprained her ankle during tryouts for the cheerleading team and was rejected, not making the squad. Knowing that opioids will alleviate the pain caused by the sprained ankle, Laura takes an opioid analgesic pill she knows is in the bathroom medicine cabinet, left over from her older brother's prescription for a wisdom tooth extraction. The opioid pill takes away Laura's pain from the sprained ankle and makes her feel relaxed (mellow and with elevated mood). She is less bothered about the tryout rejection while under the influence of the opioid. The next day, Laura decides to take another pill as the drug effects have worn off and both the ankle (physical) pain and the (psychological) pain of rejection have returned and are just as bad as the day before. The taking of the opioid medication again relieves the physical pain and takes her mind off her failure. Over the next 10 days, the drug taking cycle continues, but each day the pain-relieving effects are less (development of tolerance) and she develops a modest anxiogenic state prior to taking the drug each day (early signs of withdrawal). As the pain from the ankle sprain subsides, the association of drug taking becomes the relief of the anxious and dysphoric states that emerge as the drug wears off. On day 10, the supply of drugs from the medicine cabinet is exhausted and withdrawal sets in. Drug seeking would be a consequence during this withdrawal phase given the learned association that the aversive symptoms she is experiencing could be effectively relieved by taking the opioid (negative reinforcement). Although unlikely, given the ready access to illicit opioids throughout society, let's assume that at this point in Laura's life, she doesn't actively seek out prescription or other opioids. The teen goes through physical withdrawal (over days) with protracted mood disturbances accompanied by drug craving that is triggered by memories that she could relieve dysphoric states by taking opioids. She doesn't continue to seek opioids at this stage because the pain of rejection has dissipated and she is now engaged with other activities in her peer group. Over a year later, her boyfriend breaks up with her, her grades are not stellar, and she may not get accepted to the colleges she wants to go to. Laura becomes extremely stressed and feels unable to cope. The emergence of these negative symptoms initiates a craving for opioids that is triggered by learned associative memories. At this point in her life, she has the resources to access heroin and relapses. Laura's relapse reinforces the associative memories that aversive states are relieved by opioids and the foundation for further addictive behaviors is strengthened. In this example, drug dependence and withdrawal contribute to addiction vulnerability, and the learned associative memories are of the negative reinforcement provided by opioid relief of anxiogenic and dysphoric states, and are triggered by a different stressor. *(Taken with permission from Evans CJ, Cahill CM. Neurobiology of opioid dependence in creating addiction vulnerability. F1000Research 2016;5. F1000 Faculty Rev-1748 [217].)*

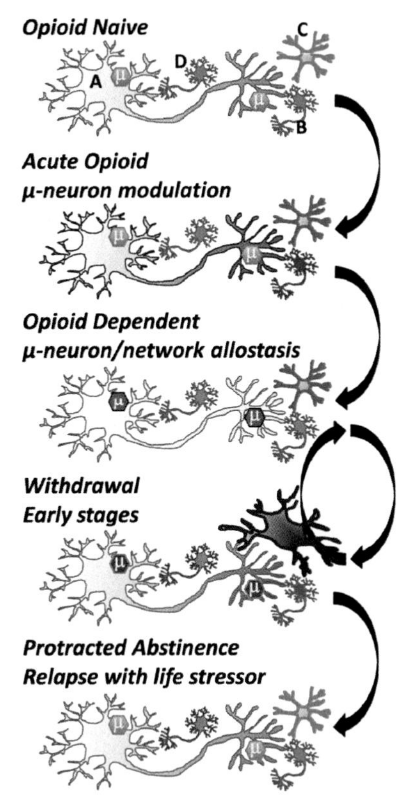

Figure 21 ***Cellular adaptations to opioids [217].*** In the "Opioid Naïve" state, μ opioid receptor-expressing cells (e.g., GABAergic neuron; top panel, neuron A) modulate reward circuitry and many other neurons (top panel, neuron B) in the brain and periphery. Brain microglia (top panel, cell C) are normally in a quiescent surveillance state. Acute opioid administration activates μ opioid receptors, which couple to inhibitory G proteins and generate an active μ-signalosome (μ) that inhibits cell activity and neurotransmitter release. In the brain, mu opioid receptors are often on GABAergic inhibitory neurons and thus can activate adjacent neurons (B) via disinhibition. Microglia may also be activated directly by opioid drugs, although this is debated. In the "Opioid Dependent" state, opioid receptor-expressing neurons adapt to the continued presence of drug. Modifications occur in the μ-signalosome, neuronal proteome, and transcriptome, alongside modifications to neuronal morphology (e.g., spine density and dendritic arborization). Maintained activation of μ-opioid receptors begins a process of cellular, network, and system adaptations. Upon drug cessation, withdrawal is triggered, and the adaptive (allostatic) changes that occur in the drug-dependent state rebound in a temporally (hours to years) orchestrated resetting of neurons and networks. Withdrawal symptoms are opposing the acute actions of opioids whereby neurons inhibited by opioid activation become excited during withdrawal. Other cells are engaged during withdrawal, including microglia (cell C) and neurons within the anxiogenic learning circuitry (cell D). Eventually, weeks to months after drug cessation, many networks re-establish a close-to-normal state, but neurons that encode memory circuits of withdrawal (cell D) and associated memories that drugs relieve aversive states can be triggered by stressful or aversive life events following "Protracted Abstinence". *Green*: resting state, *blue*: inhibited state, *yellow*: near-normal dependent state, *red*: activated state. *Arrows* denote transition between states. *(Taken with permission from Evans CJ, Cahill CM. Neurobiology of opioid dependence in creating addiction vulnerability. F1000Research 2016;5. F1000 Faculty Rev-1748 [217].)*

naloxone-precipitated withdrawal produced motivational withdrawal signs (i.e., elevation of reward thresholds or hypohedonia), somatic symptoms, and an increase in heroin intake [219]. Using place aversion, one hypothesized substrate for conditioned withdrawal in rodents is the extended amygdala [220].

4.6 Opioids, pain, the brain, and hyperkatifeia: a framework for the rational use of opioids for pain and its relevance to addiction

Shurman J, Koob GF, Gutstein HB. Opioids, pain, the brain, and hyperkatifeia: a framework for the rational use of opioids for pain. Pain Medicine 2010;11:1092–8 [213].

In humans, withdrawal from opioids and alcohol can lower pain thresholds (hyperalgesia) and exacerbate pain, and pain is one of the main triggers of relapse to addiction in methadone-maintained individuals. Such a hyperalgesic state can persist for up to 5 months in abstinent individuals with opioid addiction, and addicted individuals with more pain sensitivity also exhibit greater cue-induced craving at this time point [221]. Indeed, even acute opioid administration can produce hyperalgesia in humans [222]. Heighted pain perception has also been observed during alcohol withdrawal in humans [223]. In animal models, withdrawal from the chronic self-administration of opioids and alcohol produces hyperalgesia (i.e., lower pain thresholds; [223]). Withdrawal from opioids in humans produces the well known symptoms of physical withdrawal, such as sympathetic activation, goosebumps, and severe abdominal distress, and a severe negative emotional state that is characterized by anxiety, irritability, emotional pain, and pronounced dysphoria. Animals exhibit behavioral changes during opioid and alcohol withdrawal that reflect negative emotional-like states, such as elevations of reward thresholds (hypohedonia), anxiety-like responses, depressive-like responses, and aversive responses (place aversions).

The term "hyperkatifeia" was devised to describe the exaggerated negative emotional state that is associated with drug withdrawal [213]. Hyperkatifeia was defined as a greater intensity of the constellation of negative emotional/motivational signs and symptoms that are observed during withdrawal from drugs of abuse (derived from the Greek "katifeia" for dejection or negative emotional state). Additionally, parallels were drawn between the neural mechanisms that are responsible for hyperkatifeia and opioid-induced hyperalgesia [213].

For example, evidence suggests that the neural substrates of stress system neuroadaptations that are associated with addiction may overlap with substrates of emotional aspects of pain processing in such areas as the amygdala [224]. The spino (trigemino)-ponto-amygdaloid pathway projects from the dorsal horn of the spinal cord to the mesencephalic parabrachial area and then to the central nucleus of the amygdala. This pathway has been implicated in processing emotional components of pain perception ([225,226]; Fig. 22).

This hypothesis may explain the crosstalk between opioid addiction and chronic pain syndromes, in which some patients may be more prone to the development of hyperkatifeia during withdrawal. An allostatic view would suggest that opioid-induced hyperalgesia and hyperkatifeia would be much

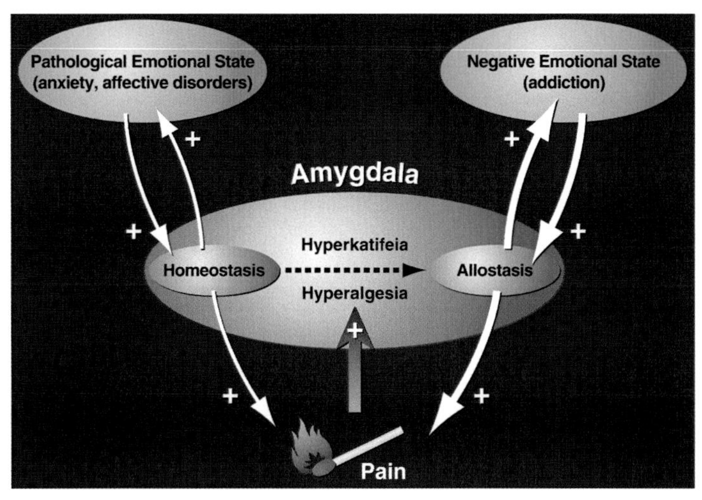

Figure 22 ***Schematic diagram summarizing the hypothesized relationship between addiction and pain [213].*** Pathological emotional states are known to exacerbate pain. We hypothesize that, in parallel, the negative emotional state of drug withdrawal and protracted abstinence can also exacerbate pain; conversely, pain can exacerbate both pathological emotional states and addiction. Hyperalgesia, an increased sensitivity to pain, caused by opioid treatment could indicate the parallel development of hyperkatifeia, or increased sensitivity to negative emotions. Hypothetically, the converse could occur in addiction (hyperkatifeia reflecting underlying hyperalgesia). The conceptual framework for such changes involves a break from emotional homeostasis termed allostasis (stability through change) in neurobiological mechanisms in the extended amygdala. *(Taken with permission from Shurman J, Koob GF, Gutstein HB. Opioids, pain, the brain, and hyperkatifeia: a framework for the rational use of opioids for pain. Pain Medicine 2010;11:1092–8 [213].)*

more likely to occur during chronic opioid administration if excessive opioids are administered, although this remains an untested hypothesis in humans. Shurman et al. [213] argued that because of overdosing, rapid escalation (overshooting), pharmacokinetic variables, or genetic sensitivity, the body will react to that perturbation by engaging opponent processes of hyperalgesia and hyperkatifeia that are mediated by significant crosstalk in such brain structures as the central nucleus of the amygdala. The repeated engagement of opponent processes without time for the system to reestablish homeostasis may engage such allostatic mechanisms. Such a framework suggests that the manifestation of opioid-induced hyperalgesia has important clinical implications: (i) the opioid has exceeded the amount that is effective for pain control, and (ii) susceptible individuals are at risk for developing hyperkatifeia, the unstable emotional and behavioral state that underlies addiction [213].

The neural substrates that underlie allostatic emotional changes that are seen in addiction include decreases in reward function that are mediated by neurochemical changes in the ventral striatum (molecular neuroadaptations in medium spiny neurons and loss of function of the dopamine system) and increases in brain stress system function that are mediated by neurochemical changes in the extended amygdala (recruitment of CRF, dynorphin, and norepinephrine; [227]). To this point, pain-responsive neurons are also abundant in the lateral part of the central nucleus of the amygdala

[228], an area that may also be responsible for negative emotional responses to abused drugs. Indeed, systemic or intraamygdalar (i.e., in the central nucleus of the amygdala) administration of a selective CRF_1 receptor antagonist inhibited anxiety-like behavior and nocifensive pain responses that were produced in an animal model of arthritis [229]. Perhaps even more compelling, opioid withdrawal and alcohol withdrawal in animal models of compulsive-like self-administration produce greater anxiety-like responses and hyperalgesia, both of which are blocked by CRF receptor antagonists [230,231].

4.7 When a good taste turns bad: neural mechanisms underlying the emergence of negative affect and associated natural reward devaluation by cocaine

Carelli RM, West EA. When a good taste turns bad: neural mechanisms underlying the emergence of negative affect and associated natural reward devaluation by cocaine. Neuropharmacology 2014;76:360–9 [232].

The authors here explore the neurobiological mechanisms that may underlie the processes by which natural rewards lose their value in cocaine addiction and provide a framework for understanding the motivating properties of negative affect-driven compulsive drug use. Using the animal model of oromotor responses (taste reactivity) in response to the intraoral infusion of a sweet (e.g., saccharin) solution, the authors showed that rats that self-administered cocaine for 2 h per day exhibited aversive taste reactivity (i.e., gapes/rejection responses) during infusion of the sweet solution that was paired with impending cocaine, similar to aversive responses during the infusion of quinine, a bitter tastant. The expression of this pronounced aversion to the sweet solution predicted the subsequent motivation to self-administer cocaine. Here, an experimental design was employed in which rats were intraorally infused with a distinctly flavored saccharin solution during daily conditioning sessions that predicted the opportunity to self-administer cocaine or saline [233]. Inherent in their design was that the rats had to wait an extended period of time (30 min) to gain access to the drug. The authors hypothesized that the "drug waiting" period allowed a strong association to develop between the tastant that was paired with delayed cocaine and enabled the emergence of a negative affective state as measured by taste reactivity [232]. The rats developed a strong aversion to the tastant that predicted cocaine (but not saline), reflected by changes in orofacial expressions during tastant delivery (gapes instead of licking and later tongue protrusions).

Using single-neuron recording by the discrimination of individual waveforms with the neurophysiological software system, neurons in the nucleus accumbens that previously responded with inhibition during infusion of the palatable sweet solution shifted to excitatory activity during infusion of the cocaine-devalued tastant [233]. Again, in parallel with the behavioral studies, the excitatory profile of neuronal firing was the same excitatory response profile that was typically observed during the infusion of quinine. The authors suggested that these results show that the once-palatable sweet becomes aversive following its association with impending but delayed cocaine, and nucleus accumbens neurons encode this aversive state. Parallel electrochemical studies that used fast scan cyclic voltammetry to measure rapid dopamine (subsecond) release in response to a tastant showed a shift (from increase to decrease) in rapid nucleus accumbens dopamine release during infusion of the cocaine-paired tastant as the aversive state developed, again resulting in responses that were similar to quinine infusion [234]. The authors also showed dampened dopamine release in the nucleus accumbens and a parallel elevation of intracranial self-stimulation thresholds during the 45 min period before cocaine availability (Fig. 23). This reward devaluation by cocaine was exacerbated by experimentally induced abstinence (30 days), and this effect was correlated with a greater motivation to lever press during extinction [232]. Altogether, these findings suggest that cocaine-conditioned cues elicit a cocaine-need state that is aversive and encoded by a distinct subset of nucleus accumbens neurons and rapid decrease in dopamine signaling, and these changes ultimately promote cocaine-seeking behavior.

In these studies, Carelli and West showed that there is a learned aversive state that occurs when a sweet signals impending but delayed drug availability, and this state extends into withdrawal and protracted abstinence. The authors hypothesize that the greater motivated behavior toward cocaine following this learned association may be a consequence of the development of a negative affective state that is a consequence of delayed drug availability [232]. Such a heightened aversive state following cocaine abstinence is consistent with the negative reinforcement model that is outlined by Koob and associates. Indeed, Carelli and West have provided compelling evidence that the tastant that predicts impending but delayed cocaine availability induces a conditioned negative affective state that is reflected by increases in the activity of distinct subsets of nucleus accumbens neurons [233,235] and decreases in rapid dopamine release [234] and is corrected by cocaine self-administration.

Earlier work established that the negative affective state that is associated with cocaine, measured by intracranial self-stimulation thresholds, can actually begin within one session of extended access to intravenous cocaine self-administration [114], suggesting that the opponent processes that are elegantly characterized in the tastant studies outlined above are an early neuroadaptation to the overactivation of dopamine release and activation of the nucleus accumbens that are induced by cocaine. Note that others have observed a decrease in dopamine release with extended access to cocaine self administration [131]. Here, rats were implanted with indwelling intravenous catheters over the course of 3 weeks of cocaine self-administration. Dopamine signaling was measured using fast-scan cyclic voltammetry simultaneously in the ventral medial striatum and dorsal lateral striatum in rats. The results showed that dopamine signaling in the ventral medial striatum declined while phasic dopamine release in the dorsal lateral striatum progressively emerged during drug taking over the course of weeks [131]. A parallel to this study is that such negative affective states that are associated with precipitated withdrawal from opioids can be readily paired with previous neutral cues, produce conditioned withdrawal, and have motivating properties to promote opioid seeking [219]. Support for this hypothesis was direct evidence that a drug-paired taste cue that was paired with naloxone-precipitated withdrawal elicited withdrawal and predicted cocaine self-administration in rats [236].

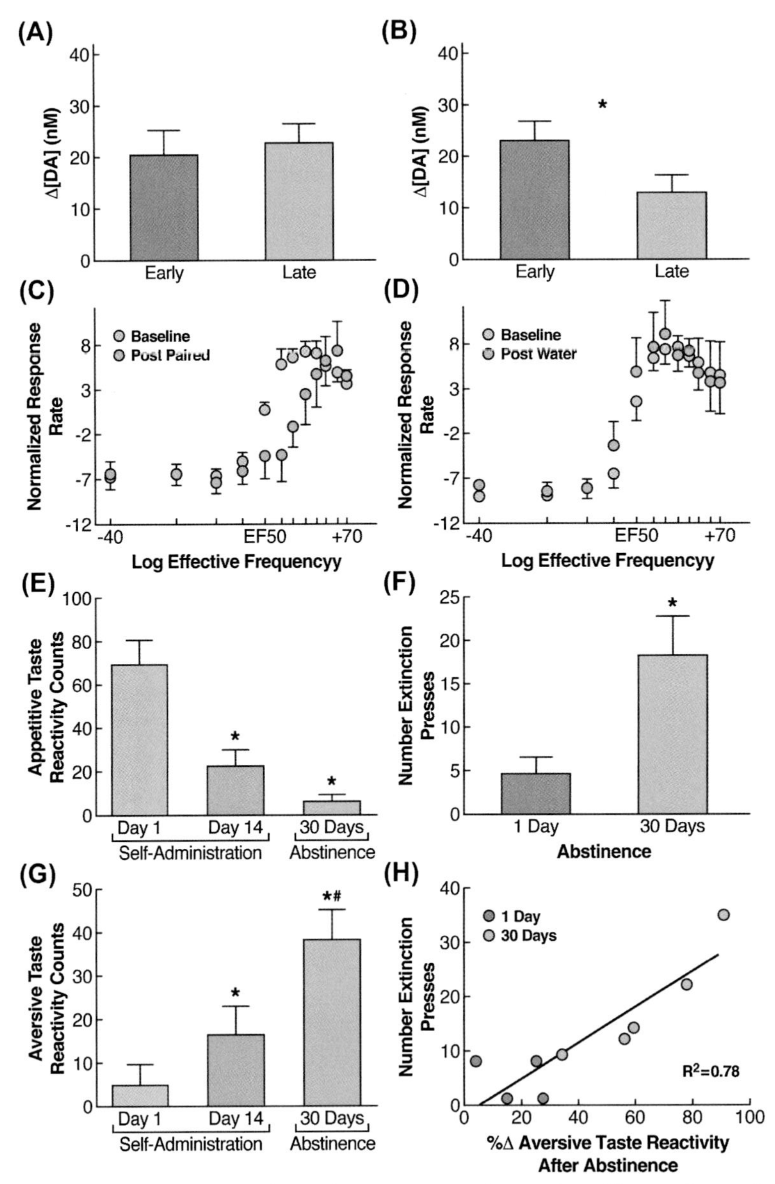

Figure 23 ***Neural mechanisms underlying the emergence of negative affect and associated natural reward devaluation by cocaine [232].*** (A–D) Dampened dopamine release during the 45-min period before cocaine availability, revealed by intracranial self-stimulation. (A) Average dopamine concentration (mean ± SEM) in the baseline period for all rats receiving the unpaired tastant is stable across trials (early versus late), $p < 0.05$. (B) Dopamine concentration decreased significantly across trials for rats receiving the cocaine-predictive taste cue. *Significant difference $p < 0.05$. (C) Baseline ICSS threshold curves were established (open circles, mean ± SEM), then rats received infusions of the cocaine-paired tastant, and the curves were redetermined (*closed circles*). The cocaine-paired tastant

4.8 Gene transcription factor CREB: role in positive and negative affective states of alcohol addiction

Pandey SC. The gene transcription factor cyclic AMP-responsive element binding protein: role in positive and negative affective states of alcohol addiction. Pharmacology & Therapeutics 2004; 104:47–58 [237].

The hypothesis here is that signaling of the gene transcription factor cyclic adenosine monophosphate (cAMP) response element binding protein (CREB) in the extended amygdala plays a role in the positive and negative affective states that are associated with alcohol addiction ([237]; Fig. 24). In this conceptual framework, increases in positive affective states and negative affective states promote and maintain alcohol addiction and form a key element in the hedonic dysregulation that is associated with compulsive drinking.

The cAMP-protein kinase A second-messenger system has long been implicated in alcohol tolerance and dependence [238,239] but in an opposite direction from opioids. Brief exposures to alcohol potentiate receptor-activated cAMP production, and chronic exposure to alcohol decreases receptor-stimulated cAMP production [240]. Lower cAMP production has been considered a cellular model of alcohol dependence. Stimulated cAMP levels are abnormally low in NG108-15 neural cell cultures after alcohol withdrawal but return to normal levels when alcohol is reintroduced [241]. CREB phosphorylation has been shown to be lower in this cortex during alcohol withdrawal [242,243]. Lower CREB phosphorylation and decreases in Ca^{2+}/calmodulin-dependent protein kinase IV (CaMKIV) protein levels have also been observed in the central nucleus of the amygdala during alcohol withdrawal and have been linked to the increase in

right-shifted ICSS thresholds significantly. *Significant differences in ICSS response rates. (D) Baseline ICSS threshold curves were established (*open circles*), and then all rats received infusions of water, and the curves were redetermined (*closed circles*). Water infusions did not change ICSS response rates. (E–H) Abstinence enhances the devaluation of saccharin that predicts impending but delayed cocaine availability during the first 15 trials of saccharin delivery. On day 1 of training, rats exhibited predominantly appetitive taste reactivity counts during saccharin infusion (E and F, left columns) but aversive counts after saccharin was paired with cocaine for 14 days (E and F, middle columns). Following 30 days of abstinence, the increase in aversive taste reactivity counts was exacerbated (E and F, right columns). $^*p < 0.05$, significantly different from day 1; $^\#p < 0.05$, significantly different from day 14;$\hat{}$, trend toward significantly different from day 14. (G) Experimenter-controlled abstinence (30 days) led to an increase in the number of lever presses under extinction compared to only 1 day of abstinence. $^*p < 0.05$, significant difference in lever presses. (H) The number of lever presses during extinction was positively correlated with the percent change in cocaine-induced devaluation following abstinence. *(Panels A–D modified from Wheeler RA, Aragona BJ, Fuhrmann KA, Jones JL, Day JJ, Cacciapaglia F, Wightman RM, Carelli RM. Cocaine cues drive opposing context dependent shifts in reward processing and emotional state. Biological Psychiatry 2011;69:1067–74 [234]. Panels E–H taken with permission from Carelli RM, West EA. When a good taste turns bad: neural mechanisms underlying the emergence of negative affect and associated natural reward devaluation by cocaine. Neuropharmacology 2014;76:360–9 [232].)*

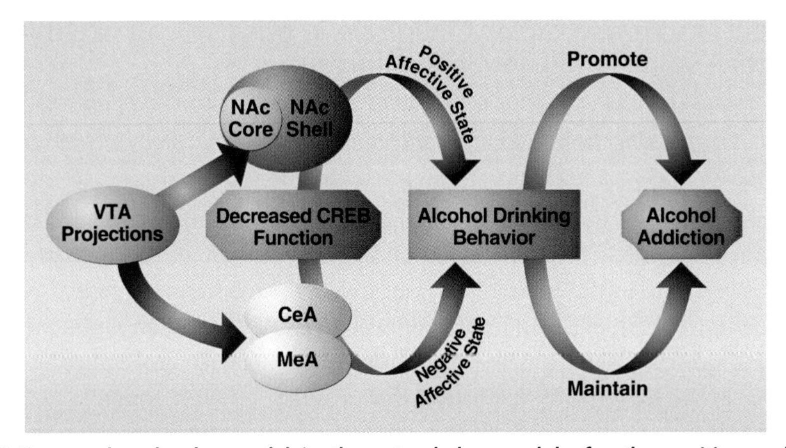

Figure 24 ***Proposed molecular model in the extended amygdala for the positive and negative affective states of alcohol addiction [237].*** The decrease in cyclic AMP-responsive element binding protein (CREB) function in the shell of nucleus accumbens (NAc), which receives dopaminergic projections from ventral tegmental area (VTA), might be associated with euphoric, rewarding (positive affective state) properties of alcohol use. The decrease in CREB function in the central and medial nucleus of amygdala (CeA and MeA, respectively) might be involved in a negative affective state related to alcohol withdrawal or as a result of pre-existing conditions. *(Taken with permission from Pandey SC. The gene transcription factor cyclic AMP-responsive element binding protein: role in positive and negative affective states of alcohol addiction. Pharmacology and Therapeutics 2004;104:47–58 [237].)*

anxiety-like responses that is associated with acute alcohol withdrawal [244]. No changes were observed during alcohol exposure *per se*. The infusion of a protein kinase A activator in the central nucleus of the amygdala normalized CREB phosphorylation and prevented the development of anxiety-like responses during alcohol withdrawal [244]. Others have shown that CREB phosphorylation decreases in the nucleus accumbens during voluntary alcohol exposure and remains lower up to 72 h into withdrawal [245]. Consistent with these results, lower CREB phosphorylation after the administration of a protein kinase A inhibitor in the central nucleus of the amygdala produced anxiety-like responses [246], and CREB mutant mice, which are deficient in transcriptionally active α and β CREB isoforms, exhibit increases in anxiety-like responses [247–249].

Preclinical and clinical studies suggest a relationship between high anxiety-like states and greater alcohol intake (see Volume 3, *Alcohol*). As such, lower CREB phosphorylation in the central nucleus of the amygdala may be responsible for the increases in anxiety-like responses in alcohol-dependent animals and thus the increase in drinking in dependent animals, possibly through actions on NPY systems, which is a CREB target gene [250]. The normalization of CREB phosphorylation by infusing the protein kinase A activator S_p-cAMPS in the central nucleus of the amygdala prevented anxiety-like effects in rats during alcohol withdrawal [244] and also reversed the decrease in the expression of NPY during alcohol withdrawal [246]. CREB regulates the expression of the gene that encodes NPY, and the decrease in NPY is associated with anxiety and excessive drinking. The decrease in the function of CREB in the central nucleus of the amygdala may regulate both anxiety and alcohol intake via the lower expression of NPY [251]. Similar results have been observed in the shell of the nucleus accumbens during acute nicotine withdrawal, in which there were reductions of total CREB and phosphorylated CREB [252].

What has been difficult to resolve is the way in which lower CREB phosphorylation in the central nucleus of the amygdala yields a decrease in NPY, hence resulting in increases in anxiety and drinking, whereas an increase in CREB phosphorylation in the nucleus accumbens with cocaine and opioids leads to a gain of dynorphin that leads to dysphoria and greater cocaine and morphine intake [253,254]. One possibility is that there are simply differences between different drugs of abuse. Although CREB alterations occur during neuroadaptations to all drugs of abuse, the valence of these changes may differ between drugs. Another possibility is that differential molecular changes occur in different parts of extended amygdala circuitry. Another possibility is that the anxiety-like responses that are associated with greater alcohol drinking during acute withdrawal reflect lower CREB function in the central nucleus of the amygdala [244], but greater CREB function in the nucleus accumbens occurs during active drug taking to decrease sensitivity to the rewarding effects of the drug and the development of tolerance [254,255]. Clearly, the relationship between these molecular changes and the motivational effects of drug addiction requires further research.

In summary, Pandey argued that CREB underexpression in the central nucleus of the amygdala promotes anxiety-like responses, possibly through a decrease in the function of NPY that in turn promotes excessive drinking, but lower CREB function in the nucleus accumbens promotes an increase in the intake of drugs to achieve an increase in reward.

4.9 Epigenetic basis of the dark side of alcohol addiction

Pandey SC, Kyzar EJ, Zhang H. Epigenetic basis of the dark side of alcohol addiction. Neuropharmacology 2017;122:74–84 [256].

The hypothesis argued here is that epigenetic modifications in the amygdala provide a molecular basis of negative affective symptoms, also known as the dark side of addiction, and such epigenetic modifications are engaged by repeated alcohol use, genetic vulnerability, and alcohol exposure during crucial developmental periods. Data from rodent models show that acute alcohol exposure produces an anxiolytic-like response that is mediated at least partially by the inhibition of histone deacetylase (HDAC), an increase in histone acetylation, an increase in the opening of chromatin, an increase in synaptic plasticity-related genes (e.g., brain-derived neurotrophic factor [BDNF]), and an increase in activity-related cytoskeleton-associated protein (ARC) and NPY in the amygdala ([251,257,258]; Fig. 25; see *Section 4.8. Gene transcription factor CREB: role in positive and negative affective states of alcohol addiction*). However, following chronic alcohol exposure and subsequent withdrawal, the opposite effects occur, with decreases in histone acetylation through an increase in HDAC activity and decreases in the expression of BDNF, ARC, and NPY in amygdala circuitry that produce anxiety-like responses.

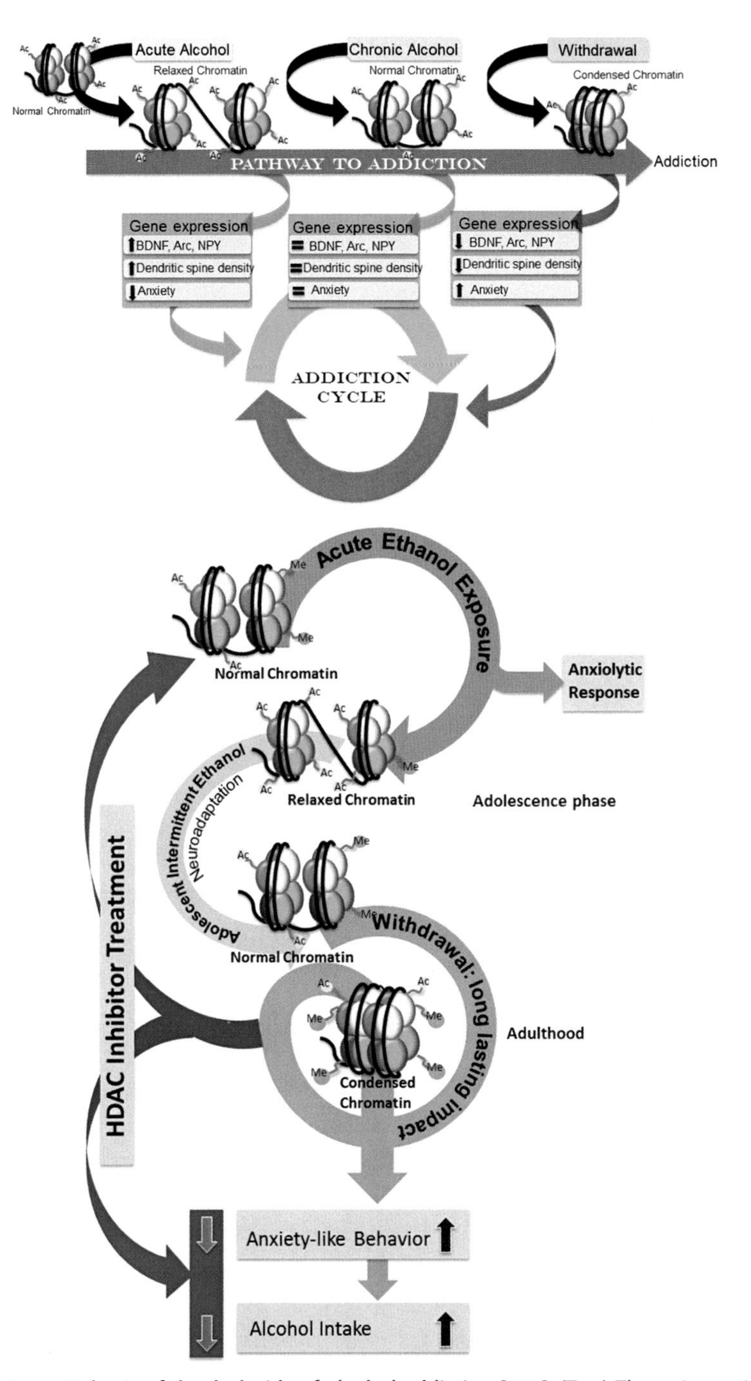

Figure 25 ***Epigenetic basis of the dark side of alcohol addiction [256].*** (Top) The epigenetic allostasis model of the dark side of alcohol addiction. The epigenome is dynamically altered by acute alcohol, chronic alcohol, and alcohol withdrawal in the amygdala and is crucial for the progression from casual

From a genetic vulnerability perspective, data have shown that innately higher expression of the HDAC2 isoform and adolescent exposure to alcohol both lead to a deficit in global and gene-specific histone acetylation in the amygdala that is associated with a decrease in the expression of several synaptic plasticity-associated genes and contributes to heightened anxiety-like responses and excessive alcohol intake [259]. Altogether, these studies demonstrate that the epigenome can undergo allostatic reprogramming in amygdaloid circuitry during various stages of excessive alcohol exposure. The authors hypothesize that chromatin opening using pharmacological or genetic manipulations to inhibit HDACs may present opportunities for novel drug development and treatment options for alcohol use disorders [260].

The authors have generated a body of work that provides insights into the molecular events that drive the dark side of addiction, specifically in the amygdala. Much work has supported the hypothesis that pro-stress neuropeptides, such as CRF and dynorphin, are engaged during alcohol and drug withdrawal, whereas others, such as NPY, have been hypothesized to buffer such pro-stress activation [261]. Pandey and colleagues outlined a compelling molecular framework for such functional changes that has a key epigenetic component. Such an epigenetic component provides a mechanism by which

use to addiction. Acute alcohol exposure opens the chromatin by inhibiting histone deacetylases (HDACs), thereby increasing histone acetylation and the expression of crucial synaptic plasticity-related genes, such as brain-derived neurotrophic factor (Bdnf), activity regulated cytoskeleton-associated protein (Arc), and neuropeptide Y (Npy) in the amygdala, leading to anxiolysis. However, these biological parameters normalize (=) with continued alcohol exposure. During alcohol withdrawal, HDAC activity is increased, leading to condensed chromatin structure and decreased expression of these genes as well as dendritic spine density. As the amygdala controls negative affective states, amygdalar chromatin conformation is critical for the development of anxiety that is seen in alcohol withdrawn animals. Additionally, the condensed chromatin state is associated with greater alcohol preference and self-administration, ostensibly to relieve negative affective states. This switch from consuming alcohol for its pro-social, anxiolytic effects to drinking as a means to relieve negative affective states is critical to the addiction cycle and is reflected by the underlying epigenome of the amygdala. (Bottom) Alcohol exposure during development leads to persistent changes in epigenetic mechanisms in the amygdala and produces long-lasting anxiety and excessive alcohol intake in adulthood. Exposure to adolescent intermittent ethanol (AIE) during this critical developmental period causes increased HDAC2 in the amygdala, leading to lower global and Bdnf- and Arc-specific histone acetylation. Lower BDNF and Arc expression in the adult amygdala of AIE animals is accompanied by greater alcohol preference and anxiety-like behavior. Aside from histone acetylation, lysine demethylase 1 (LSD1) and particularly the neuron-specific splice variant Lsd1+8a decrease in the AIE adult amygdala. This decrease is accompanied by a resultant increase in histone H3K9 dimethylation (H3K9me2) both globally and at the Bdnf exon IV promoter, possibly explaining the decrease in BDNF expression that is seen in these rats. Thus, higher HDAC2 and lower LSD1 that are associated with AIE-induced chromatin remodeling may provide novel targets for intervention in alcohol addiction and comorbid anxiety. Interestingly, HDAC inhibitor treatment in adulthood attenuates anxiety-like and alcohol drinking behaviors and also normalizes deficits in the histone acetylation of Bdnf and Arc gene promoters in the amygdala. *(Taken with permission from Pandey SC, Kyzar EJ, Zhang H. Epigentic basis of the dark side of alcohol addiction. Neuropharmacology 2017;122:74–84 [256].)*

excessive alcohol exposure causes an increase in negative emotional states and lays a foundation whereby developmental and genetic vulnerabilities interface with such exposure to drive addiction. The challenge will be to discover a means to reverse such molecular loadings to protect against vulnerability to addiction and reverse changes that have already occurred. A subpart of this challenge will be to identify a means of selectively modifying such allostatic changes without globally affecting gene expression.

5. Theories of the preoccupation/anticipation stage

5.1 Motive circuits: prefrontal cortex/ventral striatal hypotheses of addiction

Jentsch JD, Taylor JR. Impulsivity resulting from frontostriatal dysfunction in drug abuse: implications for the control of behavior by reward-related stimuli. Psychopharmacology 1999; 146:373–90 [262].

The thesis of this conceptual framework was that regions of the frontal cortex are involved in inhibitory response control that is directly affected by long-term exposure to drugs of abuse. The resulting frontal cortical cognitive dysfunction produces an inability to inhibit unconditioned or conditioned responses that are elicited by drugs [262]. Drug-seeking behavior was hypothesized to be attributable to two related phenomena: an increase in incentive motivational qualities of the drug and drug-associated stimuli through limbic/amygdalar dysfunction and impairments in inhibitory control through frontal cortical dysfunction.

The authors postulated that drugs and drug-associated stimuli rely on dopaminergic function in the nucleus accumbens, particularly the shell of the nucleus accumbens, to modulate behavioral output. They further postulated that the amygdala and prefrontal cortex contribute to learning associations between drugs and external and internal cues and also thus may contribute to impulsivity ([262]; Fig. 26). The amygdala is linked to the control of behavior by reward-related stimuli, in which the amygdala plays a role in mediating incentive learning and the incentive value of conditioned stimuli. Excitotoxic lesions of the central nucleus of the amygdala blocked the amphetamine-induced potentiation of conditioned reward [75], blunted the suppression of behavior that is produced by a fear stimulus [263], and prevented the development of autonomic responses to primary or secondary conditioned stimuli [264–267]. In contrast, excitotoxic lesions of the basolateral amygdala produced impairments in the ability of stimuli to effect instrumental responding [263]. Excitotoxic lesions of the basolateral amygdala but not central nucleus of the amygdala blocked cocaine-seeking behavior under a second-order schedule [74] and also blocked the cue-induced reinstatement of drug responding in rats ([268]; see Table 3). However, excitotoxic lesions of the basolateral amygdala did not block the acquisition of heroin-seeking behavior under a second-order schedule [269].

Many of these effects of drugs of abuse in the amygdala were hypothesized by the authors to be facilitated by the activation of dopaminergic substrates within the amygdala. For example, the intracerebral administration of amphetamine in the amygdala facilitated

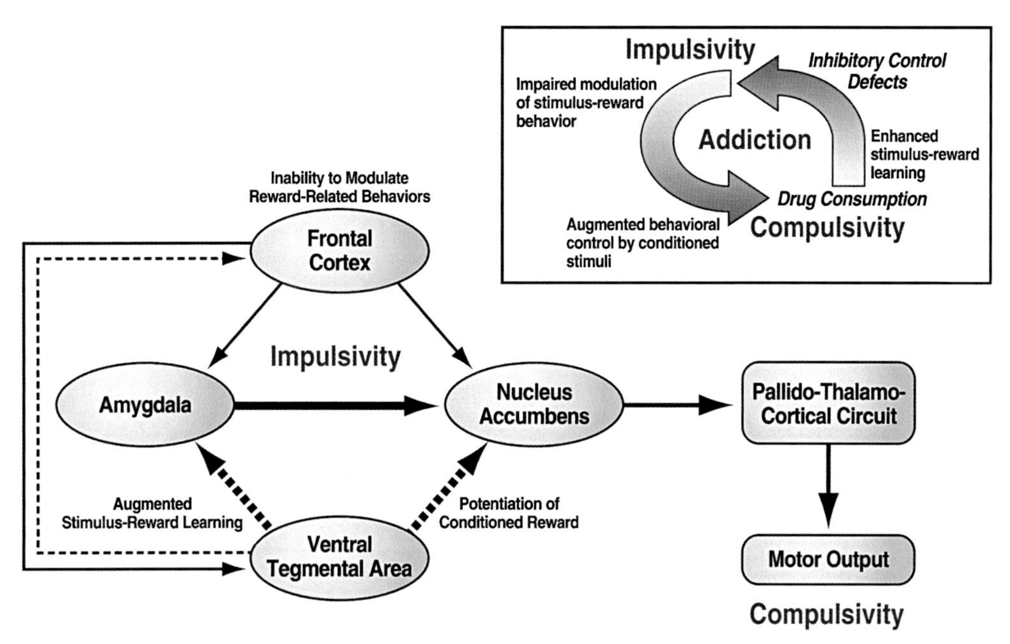

Figure 26 ***Model for functional synergism between augmented conditioned reward and deficits in the ability to modulate reward-related behavior at the cognitive level in drug addiction [262].*** Increased stimulus-reward learning and conditioned reward, produced by subcortical dopaminergic hyperactivity (*bold dashed line arrows*), may result in an augmented impulse to seek drugs. Drug-induced impulsivity and cue-elicited drug seeking may be mutually reinforcing, in which repeated drug consumption may progressively augment impulsivity, resulting in greater susceptibility to subsequent relapse. Moreover, the inhibitory control functions of the frontostriatal system that act to modulate reward-related behaviors likewise may be impaired with chronic drug use because of alterations of frontal cortical dopaminergic function (*gray dashed-line arrow*) and impairment in the frontal cortical modulation of subcortical systems by descending corticostriatal projections (*gray single-line arrows*). This theoretical model (see inset to figure) predicts that compulsive drug-seeking behavior may be seen as a functional synergism between augmented conditioned reward and deficits in the ability to modulate reward-related behaviors at a cognitive level. *(Taken with permission from Jentsch JD, Taylor JR. Impulsivity resulting from frontostriatal dysfunction in drug abuse: implications for the control of behavior by reward-related stimuli. Psychopharmacology 1999;146:373–90 [262].)*

the acquisition of stimulus-reward associations [270], and psychostimulants in general facilitated memory consolidation through actions in the amygdala [271]. Thus, Jentsch and Taylor hypothesized that during drug self-administration, the synaptic release of dopamine increases in the amygdala and helps form associations between the rewarding qualities of the drug and exteroceptive stimuli [262]. In the context of drug addiction, an increase in dopamine release in the nucleus accumbens that is produced by the repeated administration of a drug of abuse results in an increase in responding for a conditioned reinforcer, and the acquisition of stimulus-reward associations is facilitated by drug-induced neuroadaptations in the amygdala [262].

Notably, Jentsch and Taylor [262] assigned an equally important role, or a possibly primary role, to the frontal cortex in the impulsivity of drug dependence. They argued that internal motivational states to seek food, water, sex, and other primary reinforcers are regulated by an active inhibitory control mechanism in frontal striatal systems. This inhibitory control mechanism would transiently suppress rapid conditioned responses and reflexes so that slower cognitive processes can guide behavior. Lesions of the frontal cortex can lead to marked cognitive impairments, including disinhibition [272], and a preferential response for immediate small rewards over delayed rewards, in which the calculation of future outcomes is not possible [273]. Lesions of the dorsolateral prefrontal, lateral orbitofrontal, or ventromedial frontal cortex resulted in greater perseveration and deficits in inhibition [274,275]. Additionally, there is strong evidence of a role for dopamine in mediating some of these same cognitive functions in the frontal cortex [276–279]. Thus, an impairment in inhibitory control, combined with progressive enhancement of the conditioned reinforcing effects of drug-associated stimuli, would represent a state whereby reward-related stimuli could dominate responding.

Chronic high-dose administration of amphetamine, cocaine, and even cannabinoids can lead to reductions of basal frontal cortex dopamine transmission [280–283]. Similar effects have been observed with phencyclidine and have shown a correlation between performance impairments in cognitive function and dopaminergic hypofunction in the dorsolateral, prefrontal, and ventromedial frontal cortices [284]. Consistent with these results, one study showed tolerance to the footshock-enhanced release of dopamine in the medial prefrontal cortex 1 week after the chronic administration of cocaine [285]. However, others have observed greater dopamine utilization in the frontal cortex following the chronic administration of amphetamine [286] and an increase in the footshock-induced release of dopamine in the medial prefrontal cortex 1 week after the chronic administration of D-amphetamine [287]. The dysregulation of dopaminergic neurotransmission in the prefrontal cortex was hypothesized to underlie the loss of inhibition of reward-seeking behavior and the greater susceptibility to drug-induced relapse [262]. Finally, Jentsch and Taylor [262] hypothesized that the inhibitory modulation of reward-seeking behavior may critically depend on corticostriatal projections from the medial frontal cortex to the caudate nucleus and nucleus accumbens core and shell. Dysfunction of the frontal cortex or the hypofunction of dopamine activity in the frontal cortex can activate subcortical dopamine systems [288,289] (for review, see Ref. [290]). A decrease in dopamine in the prefrontal cortex induced locomotor activation through the activation of dopamine in the nucleus accumbens [291]. This functional interaction between the prefrontal cortex and nucleus accumbens is highlighted by the observation that animals that were more vulnerable to acquiring intravenous drug self-administration exhibited lower dopaminergic activity in the prefrontal cortex [289]. Such hypoactivity was hypothesized to contribute to locomotor sensitization [292].

In summary, in the Jentsch and Taylor model, compulsive drug-seeking behavior reflects functional synergism between deficits in the frontal striatal system that are driven by cortical dopamine hypofunction and augmented conditioned reward, presumably driven by subcortical (nucleus accumbens and amygdala) dopamine activation. The primary focus of this hypothesis was on changes in function in the prefrontal cortex that may underlie the loss of inhibition of reward-seeking behavior.

5.2 Frontal cortex dysfunction, cognitive performance, and executive function: disruption of frontocerebellar circuitry and function in alcoholism

Pfefferbaum A, Sullivan EV, Mathalon DH, Lim KO. Frontal lobe volume loss observed with magnetic resonance imaging in older chronic alcoholics. Alcoholism: Clinical and Experimental Research 1997;21:521–9 [293].

Sullivan EV, Harding AJ, Pentney R, Dlugos C, Martin PR, Parks MH, Desmond JE, Chen SH, Pryor MR, De Rosa E, Pfefferbaum A. Disruption of frontocerebellar circuitry and function in alcoholism. Alcoholism: Clinical and Experimental Research 2003;27:301–9 [294].

Classic neuropsychological behaviors that are typical of frontal lobe dysfunction characterize individuals with alcohol use disorder and include impaired judgment, blunted affect, poor insight, social withdrawal, reduced motivation, and attentional deficits (for review, see Refs. [295–297]). Neuroimaging and neuropathological studies have provided significant evidence of alcohol-induced abnormalities in the frontal lobes and cerebellum that are particularly relevant to these alcohol-related impairments in cognitive function [293,298,299]. More formal testing has raised the hypothesis that the frontal lobe component of these deficits can be impairments in executive function and short-term or working memory [295,300,301]. Postmortem studies [298] and *in vivo* magnetic resonance imaging (MRI) studies have shown frontal cortex abnormalities in individuals with alcohol use disorder ([293]; Fig. 27). Deficits in balance and gait in individuals with alcohol use disorder have been linked to abnormalities in cerebellar function [302,303], and the cerebellum may be involved in some classic frontal lobe functions, raising the hypothesis of an overall frontocerebellar circuitry deficit in alcohol use disorder [294].

The link between cognitive impairments and abnormalities in the frontal cortex were predicted by neuropsychological studies that revealed alcohol-dependent changes in frontal lobe function [295,303,304]. These performance deficits correlated with resting frontal lobe metabolism [305]. Perhaps more compelling were data that showed that tasks that produce robust frontal lobe activation may be especially sensitive to alcohol-related changes in brain function. Different patterns of frontal lobe activation were observed in individuals with alcohol use disorder compared with controls while performing visual attentional and working memory tasks [293]. Similarly, in a functional MRI (fMRI) study, individuals with alcohol use disorder and control subjects showed no difference in performance on a working memory task, but fMRI activation was greater in the alcohol group in the high load condition, suggesting that individuals with alcohol use disorder may require more extensive activation of frontal lobe function to maintain normal performance in such cognitive tasks [306].

5.3 Brain circuitry for addiction from brain imaging studies

Volkow ND, Fowler JS, Wang GJ. The addicted human brain: insights from imaging studies. Journal of Clinical Investigation 2003;111:1444–51 [92].

Brain imaging studies of drug addiction have largely explored addiction in humans using such imaging technologies as positron emission tomography (PET) and fMRI.

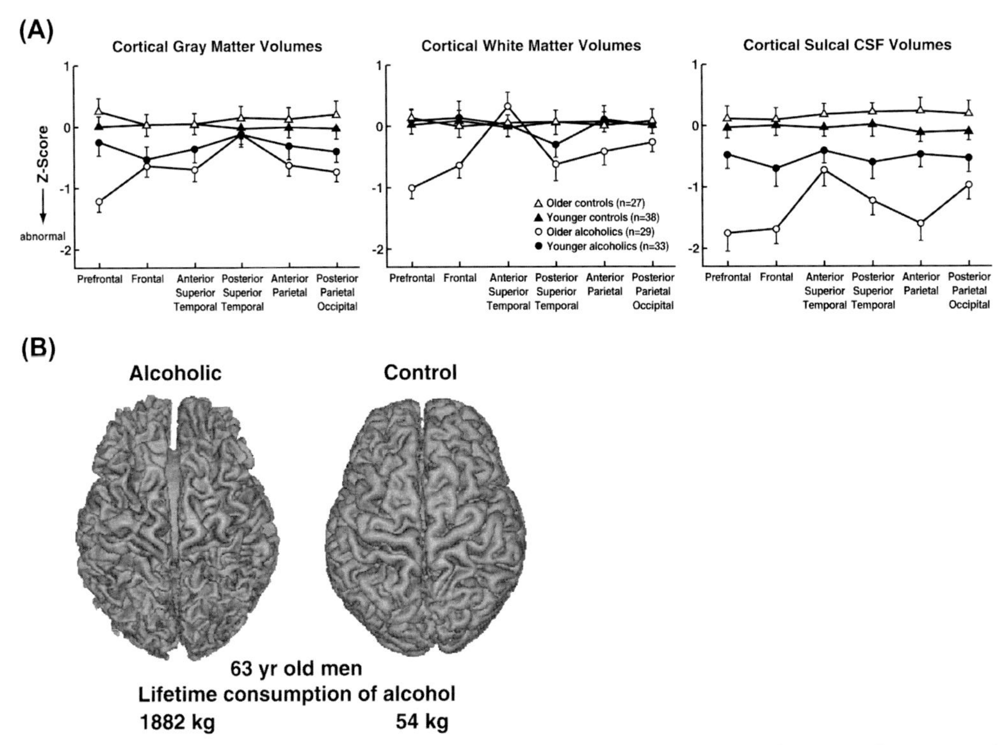

Figure 27 ***Frontal cortex dysfunction, cognitive performance, and executive function: disruption of frontocerebellar circuitry and function in alcoholism [543].*** (A) Volume profiles of (left) cortical gray matter, (middle) cortical white matter, and (right) cortical sulcal cerebrospinal fluid in two control and two alcoholic groups. For all measures, lower scores reflect greater abnormality. The two alcoholic groups had distinctly different profiles on three magnetic resonance imaging measures. Only the older alcoholic group had a cortical gray matter deficit in the prefrontal region. The white matter deficit of the older alcoholic group was greatest in the prefrontal, frontal, and posterior superior temporal regions. Both alcoholic groups had widespread regional sulcal enlargement. The enlargement of the older group was especially pronounced in the prefrontal, frontal, and parietal regions. (B) Three-dimensional renderings of the brains of two men at age 63. To produce these images, the scalp, skull, and cerebrospinal fluid were digitally peeled away to reveal the grooves (i.e., sulci) and ridges (i.e., gyri) that mark each brain's external surface. These images of living brains are comparable to postmortem photographs of brains for pathological study. The control brain is from a healthy social drinker who has an estimated lifetime consumption of 54 kg of pure alcohol. The alcoholic brain is from a male alcoholic who reported a history of heavy drinking over the past 32 years, with an estimated lifetime consumption of 1882 kg of pure alcohol. The shriveled appearance (i.e., wider sulci and narrower gyri) of the alcoholic brain sharply contrasts with the control brain's relatively plump appearance (i.e., well-filled gyri and narrower sulci) and reflects the tissue shrinkage associated with heavy drinking. Tissue shrinkage has widened the interhemispheric fissure in the alcoholic brain, exposing the bundle of fibers connecting the two hemispheres (i.e., corpus callosum). In contrast, the corpus callosum can barely be seen in the control brain. *(Taken with permission from Pfefferbaum A, Sullivan EV, Mathalon DH, Shear PK, Rosenbloom MJ, Lim KO. Longitudinal changes in magnetic resonance imaging brain volumes in abstinent and relapsed alcoholics. Alcohol Clin Exp Res 1995;19:1177–91 [543].)*

Positron emission tomography is based on the use of radiotracers that are labeled with isotopes that can be measured in the brain at very low concentrations. Positron emission tomography can measure labeled compounds that selectively bind to receptors, transporters, or enzymes. fMRI is based on changes in magnetic properties in brain tissue. During brain activation, there is an excess of arterial blood that is delivered to a given region and a change in the ratio of deoxyhemoglobin to oxyhemoglobin, which have different magnetic properties.

Based on the results of brain imaging studies, Volkow and colleagues proposed four circuits that are disrupted in drug addiction: (*i*) reward, localized to the nucleus accumbens and ventral pallidum, (*ii*) motivation/drive, localized to the orbitofrontal cortex and subcallosal cortex, (*iii*) memory and learning, localized to the amygdala and hippocampus, and (*iv*) control, localized to the prefrontal cortex and anterior cingulate gyrus ([92]; Fig. 28). In the reward circuit, imaging studies of subjects who were drug abusers and subjects who were not drug abusers showed a normal acute response of the dopamine system to drugs of abuse. Both groups presented increases in extracellular concentrations of dopamine in the striatum, and the subjects who had the greatest increases in dopamine were those who experienced the subjective effects that were associated with "euphoria" most intensely [307–309]. The route of administration that provided the fastest drug uptake produced the greatest subjective effects [309,310]. However, subjects who became drug abusers or became drug-addicted presented long-lasting decreases in the number of dopamine D_2 receptors in the striatum compared with controls, and cocaine abusers had

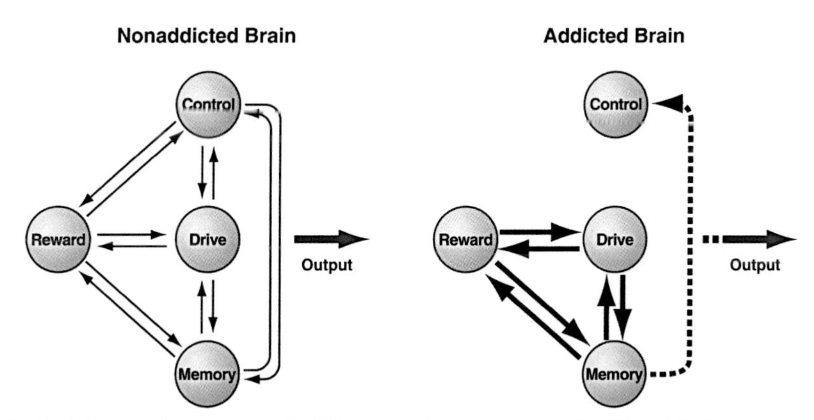

Figure 28 ***Model proposing a network of four circuits that are involved in addiction, reward, motivation/drive, memory, and control [92].*** These circuits work together and change with experience. Each is linked to an important concept: saliency (reward), internal state (motivation/drive), learned associations (memory), and conflict resolution (control). During addiction, the enhanced value of the drug in the reward, motivation, and memory circuits overcomes the inhibitory control that is exerted by the prefrontal cortex, thereby favoring a positive-feedback loop that is initiated by the consumption of the drug and perpetuated by the enhanced activation of the motivation/drive and memory circuits. *(Taken with permission from Volkow ND, Fowler JS, Wang GJ. The addicted human brain: insights from imaging studies. Journal of Clinical Investigation 2003;111:1444–51 [92].)*

lower dopamine release in response to a pharmacological challenge [311,312]. The authors concluded that the decrease in D_2 receptors and decreases in dopamine system activity would result in the lower sensitivity of reward circuits to stimulation by natural reinforcers [92], which would put subjects at greater risk for seeking drug stimulation to temporarily activate these reward circuits [311]. The hypothesis of a decrease in reward processing in addiction is supported by imaging studies in opioid-, cocaine-, and tobacco-dependent subjects [312–314].

Imaging studies have also shown disruption of the motivation/drive circuit in addiction [92]. The orbitofrontal cortex is hyperactive in active cocaine abusers [315], during intoxication [316], during the presentation of cocaine-associated cues [316–318], and during the presentation of cigarette-associated cues [319] but hypoactive during withdrawal in addicted subjects [320,321]. An increase in orbitofrontal activation is associated with obsessive compulsive disorder, and the activation of this structure may contribute to the compulsive nature of drug intake [92]. Amygdala/prefrontal (anterior cingulate) systems may be involved in conditioning or the way in which neutral stimuli that are paired with drug taking acquire reinforcing properties and motivational salience. Such structures as the amygdala and anterior cingulate have been shown to be activated during intoxication and during craving that is induced by drug exposure or drug-related videos [317,322,323].

In summary, Volkow and colleagues hypothesized a control circuit in drug addiction based on one of the most robust findings from imaging studies of individuals with substance use disorder, showing abnormalities in the prefrontal cortex and anterior cingulate gyrus. Disruption of the prefrontal cortex would impair inhibitory control and decision making, leading individuals with substance use disorder to choose immediate rewards over delayed rewards and contributing to the loss of control over intake [42].

5.4 Impaired response inhibition and salience attribution syndrome of addiction

Goldstein RZ, Volkow ND. Drug addiction and its underlying neurobiological basis: neuroimaging evidence for the involvement of the frontal cortex. American Journal of Psychiatry 2002;159:1642–52 [42].

In an elaboration of the role of the frontal cortex and orbitofrontal cortex in the motivational effects of drug addiction, Goldstein and Volkow [42] conceptualized drug addiction as a syndrome of impairments in response inhibition and salience attribution. The authors argued that four clusters of behaviors are interconnected in a positive feedback loop (drug reinforcement, craving, bingeing, and withdrawal), and activity of the prefrontal circuits and the subcortical reward pathway are differentially represented at each stage ([42]; Fig. 29). During the drug intoxication stage, strong positive and negative reinforcement is strengthened through repeated self-administration of the drug, giving more attribution to incentive salience of the drug at the expense of less powerful reinforcers. During this stage, the prefrontal and orbitofrontal cortices are activated in individuals with substance use disorder who are challenged with drug [324–327]. During the relapse and bingeing stage, high levels of brain activation have also been observed in the prefrontal cortex

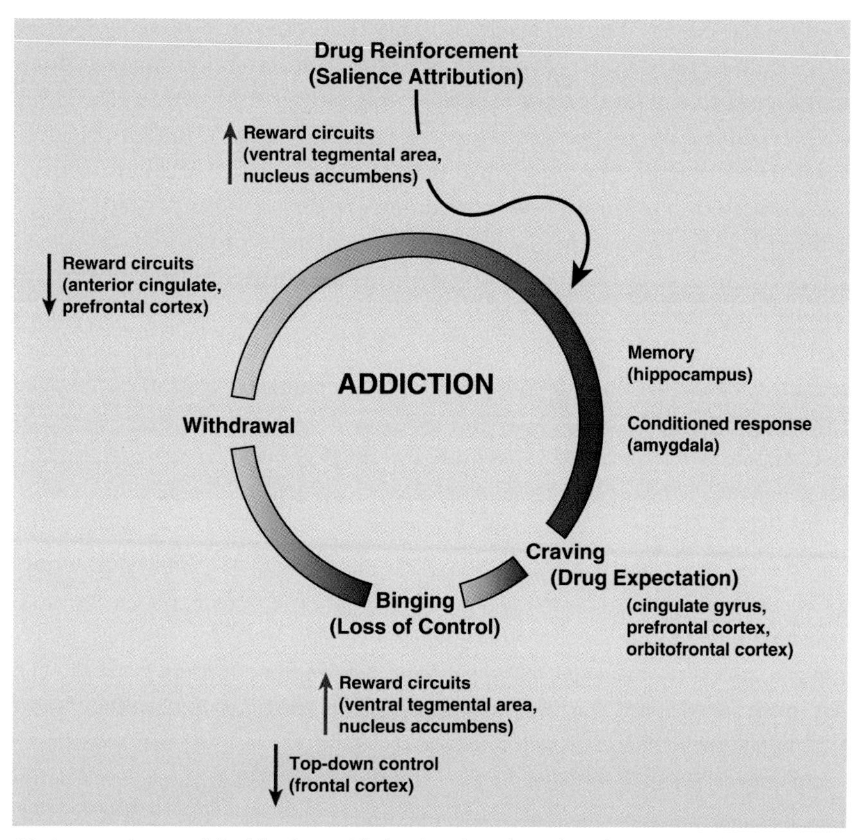

Figure 29 ***Integrative model of brain and behavior that describes the Impaired Response Inhibition and Salience Attribution (I-RISA) syndrome of drug addiction [42].*** The mesolimbic dopamine circuit, which includes the nucleus accumbens, amygdala, and hippocampus, has been traditionally associated with the acute reinforcing effects of a drug and with the memory and conditioned responses that have been linked to craving. It is also likely to be involved in the emotional and motivational changes that are seen in drug abusers during withdrawal. The mesocortical dopamine circuit, which includes the prefrontal cortex, orbitofrontal cortex, and anterior cingulate, is likely to be involved in the conscious experience of drug intoxication, drug incentive salience, drug expectation/craving, and compulsive drug administration. Note the circular nature of this interaction: the attribution of salience to a given stimulus, which is a function of the orbitofrontal cortex, depends on the relative value of a reinforcer compared with simultaneously available reinforcers, which requires knowledge of the strength of the stimulus as a reinforcer, a function of the hippocampus and amygdala. Consumption of the drug in turn will further activate cortical circuits (orbitofrontal cortex and anterior cingulate) in proportion to the dopamine stimulation by favoring the target response and decreasing non-target-related background activity. The activation of these interacting circuits may be indispensable for maintaining the compulsive drug administration that is observed during bingeing and to the vicious circle of addiction. *(Taken with permission from Goldstein RZ, Volkow ND. Drug addiction and its underlying neurobiological basis: neuroimaging evidence for the involvement of the frontal cortex. American Journal of Psychiatry 2002;159:1642–52 [42].)*

and orbitofrontal cortex. Higher levels of brain activation, measured by glucose metabolism or cerebral blood flow, have also been seen in frontolimbic areas in drug abusers who are exposed to stimuli that are associated with drug [317,318,322,328–330].

However, during the relapse and bingeing stage, the response inhibition system is hypothesized to be impaired because of impairments in salience attribution that lead to response disinhibition or impulsive responding, again to immediately salient drug-related rewards [42]. During drug withdrawal, particularly protracted abstinence, brain metabolism is lower in the orbitofrontal cortex, frontal cortex, and anterior cingulate gyrus in drug abusers/addicts than in normal controls [321,331,332]. Dysthymia is a key symptom of abstinence and is hypothesized to be produced by lower sensitivity of the brain reward systems that is induced by adaptational changes in response to repeated activation by drugs of abuse.

The Goldstein and Volkow [42] and Volkow et al. [92] models conceptualize addiction as a state that is initiated by taking drugs that are highly effective rewards compared with other stimuli, and this state triggers a series of adaptations in the reward, motivation/drive, memory, and control circuits of the brain [92]. As a result, the drug has greater saliency value, superimposed on a reward deficit for non-drug rewards and the loss of inhibitory control, thus favoring the emergence of compulsive drug administration. The authors further argued that their model leads to suggestions for treatment strategies, such as decreasing the rewarding value of drugs while increasing the value of non-drug reinforcers, weakening learned drug associations, and strengthening cognitive function specifically in the frontal control circuit [92].

In summary, addiction is conceptualized as the disruption of incentive salience attribution for normal rewards, with a redirection of salience to drug rewards and accompanied by deficits in response inhibition and the lower sensitivity of reward function during withdrawal.

5.5 Competing neurobehavioral decision systems

Bickel WK, Jarmolowicz DP, Mueller ET, Gatchalian KM, McClure SM. Are executive function and impulsivity antipodes? A conceptual reconstruction with special reference to addiction. Psychopharmacology 2012a;221(3):361–87 [333].

These authors developed a hypothesis, termed the competing neurobehavioral decision systems (CNDS) theory, to account for self-control failure and to integrate the neuroscience of addiction with developmental processes, socioeconomic status, and comorbidities [334–337]. This hypothesis, consistent with a broad array of dual-systems theories, suggests that choices for drugs result from the interaction between the two decision systems and that those who experience addiction suffer from an imbalance or dysregulation between these two systems [338].

The CNDS theory, dual-systems model, places an emphasis on relative control between impulsive and executive decision systems. The impulsive system (composed of limbic and paralimbic brain regions) and executive system (composed of prefrontal

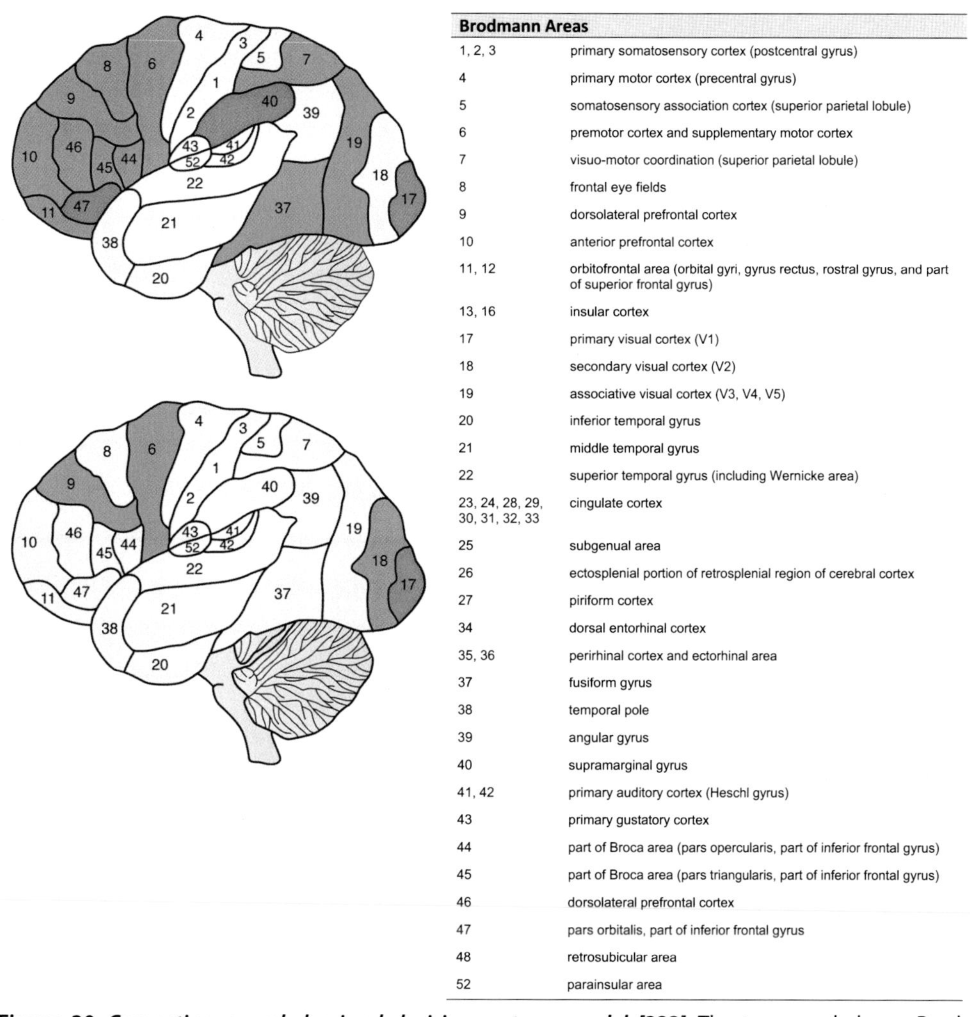

Brodmann Areas	
1, 2, 3	primary somatosensory cortex (postcentral gyrus)
4	primary motor cortex (precentral gyrus)
5	somatosensory association cortex (superior parietal lobule)
6	premotor cortex and supplementary motor cortex
7	visuo-motor coordination (superior parietal lobule)
8	frontal eye fields
9	dorsolateral prefrontal cortex
10	anterior prefrontal cortex
11, 12	orbitofrontal area (orbital gyri, gyrus rectus, rostral gyrus, and part of superior frontal gyrus)
13, 16	insular cortex
17	primary visual cortex (V1)
18	secondary visual cortex (V2)
19	associative visual cortex (V3, V4, V5)
20	inferior temporal gyrus
21	middle temporal gyrus
22	superior temporal gyrus (including Wernicke area)
23, 24, 28, 29, 30, 31, 32, 33	cingulate cortex
25	subgenual area
26	ectosplenial portion of retrosplenial region of cerebral cortex
27	piriform cortex
34	dorsal entorhinal cortex
35, 36	perirhinal cortex and ectorhinal area
37	fusiform gyrus
38	temporal pole
39	angular gyrus
40	supramarginal gyrus
41, 42	primary auditory cortex (Heschl gyrus)
43	primary gustatory cortex
44	part of Broca area (pars opercularis, part of inferior frontal gyrus)
45	part of Broca area (pars triangularis, part of inferior frontal gyrus)
46	dorsolateral prefrontal cortex
47	pars orbitalis, part of inferior frontal gyrus
48	retrosubicular area
52	parainsular area

Figure 30 ***Competing neurobehavioral decision systems model [333].*** The top panel shows Brodmann's areas color-coded to show areas wherein lower levels of activation are associated with executive dysfunction (*blue*), impulsivity (*red*), or both executive dysfunction and impulsivity (purple). The bottom panel shows Brodmann's areas color-coded to show areas wherein higher levels of activation are associated with executive dysfunction (*blue*), impulsivity (*red*), or both executive dysfunction and impulsivity (*purple*). *(Taken with permission from Bickel WK, Jarmolowicz DP, Mueller ET, Gatchalian KM, McClure SM. Are executive function and impulsivity antipodes? A conceptual reconstruction with special reference to addiction. Psychopharmacology 2012a;221(3):361–87 [333].)*

and parietal cortices) are hypothesized to be interdependent and compete for relative control during decision-making ([333]; Fig. 30). Normal functioning results when the systems are in regulatory balance. Pathology can result when the two systems are not in regulatory balance [339].

In the CNDS model, delayed discounting is further argued to be the behavioral measure of self-control that indicates the relative strength of the competing decision-making systems [335]. Neurobiological evidence clearly supports the hypothesis that the impulsive decision system is mediated by what these authors term the "limbic" system (midbrain, amygdala, habenular commissure, and striatum) and paralimbic system (insula and nucleus accumbens; [336]). In contrast, the executive function system is argued to be mediated by the parietal lobes and prefrontal cortex [336]. The authors of the CNDS theory argue that there is hyperactivation of the impulsive system and hypoactivation of the executive function system in stimulant addiction. They further argue that CNDS provides a framework for differential development of the CNDS across the lifespan that can explain developmental vulnerabilities to addiction, socioeconomic vulnerabilities, and comorbidities with other substance use disorders, as well as a framework for the innovative treatment of addictive disorders.

The strengths of this theory are that it provides an endophenotype of impairments in executive function for significant neurobiological neurocircuitry changes in addiction that globally involve the prefrontal cortex and allocortex. As such, the CNDS theory may be a boon to identifying the vulnerability to addiction and recovery from addiction. Studies that are currently underway in the field will certainly test this theory in the coming years.

One weakness of the CNDS theory is that it discounts (pun intended) the role of the dark side. There is strong evidence that negative emotional states can drive compulsivity via negative reinforcement [340,341], and there is a neurobiological basis for such impairment in the prefrontal control of the amygdala and brainstem structures, such as the periaqueductal gray. Perhaps even more intriguingly, the dark side of addiction can drive impulsivity via negative urgency [342]. Here, negative urgency is conceived as an emotion-based trait, referring to acting rashly and impulsively when in extreme distress and involving impairments in inhibitory control [343]. Negative urgency has been implicated in compulsive substance use, including tobacco [344] and alcohol [345], and compulsive eating [346].

5.6 Somatic marker theory of addiction

Verdejo-Garcıa A, Bechara A. A somatic marker theory of addiction. Neuropharmacology 2009; 56:48–62 [347].

Verdejo-Garcia and Bechara sought to apply a "somatic marker" model of addiction to explain "myopia for the future," manifested in the behavioral decisions of many individuals with a history of chronic drug use. The somatic marker hypothesis was originally proposed by Damasio [348]. It provides a systems-level neuroanatomical and cognitive framework for decision-making and for choosing according to long-term outcomes rather than short-term ones. Here, the term "somatic markers" refers to the collection of body- and brain-related responses that have been connected by learning anticipated future outcomes of certain scenarios. A negative somatic marker is linked to a particular future outcome and functions as an alarm bell. A positive somatic marker is linked to a future outcome that functions as an incentive [348]. Verdejo-Garcia and Bechara [347] argued that there are at least two underlying types of dysfunction in addiction, in which emotional signals (somatic markers) turn in favor of immediate outcomes: (*i*) hyperactivity in the amygdala or impulsive system, which exaggerates the rewarding impact of available incentives, and (*ii*) hypoactivity in the prefrontal cortex or reflective system [347]. They further argued that hypoactivity in the prefrontal cortex forecasts the long-term consequences of a given action.

The neurobiological basis for the somatic marker theory of addiction is hypothesized to involve dysfunction in the ventromedial prefrontal cortex system and hyperactivity of the amygdala ([347]; Fig. 31). The ventromedial prefrontal cortex is critical for processing emotional (somatic) states from secondary inducers and has been termed a "reflective" system, in which its dysfunction causes loss of the ability to process and trigger somatic signals that are associated with future prospects [347]. The amygdala is critical for processing emotional (somatic) states from primary inducers and is hypothesized to be an "impulsive" system. In such a functional context, the amygdala is hypothesized to become altered such that it diminishes the emotional/somatic impact of natural reinforcers, but it exaggerates the emotional/somatic impact of the immediate prospects of obtaining drugs [347]. Thus, these authors hypothesized that drug cues acquire properties that trigger bottom-up, automatic, and involuntary somatic states through the amygdala, and this bottom-up somatic bias can modulate top-down cognitive mechanisms that are normally under control of the ventromedial prefrontal cortex. If the bias is sufficiently strong, then drugs can trigger bottom-up, involuntary signals through the amygdala that modulate and compromise top-down, goal-driven attentional resources that are needed for the normal operation of the reflective system and are critical for enabling an individual to resist the temptation to seek the drug.

Data that support this hypothesis include numerous studies that showed impairments in decision-making performance among individuals with alcohol, cannabis, cocaine, opioid, methylenedioxy-methamphetamine, and methamphetamine abuse [347]. The experimental task that is often used is a variant of the Iowa Gambling Task [349], which measures several factors, including immediate rewards and delayed punishments, risk, and the uncertainty of outcomes. Brain imaging patterns in methamphetamine-using individuals while performing another task, a two-choice prediction task that measures decision-making function under conditions of uncertainty, showed lower activation of the orbitofrontal, dorsolateral, insular, and inferior parietal cortices. Orbitofrontal activation was inversely correlated with the duration of methamphetamine use [350]. Stroke that involved the insula was associated with a greater likelihood of eliminating craving cigarettes [351]. Most studies of emotional perception have focused on analyzing alterations of the processing of emotional facial expressions in subjects with long-term substance abuse. In alcohol-dependent individuals, several studies have observed significant alterations of the processing of facial expressions, although the range of emotions that are affected is still controversial. One study showed that alcohol-dependent individuals presented specific overestimation of the intensity of emotion in facial expressions that were mainly related to

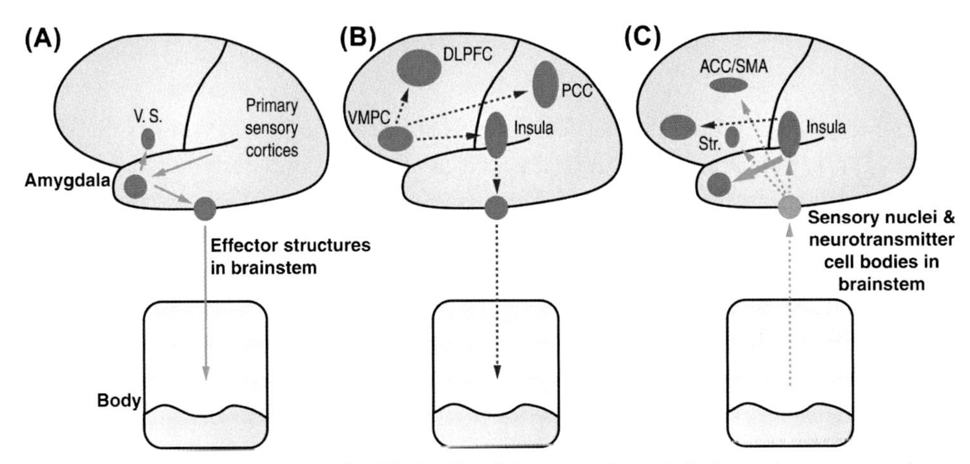

Figure 31 ***Somatic marker theory of addiction [347].*** A schematic model of somatic state activation and decision-making. (A) This is the neural circuitry (in *green*) that represents the impulsive system, in which the amygdala is a trigger structure for emotional (somatic) states from primary inducers. It couples the features of primary inducers, which can be processed subliminally (e.g., via the thalamus) or explicitly (e.g., via primary sensory cortices), with effector structures that trigger the emotional/somatic response. However, the amygdala is also directly connected to the ventral striatum (V.S.), and its trigger can also activate classic motivational systems that are associated with the approach of drug-related cues. (B) This is the neural circuitry (in *red*) that represents the reflective system, in which the ventromedial prefrontal cortex (VMPC) is a trigger structure for emotional (somatic) states from secondary inducers. It couples systems that are involved in memory to systems that are involved in processing emotions, so that memories or thoughts about drug cues are linked to their emotional attributes. Memories, information, or knowledge that are held temporarily in working memory and manipulated by the executive processes of working memory are dependent on the dorsolateral prefrontal cortex (DLPFC), but these working processes also include the ventrolateral prefrontal cortex and lateral region of the orbitofrontal cortex. The hippocampus is also engaged, at least in situations in which the memory of a scenario exceeds 40 s, which is the maximum capacity of short-term memory that is mediated by the DLPFC. Structures that are involved in representing previous feeling states include the insula and surrounding somatosensory cortices, as well as the posterior cingulate cortex (PCC) and precuneate cortex. This coupling that is mediated by the VMPC can also lead to the triggering of somatic states (*red lines*). (C) Somatic signals that are triggered simultaneously by the impulsive system and reflective system compete. This competition is argued to actually occur in the body proper, although some authors have challenged the validity of this peripheral link. Even if this peripheral link proves to be invalid, neuroanatomical evidence supports neural competition at the next link of the circuitry, the neurotransmitter cell bodies in the brainstem. Whether the competition of hailing somatic signals via the impulsive and reflective systems occurs in the body or the brainstem, a "winner takes all" somatic signal emerges, either positive or negative, and this resultant ascending feedback somatic signal (*blue lines*) participates in two functions: in one, it provides a substrate for feeling the emotional state through the insula and surrounding somatosensory cortices; in the other, it provides a substrate for biasing decisions through motor effector structures, such as the striatum (Str.), the anterior cingulate cortex (ACC), and adjacent supplementary motor area (SMA). Another highly candidate region that is involved in this complex motor preparation is the cerebellum (not shown in this diagram). In light of more recent evidence on the role of the insula in cigarette smoking, we propose that the insula has a more specific role. Especially during withdrawal or drug deprivation, homeostatic signals that arise in the body are perceived in the insula as feelings of urges, which in turn act on the impulsive system and sensitize it (*green line*), whereas its action on the reflective system is to inhibit it or "hijack" it. The mechanism by which insular activity sensitizes the impulsive system remains to be determined. One possibility, for example, is that neuronal signals from the insula activate the ventral tegmental area and increase the mesolimbic dopamine surge, thereby heightening the motivation to seek drugs. Such a possible mechanism can connect the insula to the classic neural systems that have been implicated in addiction (i.e., the mesolimbic dopamine system). *(Taken with permission from Verdejo-Garcıa A, Bechara, A. A somatic marker theory of addiction. Neuropharmacology 2009;56:48–62 [347].)*

expressions of fear [352], but others have reported impairments in recognizing facial expressions of happiness and anger [353].

The somatic marker hypothesis provides an interesting link between the CNDS hypothesis of Bickel and colleagues ([333]; see above) and the dark side of Koob and Le Moal [340,341]. Particularly intriguing is the hypothesis that overactivity of the amygdala, combined with the hypoactivity of top-down control, can trigger impulsivity as outlined by Verdejo-Garcia and Bechara [347]. Indeed, one could argue that the somatic marker hypothesis lays the foundation for a neurobiological circuit for the phenomenon of negative urgency [342]. One take-home message is that imaging and neuropsychological studies in the addiction field should engage more of the dark side in tasks that reflect elements of negative emotional states that are associated with drug withdrawal and protracted abstinence, not only from the perspective of negative reinforcement but also from the perspective of impulsivity and decision making.

5.7 Neurocircuits for relapse to drugs of abuse or brain circuitry of reinstatement of drug-seeking

Kalivas PW, McFarland K. Brain circuitry and the reinstatement of cocaine-seeking behavior. Psychopharmacology 2003;168:44–56 [354].

Shaham Y, Shalev U, Lu L, De Wit H, Stewart J. The reinstatement model of drug relapse: history, methodology and major findings. Psychopharmacology 2003;168:3–20 [138].

The thesis of this conceptual framework, based almost exclusively on animal models, was that the reinstatement of drug-seeking behavior by the drug itself (drug priming), drug cues, or stressors converges on the medial prefrontal cortex (anterior cingulate cortex) and output through the core of the nucleus accumbens [354]. A neurochemical subhypothesis is that the effects of such priming stimuli are conveyed by a glutamatergic projection from the anterior cingulate to the nucleus accumbens core. As a result, changes in glutamate system neurotransmission may be a key component of the vulnerability to relapse. Three distinct types of stimuli were postulated to produce a related interoceptive state that increases the probability of executing drug-seeking behavior: exposure to a pharmacological stimulus that induces a common component of the drug experience, exposure to an environmental stimulus that is associated with the drug, and exposure to stressors.

The animal models that are used to explore the neurobiological bases of reinstatement typically utilize a paradigm in which rats are allowed limited access to a drug (usually cocaine) until they achieve stable responding, and then the animals are subjected to prolonged extinction procedures. Typically, the animals are trained to self-administer intravenous cocaine in daily 2 h sessions ([355]; see Chapter 2, *Animal Models of Addiction*). The animals are then subjected to extinction, in which responding on the active lever results in illumination of the stimulus light for 20 s and the infusion of saline instead of cocaine. Once a criterion of $< 10\%$ of the average responding during the maintenance of self-administration is achieved, the animals are tested for reinstatement following the passive administration of drug (cocaine-primed reinstatement), cues (discrete and/or contextual cues that are associated with drug administration), or stressors.

The dorsal prefrontal cortex, core of the nucleus accumbens, and ventral pallidum but not the basolateral amygdala appear to play a critical role in drug-induced reinstatement (such as that primed by cocaine; [356,357]). Cocaine-primed reinstatement was unaltered by the blockade of dopamine receptors in the core of the nucleus accumbens [357] but was blocked by the inactivation of D_1 receptors in the shell of the nucleus accumbens [358]. Cocaine-primed reinstatement is blocked by inhibiting ionotropic AMPA/kainate glutamate receptors but not NMDA glutamate receptors in the core of the nucleus accumbens. In fact, NMDA receptor antagonists themselves can reinstate cocaine-seeking behavior [359,360]. Both dopamine receptor agonists and glutamate receptor agonists that are administered directly in the nucleus accumbens can induce reinstatement in rats [357,361]. Thus, the activation of either dopamine in the nucleus accumbens shell or AMPA glutamate receptors in the nucleus accumbens core play a role in drug-primed reinstatement [362].

Reversible inactivation of the anterior cingulate/prelimbic region of the prefrontal cortex prevents cocaine-primed reinstatement [357], and the blockade of dopamine D_1/D_2 receptors in the dorsal prefrontal cortex prevents cocaine-primed reinstatement [357]. Consistent with the interaction of the frontal cortex with the nucleus accumbens, the reinstatement that is produced by a cocaine microinjection in the prefrontal cortex was blocked by the nucleus accumbens inhibition of AMPA receptors but not NMDA receptors [360].

In summary, cocaine-induced reinstatement involves a frontal cortex (cingulate cortex) glutamatergic projection to the nucleus accumbens core, and dopamine projections to both the prefrontal cortex and nucleus accumbens shell are involved in cocaine-induced reinstatement ([138,354]; Figs. 32 and 33).

Parts of the medial prefrontal cortex (notably the anterior cingulate/prelimbic cortices) and amygdala (rostral basolateral amygdala) appear to be particularly involved in cue-induced reinstatement. Reversible inactivation of the anterior cingulate/prelimbic region of the rat prefrontal cortex (termed "cingulate" by [354]) prevents cue-induced and drug-induced reinstatement [357]. Inactivation of the rostral parts of the basolateral amygdala also blocked cue-induced reinstatement [363]. Additionally, inactivation of the ventral tegmental area or nucleus accumbens core also reversed cocaine-seeking under a second-order schedule of reinforcement [364], and inactivation of the lateral orbitofrontal cortex also blocked cue-induced reinstatement [364,365]. The blockade of dopamine receptors in the basolateral amygdala blocked cue-induced reinstatement [366] and cocaine-seeking behavior in a second-order schedule [367]. Discriminative stimuli that predict the availability of cocaine and reinstate cocaine-seeking behavior increased dopamine and glutamate release in the amygdala and nucleus accumbens [368]. These same discriminative stimuli increased Fos activation in the basolateral amygdala and medial prefrontal cortex, and this Fos activation and reinstatement were reversed by dopamine D_1 receptor antagonism [368,369]. The blockade of ionotropic AMPA/kainate receptors

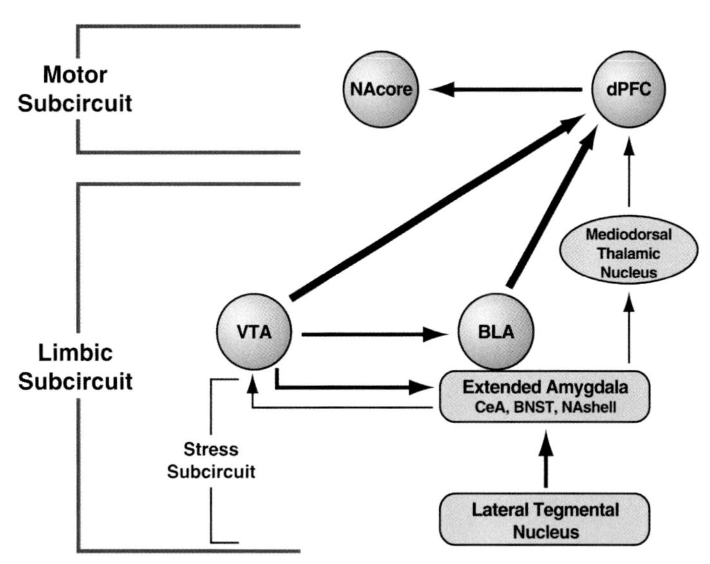

Figure 32 ***Schematic illustration of the cocaine reinstatement circuit [354].*** Schematic illustration of the hypothetical circuitry involved in primed reinstatement. It is proposed that the motor subcircuit that contains the projection from the anterior cingulate/prelimbic region of the dorsal prefrontal cortex (dPFC) to the core of the nucleus accumbens (NAcore) corresponds to a final common pathway for all priming stimuli. The remainder of the circuit is termed the limbic subcircuit and contains the connected nuclei that are shown to be distinct to a class of priming stimulus (e.g., cue, stress, and drug). This includes the ventral tegmental area (VTA) and basolateral amygdala (BLA) for drug and cue priming, and adrenergic projections from the lateral tegmental nucleus (LTN) to the extended amygdala for stress-primed reinstatement. The *lighter lines* correspond to anatomical connections that might theoretically link stress priming to the dPFC, but for which no evidence currently exists. BNST, bed nucleus of the stria terminalis, CeA, central nucleus of the amygdala; NAshell, shell of the nucleus accumbens. *(Taken with permission from Kalivas PW, McFarland K. Brain circuitry and the reinstatement of cocaine-seeking behavior. Psychopharmacology 2003;168:44–56 [354].)*

but not NMDA glutamate receptors in the core of the nucleus accumbens prevented drug-seeking behavior in response to a cocaine-associated cue in a second-order schedule [370], and this region of the nucleus accumbens is preferentially innervated by the rostral basolateral amygdala.

In summary, cue-induced reinstatement appears to be dependent on activation of the basolateral amygdala and medial prefrontal cortex (cingulate cortex) via glutamatergic innervation of the nucleus accumbens core [138,354]. Dopamine innervation of the basolateral amygdala and prefrontal cortex may also play a role in cue-induced reinstatement.

Inactivation of the central nucleus of the amygdala and bed nucleus of the stria terminalis prevented stress-induced reinstatement [186], and CRF receptor antagonists that are injected in the central nucleus of the amygdala and lateral bed nucleus of the stria terminalis also blocked stress-induced reinstatement, but injections in the basolateral amygdala did not block stress-induced reinstatement. Microinjections of noradrenergic

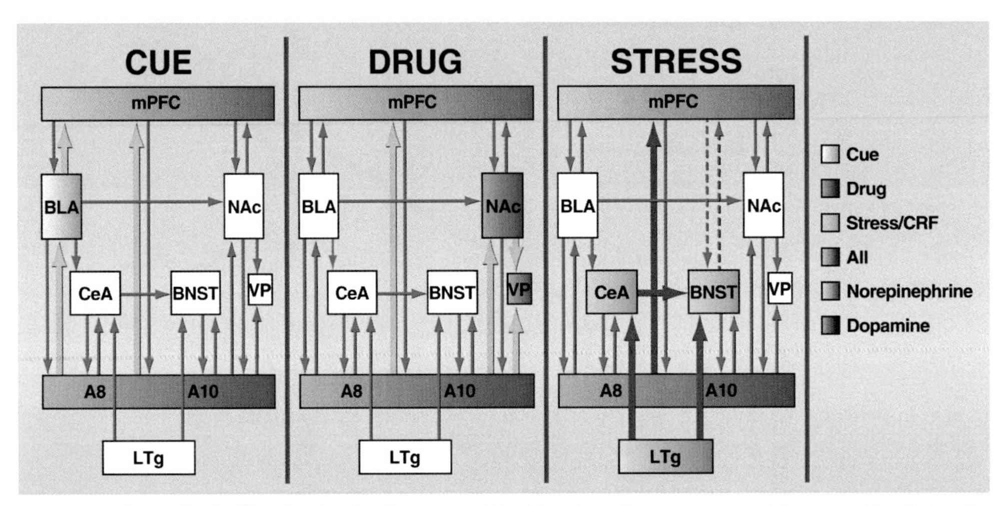

Figure 33 ***Hypothetical brain circuits that are critical for the reinstatement of drug-seeking behavior by cues, drugs, and footshock stress [138].*** Most of the data that are described [in this figure] are from studies with cocaine-trained rats. In each case, *fat arrows* indicate pathways that may be involved in reinstatement; *thin arrows* indicate some of the existing direct anatomical connections; *dashed arrows* indicate indirect connections. Cue-induced reinstatement: the effect of cocaine cues on reinstatement is blocked by systemic and intra-BLA injections of D_1-like receptor antagonists and by reversible inactivation of the BLA and dorsal mPFC. Drug-induced reinstatement: the effect of cocaine priming on reinstatement is blocked by systemic, intra-mPFC, and possibly intra-NAc shell injections of dopamine receptor antagonists. This effect of cocaine priming is also blocked by reversible inactivation of the dorsal mPFC, NAc core, VP and VTA. In contrast, the effect of cocaine on reinstatement is mimicked by systemic injections of D_2-like receptor agonists, the activation of VTA dopamine neurons with morphine or excitatory amino acid agonists, and infusions of cocaine, dopamine, and amphetamine into the NAc and mPFC. Footshock stress-induced reinstatement: the effect of footshock on reinstatement is blocked by systemic and intra-BNST, but not intra-amygdala, injections of a CRF receptor antagonist. This effect of footshock stress is also blocked by systemic and intracerebroventricular, but not intra-LC, injections of α_2-adrenoceptor agonists (which reduce noradrenergic cell firing and release), by intra-BNST and intra-amygdala infusions of α_2-adrenoceptor antagonists, and by 6-OHDA lesions of VNAB projections that arise from the LTg. Additionally, the effect of footshock on reinstatement is attenuated by disrupting CRF projections from the CeA to the BNST by infusions of a CRF antagonist into the BNST and TTX into the contralateral CeA. 6-OHDA, 6-hydroxydopamine; A8 and A10, dopamine (DA) cell groups; BLA, basolateral amygdala; BNST, bed nucleus of the stria terminalis; CeA, central nucleus of the amygdala; CRF, corticotropin-releasing factor; LC, locus coeruleus; LTg, noradrenergic cell groups of the lateral tegmental nuclei; mPFC, medial prefrontal cortex; NAc, nucleus accumbens; TTX, tetrodotoxin; VTA, ventral tegmental area; VP, ventral pallidum. *(Taken with permission from Shaham Y, Shalev U, Lu L, De Wit H, Stewart J. The reinstatement model of drug relapse: history, methodology and major findings. Psychopharmacology 2003;168:3–20 [138].)*

receptor antagonists in both the central nucleus of the amygdala and lateral bed nucleus of the stria terminalis also blocked stress-induced reinstatement [191]. Transient inactivation of the prefrontal cortex, extended amygdala, nucleus accumbens, and ventral tegmental area can also prevent stress-induced reinstatement [138,354,357,371].

In summary, an important role for the extended amygdala (central nucleus of the amygdala and lateral bed nucleus of the stria terminalis) has been hypothesized in stress-induced reinstatement, with a possible link between the extended amygdala and prefrontal cortex—accumbens projection that was hypothesized to be critical for drug- and cue-induced reinstatement.

Kalivas and McFarland divided the neurocircuitry that is involved in the drug-, cue-, and stress-induced reinstatement of cocaine-seeking behavior into three subcircuits [138,354]. The frontal cortex (anterior cingulate) projection to the core of the nucleus accumbens is referred to as the motor subcircuit. The circuitry that is distinct for each mode of priming is described as the limbic subcircuit. The stress subcircuit is embedded within the limbic subcircuit. The neurocircuitry of drug-induced reinstatement is described as a series of connections from the ventral tegmental area to prefrontal cortex (anterior cingulate) and then to the core of the nucleus accumbens. The neurocircuitry of cue-induced reinstatement is described as a series of connections from the ventral tegmental area to basolateral amygdala, then from the basolateral amygdala to prefrontal cortex (anterior cingulate), and then to the core of the nucleus accumbens. The neurocircuitry of stress-induced reinstatement is described as a series of adrenergic and CRF inputs to the extended amygdala that in turn activate the prefrontal cortex (anterior cingulate), which then projects to the core of the nucleus accumbens. One hypothesis that was postulated by the authors is that afferents from the extended amygdala modulate GABAergic projections from the ventral tegmental area to prefrontal (cingulate) cortex [354]. Another possibility may be a projection from the extended amygdala to the mediodorsal thalamus and then to the prefrontal (cingulate) cortex.

The key component of this conceptualization is the hypothesized role of the prefrontal (cingulate) cortex-to-nucleus accumbens glutamatergic projection as a critical subcircuit in driving drug-, cue-, and stress-induced reinstatement and initiating motivated behavior in general [354]. Although there is a common focus on the frontal cortex with the Kalivas and McFarland (2003) [354], Jentsch and Taylor [262], and Volkow et al. [92] conceptual frameworks, one area of significant discrepancy is that the reinstatement neurocircuitry hypothesis posits activation of the prefrontal cortex, whereas the imaging [92] and cognitive-behavioral hypotheses posit hypoactivity in the prefrontal cortex (hypofrontality). Kalivas and McFarland suggest an explanation via a greater signal-to-noise ratio, but another issue that needs exploration is that all of the reviewed animal models of reinstatement to date have involved nondependent animal models, and most of the imaging studies in humans have involved cocaine-dependent subjects. Data from reinstatement models, in which there is a history of dependence, may shed light on the actual valence of changes within these circuits that are so critical for guiding motivated behavior.

5.8 Glutamate systems in cocaine addiction

Kalivas PW, McFarland K, Bowers S, Szumlinski K, Xi ZX, Baker D. Glutamate transmission and addiction to cocaine. In: Moghaddam B, Wolf ME, editors. Glutamate and disorders of cognition and motivation. Annals of the New York Academy of Sciences, vol. 1003. New York: New York Academy of Sciences; 2003. p. 169–75 [372].

Kalivas PW. Glutamate systems in cocaine addiction. Current Opinion in Pharmacology 2004;4: 23–9 [373].

As noted above, Kalivas and colleagues and other groups have shown that reversible inhibition of the prefrontal cortex prevents the reinstatement of drug-seeking that is induced by cocaine, stress, or cocaine-associated cues [357]. Frontal cortex activation that is involved in reinstatement has been linked to a glutamatergic pathway to the core of the nucleus accumbens [362]. With this neurocircuitry as a background, Kalivas identified four neuroadaptations of potential importance in mediating the dysregulation of glutamatergic neurotransmission that were associated with the cocaine-primed reinstatement of drug seeking ([372]; see [1] in Fig. 34).

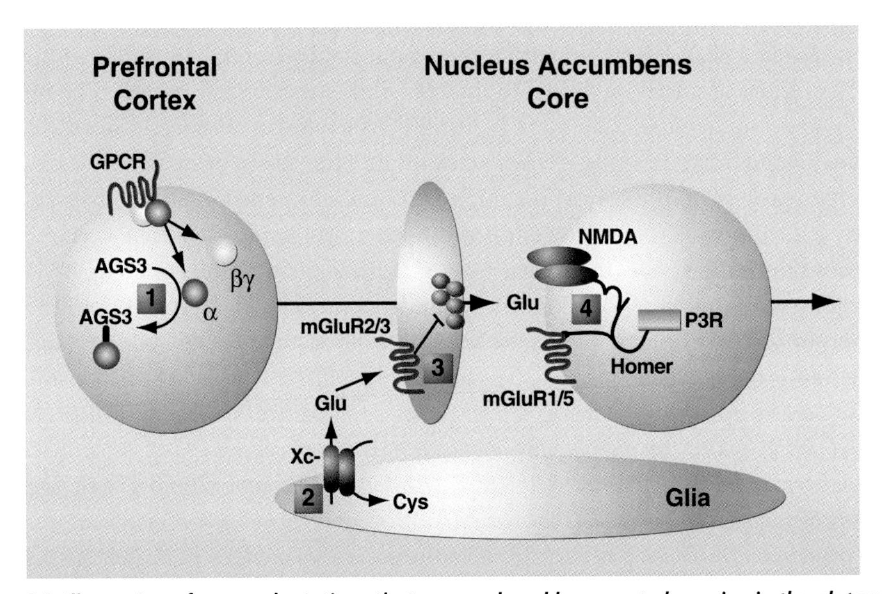

Figure 34 ***Illustration of neuroadaptations that are produced by repeated cocaine in the glutamatergic projection from the prefrontal cortex to the nucleus accumbens core [372].*** The numbers refer to specific adaptations that are described in the text. (1) Increased AGS3 results in reduced signaling through G_i protein coupled receptors (GPCR) because AGS3 competes with ß for G_i binding. (2) The rate of cystine-glutamate exchange (Xc-) is reduced in the nucleus accumbens, which decreases the basal extracellular level of glutamate. (3) Reduced extracellular glutamate, combined with desensitization of mGluR2/3, decreases inhibitory tone on synaptic glutamate release. (4) Reduced levels of Homer1bc compromise signaling in the postsynaptic density between group I mGluR (mGluR1/5), NMDA receptors, and IP3 receptors. AGS3, activator of G-protein signaling 3; Cys, cysteine; Glu, glutamate; mGluR, metabotropic glutamate receptor; NMDA, *N*-methyl-D-aspartate. *(Taken with permission from Kalivas PW, McFarland K, Bowers S, Szumlinski K, Xi ZX, Baker D, Glutamate transmission and addiction to cocaine. In: Moghaddam B, Wolf ME, editors. Glutamate and disorders of cognition and motivation. Annals of the New York Academy of Sciences, vol. 1003. New York: New York Academy of Sciences; 2003. p. 169–75 [372].)*

One neuroadaptation involves the protein regulation of a G-protein-linked receptor to increase glutamate release in prefrontal cortex neurons. AGS3 is a protein that regulates G-protein signaling by reducing the access of $G_{i\alpha}$ to receptor and effector proteins. Withdrawal from repeated cocaine administration produces an increase in AGS3 in the prefrontal cortex and nucleus accumbens core and is accompanied by a reduction of the signaling of presynaptic $G_{i\alpha}$-coupled receptors, notably metabotropic glutamate 2/3 receptors (mGluR2/3) and an increase in cocaine-primed glutamate release in the nucleus accumbens core ([372,374]; see [1] in Fig. 34). Supporting a role for enhanced glutamate release in this pathway that is associated with cocaine-induced reinstatement is the observation that infusions of antisense oligonucleotides to AGS3 in the prefrontal cortex prevented cocaine-induced reinstatement in rats that were subjected to extinction, and increasing AGS3 facilitated the ability of cocaine to elevate glutamate levels in the nucleus accumbens [375]. Thus, the cocaine-induced elevation of AGS3 in the prefrontal cortex is associated with an increase in the release of prefrontal-accumbens glutamate and reinstatement of cocaine self-administration.

A second proposed neuroadaptation of glutamate neurotransmission that is associated with the repeated administration of cocaine is a decrease in cystine-glutamate exchanger function ([372]; see [2] in Fig. 34). The cystine-glutamate exchanger is a heterodimer that is hypothesized to play a primary role in regulating extracellular levels of glutamate in the nucleus accumbens. Repeated cocaine administration decreases the rate of cystine-glutamate exchange and thus reduces extracellular basal levels of glutamate. Administration of the procysteine drug *N*-acetylcysteine normalizes extracellular levels of glutamate and blocks reinstatement by a cocaine priming injection [376].

A third proposed neuroadaptation is that lower basal extracellular glutamate levels reduce tone at mGluR2/3 autoreceptors and thus facilitates the phasic synaptic release of glutamate during cocaine-primed reinstatement ([372]; see [3] in Fig. 34).

A fourth proposed neuroadaptation of glutamate neurotransmission is a reduction of the expression of postsynaptic glutamate receptors that is caused by a reduction of Homer proteins, notably Homer 1bc and Homer 2 ([372]; see [4] in Fig. 34). Homer 2 knockout mice are more sensitive to the motor-stimulant and rewarding effects of cocaine [377,378]. Thus, a reduction of Homer proteins resulted in lower mGluR1/5 signaling, which is hypothesized to increase the responsiveness to cocaine. Notably, deletion of the mGluR5 gene or the administration of mGluR5 antagonists blocked cocaine reinforcement, suggesting that the above effects are indeed mediated by mGluR1 [373,379,380].

Thus, the Kalivas model focuses on molecular changes within a specific neurotransmitter system (glutamate) within a specific neurocircuit (prefrontal cortex-to-nucleus accumbens). Prolonged decreases in tonic glutamate in the prefrontal cortex-nucleus accumbens core pathway, combined with the desensitization of mGluR2/3, results in a loss of the feedback control of dynamic synaptic glutamate release. Lower basal levels of glutamate, combined with increases in the release of synaptic glutamate in the nucleus accumbens core, in response to the activation of prefrontal cortical afferents results in a hypothesized amplified signal, thus resulting in a greater drive for drug seeking at the behavioral level. Various postsynaptic glutamate receptor functional changes have been described, including a decrease in both NMDA receptor and mGluR2/3 function. However, the exact nature of this relationship remains unknown. For example, mGluR5 antagonists block cocaine self-administration [379], but molecular changes that are associated with prefrontal glutamate dysfunction in preclinical reinstatement models need to be reconciled with clinical observations of prefrontal cortical hypofunction in individuals with established substance use disorder. The ways in which such changes are associated with the vulnerability to drug addiction and relapse remain to be established.

In summary, the Kalivas model emphasizes molecular changes within the glutamate projection from the prefrontal cortex to the nucleus accumbens. Lower basal levels of glutamate that are caused by a decrease in cystine-glutamate exchange, combined with the augmentation of cocaine-induced glutamate release through the lower tone of mGluR2/3 autoreceptors and an increase in AGS3 protein, are hypothesized to promote cocaine reinstatement.

5.9 Nucleus accumbens: mechanisms of addiction across drug classes reflect the importance of glutamate homeostasis

Scofield MD, Heinsbroek JA, Gipson CD, Kupchik YM, Spencer S, Smith AC, Roberts-Wolfe D, Kalivas PW. The nucleus accumbens: mechanisms of addiction across drug classes reflect the importance of glutamate homeostasis. Pharmacological Reviews 2016;68:816–71 [381].

As elaborated in earlier versions of the glutamate hypothesis and in the Wolf synaptic hypothesis, chronic drug exposure and subsequent withdrawal produce numerous alterations of glutamatergic transmission in the nucleus accumbens. The authors argued that these drug-induced neuroadaptations serve as a molecular basis for relapse vulnerability. Five main elements of synaptic plasticity in the nucleus accumbens are altered after exposure to drugs of abuse and withdrawal.

First, using cocaine as a prototypic drug, during withdrawal from repeated administration of cocaine, there is a loss of the ability to induce electrically stimulated mGluR2/3 LTD, and mGluR2/3 agonists reduce reinstated cocaine seeking [382]. This may be a result of a loss in the tonic activation of mGluR2/3 after the extinction of cocaine self-administration [383], presumably because of a reduction of extracellular glutamate levels [384]. *N*-acetylcysteine normalizes extracellular glutamate levels and restores the ability to electrically induce LTD in the nucleus accumbens [382], and the effect of *N*-acetylcysteine is blocked by mGluR2/3 antagonists [383].

Second, during withdrawal from several drugs of abuse, persistent potentiation of the glutamatergic input to the nucleus accumbens occurs [381], and high-frequency stimulation of nucleus accumbens afferents leads to LTP [382]. This LTP in the nucleus accumbens requires the activation of postsynaptic NMDA receptors, the entry of Ca^{2+} into dendritic spines, the activation of several molecular-biochemical changes, and the insertion of new AMPA receptors into the postsynaptic membrane. This insertion of new AMPA receptors into the postsynaptic membrane is not an acute withdrawal phenomenon but rather a delayed effect during protracted withdrawal and may involve a "silent synapse" mechanism ([381]: see *Section 5.11. Neurobiology of cue-induced incubation ("craving"): synaptic mechanism hypothesis*).

Third, another synaptic mechanism that involves changes in dendritic spine formation is hypothesized to involve glutamatergic plasticity. In general, the formation of new spines or the enlargement of existing spines is considered a correlate of LTP, whereas the retraction or contraction of spines is associated with LTD [381]. The activation of AMPA receptors increases spine head diameter [385]. An increase in spine head diameter has been hypothesized to be the result of an increase in actin cycling and AMPA receptor trafficking to the cell surface [386]. The reinstatement of drug seeking produces increases in head diameter after contingent exposure to discrete cues or environmental contexts that are associated with the drug of abuse [381]. Chronic cocaine exposure is also associated with an increase in F-actin and actin cycling [387].

In a review of pharmacological studies of glutamatergic neurotransmission in drug seeking, a key observation was that AMPA receptor activation in the nucleus accumbens is required for the acute excitation of medium spiny neurons by glutamatergic inputs that are required to induce drug seeking ([388]; see Section 5.11. *Neurobiology of cue-induced incubation ("craving"): synaptic mechanism hypothesis*). The systemic delivery of AMPA receptor antagonists inhibits cue-induced drug seeking and conditioned place preference [381]. Extended-access cocaine self-administration, followed by the incubation of craving, significantly increases drug-seeking behavior and decreases the surface postsynaptic expression of mGluR1 in the nucleus accumbens. The restoration of mGluR1 tone with positive allosteric modulators of mGluR1 inhibited cocaine seeking [389]. mGluR5 is preferentially localized postsynaptically, and mGluR5 activation inhibits drug self-administration and cue-induced drug seeking [381]. Consistent with above, the systemic administration of mGluR2/3-selective agonists inhibits cue- and context-induced drug seeking [381]. mGluR7 activation augments glutamate release and inhibits drug seeking, possibly via glutamate activity on mGluR2/3 [390].

Another contribution to the way in which chronic drug exposure alters glutamate synaptic plasticity in the nucleus accumbens is the lower expression of glial proteins that regulate homeostatic levels of extrasynaptic glutamate through glutamate release via the glial cysteine-glutamate exchanger and glial uptake ([391]; Fig. 35). Drug-induced disruption of these processes can then affect plasticity by influencing extrasynaptic glutamate levels, leading to changes in extrasynaptic mGluRs that influence glutamatergic plasticity. An increase in the activity of glutamate transporter-1 with such drugs as ceftriaxone and *N*-acetylcysteine can block drug-, cue-, and context-induced drug seeking [381,391]. The hypothesis is that enhancing glutamate transporter-1 activity normalizes extrasynaptic glutamate levels and promotes glutamate uptake to counteract drug- or cue-induced glutamate overflow in the nucleus accumbens [381,391]. For a listing of preclinical pharmacological studies that have been conducted to date with

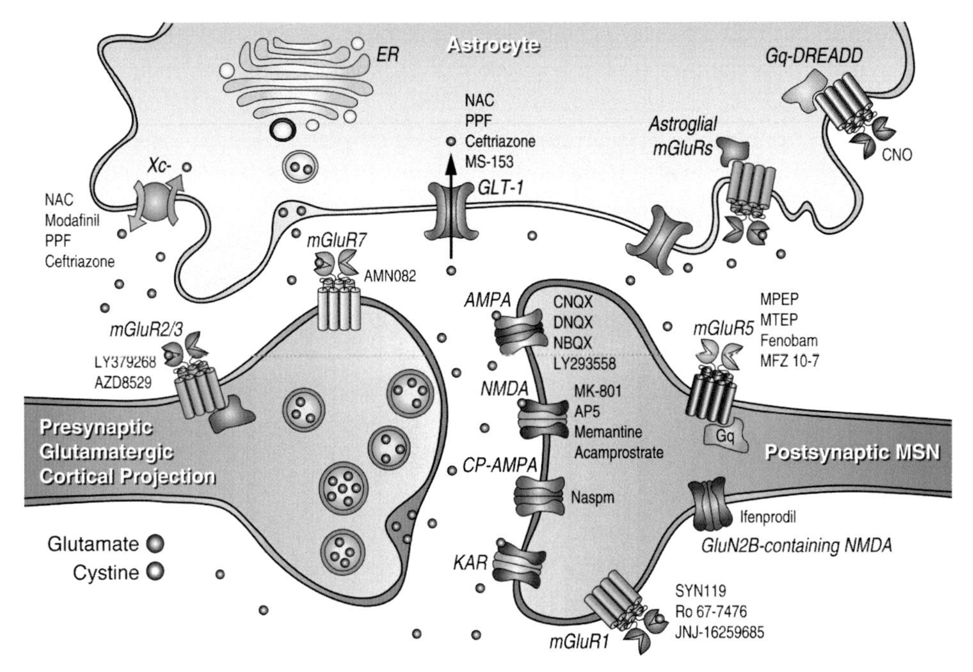

Figure 35 ***Pharmacological targets at the glutamatergic nucleus accumbens core synapse [381].*** Shown here is a schematic of a glutamate synapse in the NAC core with the pre- (*green*) and postsynaptic (*blue*) terminals as well as an astrocytic contact (*yellow*). Glutamate is depicted as *orange spheres*, and cysteine is shown as *gray spheres*. Listed next to AMPA, NMDA, mGLuR2/3, mGluR1, mGluR5, mGluR7, x_c-, and GLT-1 are the drugs that affect these proteins, which have been shown to inhibit drug seeking. *(Taken with permission from Scofield MD, Heinsbroek JA, Gipson CD, Kupchik YM, Spencer S, Smith AC, Roberts-Wolfe D, Kalivas PW. The nucleus accumbens: mechanisms of addiction across drug classes reflect the importance of glutamate homeostasis. Pharmacological Reviews 2016;68: 816–71 [381].)*

glutamate modulators in drug seeking, see Tables 2–4 in Scofield et al. [381]. For a listing of clinical pharmacological studies with glutamate modulators in addiction, see Table 5 in Scofield et al. [381].

Thus, evidence from numerous preclinical approaches and using many models of addiction in rats and mice support the hypothesis that nucleus accumbens glutamate transmission plays a key role in the neurobiology of addiction-related behaviors and the relapse to drug seeking. Most work focuses on drug-induced glutamatergic dysfunction in the cortico-nucleus accumbens circuit, a shared feature of exposure to many types of addictive drugs. Note that these persistent alterations of glutamatergic plasticity with a focus on the cortico-accumbens circuit are also the focus of the molecular basis of long-lasting relapse vulnerability that is associated with addiction (see Section 5.13. *Neuroepigenetics and addiction*).

The glutamate hypothesis has gained significant traction in the domain of the cue- and context-dependent reinstatement of drug seeking, but delayed effects are seen in animals with extended access to drugs, especially psychostimulants, suggesting that the AMPA glutamate receptor alterations may be particularly relevant in situations of protracted abstinence. One could argue that the incubation model fits the domain of protracted abstinence. This argues that drugs that restore glutamatergic homeostasis may have some therapeutic potential in the preoccupation/anticipation stage of the addiction cycle. Given the limited efficacy of glutamatergic modulating drugs in double-blind, placebo-controlled trials in addiction to date (see Table 5 in Ref. [381]), perhaps more of a focus should be placed on long-term abstinence and recovery after a period of abstinence. Note that for the glutamatergic modulator acamprosate (a Food and Drug Administration-approved medication for alcohol use disorder with modest efficacy in clinical trials), treatment efficacy was particularly high in patients who had a clearly identified goal of achieving abstinence before starting treatment [392].

5.10 Drug addiction and relapse: amygdala and corticostriatopallidal circuits

Everitt BJ, Wolf ME. Psychomotor stimulant addiction: a neural systems perspective. Journal of Neuroscience 2002;22:3312–20 [erratum: 22(16):1a] [97].

See RE, Fuchs RA, Ledford CC, McLaughlin J. Drug addiction, relapse, and the amygdala. In: Shinnick-Gallagher P, Pitkanen A, Shekhar A, Cahill L, editors. The amygdala in brain function: basic and clinical approaches. Annals of the New York Academy of Sciences, vol. 985. New York: New York Academy of Sciences; 2003. p. 294–307 [393].

Relapse to drugs of abuse has long been associated with a nebulous concept termed "craving" that involves not only memories of the positive-reinforcing effects of drugs but also drug-opposite associated effects, such as the motivational components of withdrawal [394]. Nevertheless, evidence is clear that drug-related stimuli can trigger greater motivation toward drug seeking and drug taking [395]. See and colleagues argued that through a process of associative learning, previously neutral stimuli come to acquire incentive-motivational properties after repeated pairings with the drug [393]. Everitt and Wolf further proposed that at the systems level, progressive strengthening or the "consolidation" of behavior that parallel the progression to addiction may be a form of habit learning [97]. In such habit learning, voluntary control over drug use is further hypothesized to be lost, and the propensity to relapse is high and readily precipitated by exposure to drug-associated stimuli. While studying cocaine-seeking behavior, both frameworks have identified corticostriatal pallidal systems as key components, with a focus on the basolateral amygdala for cue-induced relapse and cue-controlled cocaine seeking and a role for the prefrontal cortex in the loss of inhibitory control mechanisms that is hypothesized to reflect the development of locomotor sensitization in animals and impulsivity in humans ([97,393]; Figs. 36 and 37).

Evidence of a role for the basolateral amygdala in the cue-induced reinstatement of cocaine-seeking behavior is overwhelming. Inactivation of the basolateral amygdala with excitotoxic lesions, tetrodotoxin, or lidocaine blocked the reinstatement of cocaine

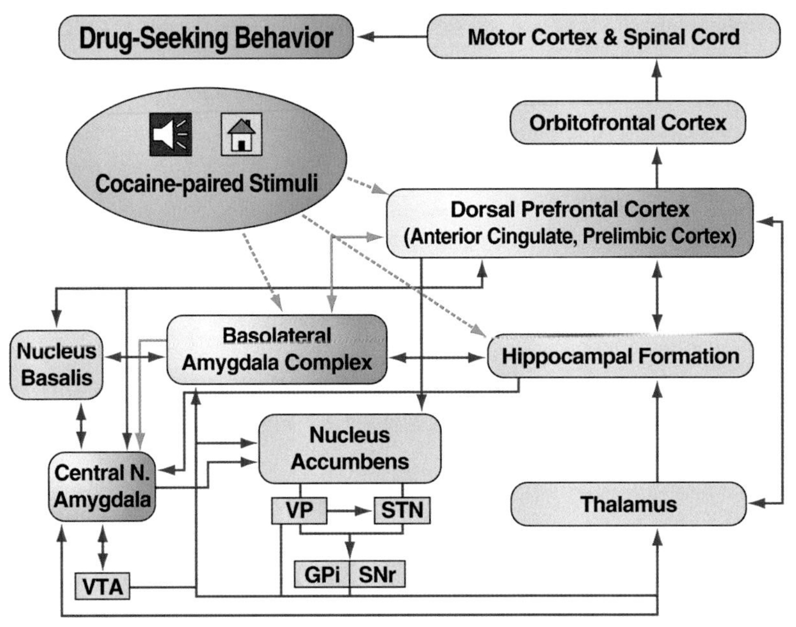

Figure 36 ***The conditioned-cued reinstatement of drug-seeking behavior neurocircuit [393].*** The motivational properties of drug-associated stimuli are processed by a network of brain structures which, based on the results of lesion and pharmacological inactivation experiments, include the basolateral amygdala complex, central nucleus of the amygdala, anterior cingulate cortex, and prelimbic cortex. Note: A number of nuclei and connections are omitted for the sake of clarity. GPi, globus pallidus internal segment; SNr, substantia nigra pars reticulata; STN, subthalamic nucleus; VP, ventral pallidum; VTA ventral tegmental area. *(Taken with permission from See RE, Fuchs RA, Ledford CC, McLaughlin J. Drug addiction, relapse, and the amygdala. In: Shinnick-Gallagher P, Pitkanen A, Shekhar A, Cahill L, editors. The amygdala in brain function: basic and clinical approaches. Annals of the New York Academy of Sciences, vol. 985. New York: New York Academy of Sciences; 2003. p. 294–307 [393].)*

seeking by cues that were paired with cocaine self-administration in animal models of cue-induced reinstatement [268,356,363]. Typically, various stimuli that were paired with drug administration in rats that were trained to self-administer cocaine are then presented in the absence of the drug after extinction, and the amount of responding for these stimuli is used as a measure of cue-induced reinstatement (see Chapter 2, *Animal Models of Addiction*). These lesions do not block subsequent cocaine self-administration, but the lesions block the ability of cocaine-paired stimuli to reinstate extinguished lever responding. Intra-basolateral amygdala administration of tetrodotoxin also blocked the acquisition of cue-induced cocaine-seeking behavior and the expression of reinstatement [396]. Central nucleus of the amygdala lesions only blocked the expression of reinstatement. The dorsomedial prefrontal cortex (also labeled the anterior cingulate cortex or prelimbic cortex; [393]) has been hypothesized to act in concert with the basolateral amygdala during the process of conditioned cued reinstatement [393]. Tetrodotoxin

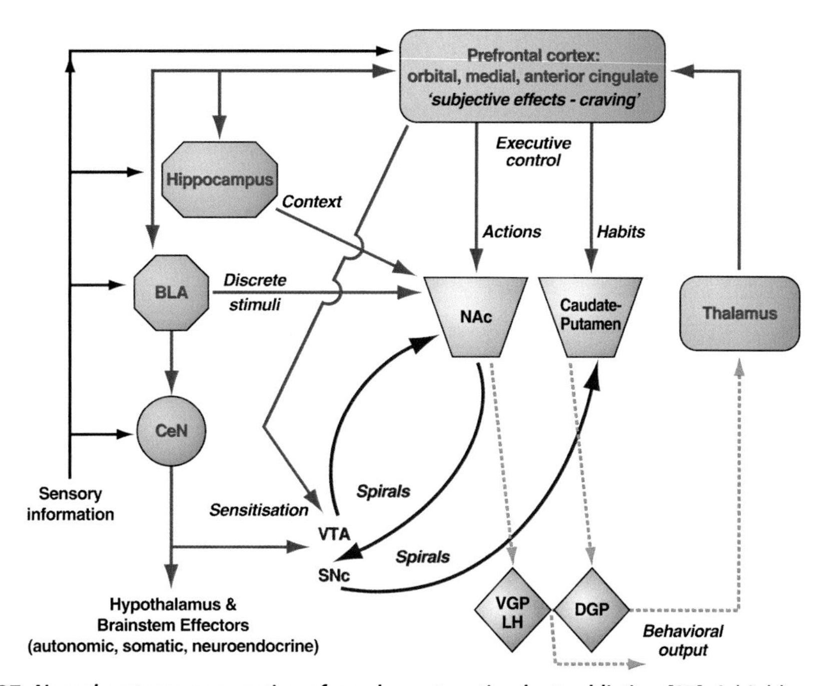

Figure 37 ***Neural systems perspective of psychomotor stimulant addiction [97].*** A highly schematic representation of limbic cortical-ventral striatopallidal circuitry that tentatively localizes particular functions: (1) sensitization—ventral tegmental area and also the nucleus accumbens, via glutamate-dopamine interactions; (2) processing of discrete and contextual drug-associated conditioned stimuli—basolateral amygdala and hippocampal formation, respectively; (3) goal-directed actions ("action-outcome" associations)—nucleus accumbens; (4) "habits" (stimulus-response learning)—dorsal striatum. Both 3 and 4 involve interactions between cortical afferents and striatal processes that are modulated by dopamine. (5) "Executive control"—prefrontal cortical areas; (6) subjective processes, such as craving, activate areas such as the orbital and anterior cingulate cortices, as well as temporal lobe structures, including the amygdala, in functional imaging studies; (7) "behavioral output" is intended to subsume ventral and dorsal striatopallidal outflow via both brainstem structures and reentrant thalamocortical loop circuitry; (8) "spirals" refers to the serial, spiraling interactions between the striatum and midbrain dopamine neurons that are organized in a ventral-to-dorsal progression [61]; (9) *blue arrows* indicate glutamatergic pathways; *yellow arrows* indicate GABAergic pathways; *red arrows* indicate dopaminergic pathways. The neurotransmitter that is used by central amygdala neurons is less certain but is probably glutamate and also possible neuropeptides. BLA, basolateral amygdala; CeN, central nucleus of the amygdala; VTA, ventral tegmental area; SNc, substantia nigra pars compacta. *(Taken with permission from Everitt BJ, Wolf ME. Psychomotor stimulant addiction: a neural systems perspective. Journal of Neuroscience 2002;22:3312–20 [erratum: 22(16):1a] [97].)*

lesions of the dorsomedial prefrontal cortex also significantly attenuated conditioned cued reinstatement that was produced by cocaine-paired stimuli [397]. The neuropharmacological basis of the cue-induced reinstatement of cocaine-seeking behavior appears to depend on dopamine D_1 receptor elements in the basolateral amygdala but not D_2 receptors or glutamate AMPA or NMDA receptors [366]. However, systemic administration of both D_1 and D_2 receptor antagonists blocked cue-induced reinstatement [398]. The presentation of a stimulus that predicted cocaine availability increased extracellular dopamine levels in the basolateral amygdala and increased basolateral amygdala Fos activity [368,369]. Finally, imaging studies in humans showed that cocaine-related cues activated the amygdala [322,399].

Consistent with the results of cue-induced reinstatement, excitotoxic lesions of the basolateral amygdala and medial prefrontal cortex also blocked the acquisition of cocaine-seeking behavior under a second-order schedule of reinforcement [74,400]. Inactivation of the nucleus accumbens core [58] and the administration of an AMPA but not NMDA receptor antagonist in the core of the nucleus accumbens [370] blocked cocaine-seeking behavior in a second-order schedule. Consistent with the results that have been reported with second-order schedules, intra-nucleus accumbens core administration of an AMPA receptor antagonist blocked psychostimulant-induced conditioned place preference [401,402]. Further reinforcing the role of corticostriatal glutamate projections are neurochemical studies that showed that discrete cocaine-related stimuli increased nucleus accumbens glutamate levels but decreased basal nucleus accumbens glutamate levels [403]. The potentiation of conditioned reinforcement that is associated with psychostimulants also critically depends on the basolateral and central nuclei of the amygdala and their interactions with the shell of the nucleus accumbens ([76]; for review, see Ref. [404]).

Notably, another glutamate projection from the ventral subiculum has been implicated in cocaine reinstatement [405]. Stimulation of the ventral subiculum induced long-lasting dopamine release in the nucleus accumbens and reinstated cocaine-seeking behavior [405]. Lesions of the ventral subiculum also blocked the facilitation of responding for conditioned reinforcers that was produced by an intra-nucleus accumbens injection of D-amphetamine [76]. Thus, glutamatergic inputs to the nucleus accumbens via the ventral subiculum may amplify information that also is provided by the basolateral amygdala [404].

However, discriminated approach to appetitive Pavlovian stimuli in an autoshaping procedure appears to depend more on the central nucleus of the amygdala and the anterior cingulate-core of the nucleus accumbens system but not the basolateral nucleus of the amygdala [406–408]. In such a procedure, the presentation of a visual stimulus is followed by food, and the animal develops a conditioned response of approaching the food-predicting conditioned stimulus before retrieving the primary reward. Similarly, the potentiation of instrumental behavior by noncontingent presentations of Pavlovian conditioned stimuli also depends on the central nucleus of the amygdala and core of the nucleus accumbens but not the basolateral amygdala ([73]; Table 5).

Such potentiation of instrumental responding by noncontingent presentations of Pavlovian conditioned stimuli is augmented by an increase in dopamine in the nucleus accumbens shell and has been linked to the sensitized conditioned salience that is associated with locomotor sensitization [95]. Locomotor sensitization also appears to depend on activation of the prefrontal cortex and basolateral amygdala. Lesions of the prefrontal

Table 5 Effects of lesions of specific forebrain sites on stimulus-reward learning.

	Stimulus-reward conditioned response (NAc amphetamine)	Stimulus-reward conditioned response	Pavlovian approach	Pavlovian instrumental transfer	Reference
Basolateral amygdala	↓	↓	–	–	[73,76,407,538]
Central amygdala	↓	–	↓	↓	[73,75,406,407]
Nucleus accumbens shell	↓	–	–	–	[72,73,408]
Nucleus accumbens core	–	–	↓	↓	[72,73,408]
Medial prefrontal cortex	–	–	↓	nt	[76,408,539]
Subiculum	↓	–	nt	nt	[76]

nt, not determined.

cortex, basolateral amygdala, and ventral tegmental area blocked the induction and expression of locomotor sensitization to the repeated administration of psychostimulant drugs [409–411]. The mechanism was hypothesized to be mediated by a glutamate projection to the ventral tegmental area [412].

Complex interactions have been proposed in an attempt to explain the way in which locomotor sensitization leads to the loss of inhibitory control mechanisms and the development of impulsivity [413]. For example, in the sensitized state, the loss of inhibitory tone in the prefrontal cortex may lead to prefrontal disinhibition of the basolateral amygdala, which in turn could enhance the basolateral amygdala-nucleus accumbens role in drug-conditioned reward [97] and impulsivity [262,413]. In contrast, compulsivity was hypothesized to arise from the consolidation of habitual stimulus-response drug seeking through the engagement of corticostriatal loops that operate through both the dorsal and ventral striatum (nucleus accumbens core; [413,414]). These changes that lead from impulsivity to compulsivity may be linked. These and other authors argued that the transition from voluntary drug seeking to a compulsive habit may also depend on the disruption of executive control that is conferred by descending prefrontal cortex influences on striatal mechanisms ([92,262]; see Section 5.1. Motive circuits: prefrontal cortex/ventral striatal hypotheses of addiction, Section 5.2. Frontal cortex dysfunction, cognitive performance, and executive function: disruption of frontocerebellar circuitry and function in alcoholism, Section 5.3. Brain circuitry for addiction from brain imaging studies, and Section 5.4. Impaired response inhibition and salience attribution syndrome of addiction).

What is somewhat difficult to reconcile is the way in which frontostriatal dysfunction—which simultaneously blocks cue-induced reinstatement, cocaine-induced reinstatement, and locomotor sensitization in animal studies — "acts synergistically with sensitization of stimulus-response mechanisms to produce compulsive drug-seeking behavior" [97]. This conundrum becomes particularly apparent when one notes that cocaine-induced locomotor sensitization is accompanied by a reduction of the magnitude of AMPA receptor-mediated quantal synaptic events and LTD in the nucleus accumbens shell and ventral tegmental area [94,106]. Interestingly, the authors ultimately suggested that lower excitability of the nucleus accumbens that is associated with repeated cocaine administration that is produced by the loss of prefrontal cortex executive control and loss of glutamate tone in the nucleus accumbens could be specifically related to "withdrawal phenomena," such as elevations of reward thresholds [415] and anhedonia or dysphoria [90] that may also contribute to persistent cocaine seeking and relapse.

In summary, Everitt and Wolf [97] and See et al. [393] presented a framework in which cue-induced reinstatement depends on a circuit that involves the basolateral amygdala, prefrontal cortex, and core of the nucleus accumbens (ventral striatal—ventral pallidal loops). Other stimulus-reward associations, measured by approach to appetitive Pavlovian stimuli and Pavlovian instrumental transfer, also involve activation of the nucleus accumbens core. The psychostimulant-induced potentiation of conditioned reinforcement involves the basolateral and central nuclei of the amygdala but via the shell of the nucleus accumbens. Impulsivity is linked to the sensitization of dopamine systems via the loss of inhibitory control from the prefrontal cortex. Compulsivity is hypothesized to develop from the consolidation of stimulus response drug seeking through the engagement of corticostriatal loops that operate through both the ventral and dorsal striatum.

5.11 Neurobiology of cue-induced incubation ("craving"): synaptic mechanism hypothesis

Wolf ME. Synaptic mechanisms underlying persistent cocaine craving. Nature Reviews Neuroscience 2016;17:351–65 [416].

Neuroimaging studies in humans have identified specific prefrontal cortex connections in cocaine craving that are elicited by environmental cues. Much work in rodents has used a face validity-type approach to animal models of craving by measuring the cue-induced reinstatement of responding. Many such studies of cocaine craving-like responding during abstinence have focused on the progressive intensification (incubation) of the cue-induced reinstatement of cocaine seeking that occurs as the duration of abstinence increases. An early observation was that adaptations occur in crucial regions of reward circuitry during this process ([416]; Fig. 38). For example, after 3–4 weeks of forced abstinence, the activation of a network that includes the nucleus accumbens, medial prefrontal cortex, and central nucleus of the amygdala is required for the expression of incubation. The nucleus accumbens has evolved as a focal point because many pathways that regulate motivated behavior converge on this structure.

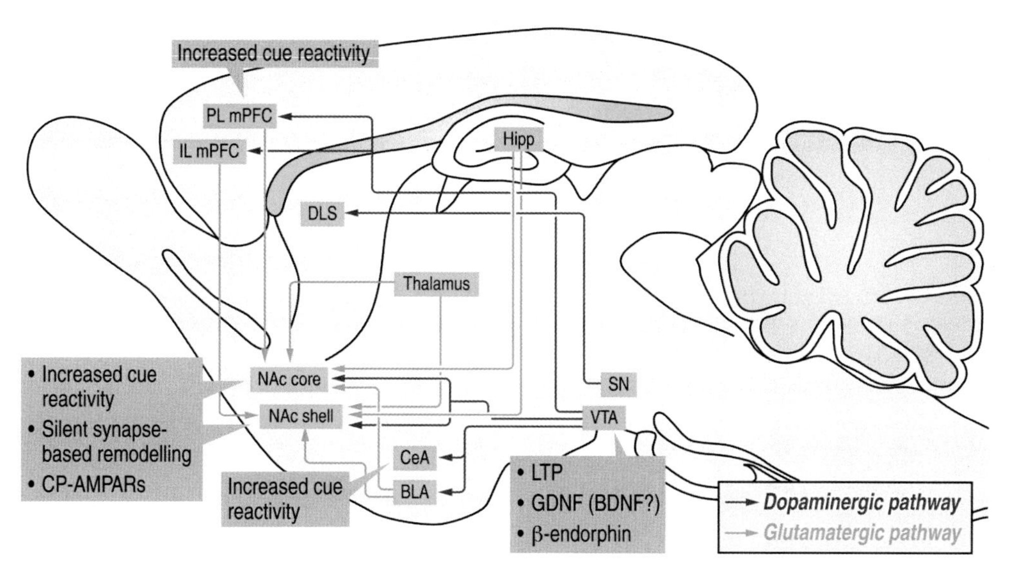

Figure 38 ***Neurobiology of cue-induced incubation ("craving"): synaptic mechanism hypothesis [416].*** Reward pathways and their roles in the incubation of cue-induced cocaine craving. In rats, glutamate inputs to medium spiny neurons (MSNs) of the nucleus accumbens (NAc) influence the behavioral response to rewards and the cues that predict them. Medium spiny neurons receive convergent inputs from multiple brain regions, and these inputs interact synaptically [493,494]. Inputs from the medial prefrontal cortex (mPFC) provide executive control (which is important for behavioral flexibility, response inhibition, and salience attribution). Basolateral amygdala (BLA) inputs relay conditioned associations and affective drive. Inputs from the hippocampus (Hipp) supply spatial and contextual information. Thalamic inputs presumably convey information about arousal and direct attention to behaviorally significant events. The central nucleus of the amygdala (CeA) does not project directly to the NAc, but the CeA (and BLA) can influence the NAc and other regions indirectly through connections with the midbrain (not shown), as well as through other routes. Note that different anatomical substrates underlie the incubation of cue-induced cocaine craving versus the reinstatement of drug seeking after extinction or even cocaine seeking upon return to the drug self-administration context [495,496]. Furthermore, differences exist between the neuronal pathways that are implicated in incubation of craving for cocaine versus other drugs of abuse [497,498]. *Gray boxes* list mechanisms or functional consequences that are associated with incubation. BDNF, brain-derived neurotrophic factor; CP-AMPARs, Ca^{2+}-permeable AMPA receptors; DLS, dorsolateral striatum; GDNF, glial cell line-derived neurotrophic factor; IL, infralimbic; LTP, long-term potentiation; PL, prelimbic; SN, substantia nigra; VTA, ventral tegmental area. *(Taken with permission from Wolf ME. Synaptic mechanisms underlying persistent cocaine craving. Nature Reviews Neuroscience 2016;17: 351–65 [416].)*

The synaptic mechanism hypothesis reflects analyses that were performed at the level of particular brain regions or major phenotypically defined cell types in small, functionally linked subsets of neurons in a particular region—neuronal ensembles—that are hypothesized to contribute to behavioral changes in animal models of addiction in largely, but not exclusively, studies of cocaine self-administration [417]. The underlying cellular mechanisms of heightened cue reactivity in rodents during the incubation of cocaine-induced reinstatement have focused on the nucleus accumbens, where excitatory transmission onto medium spiny neurons is strengthened through changes in AMPA receptor subunit composition and silent synapse-based remodeling. In the nucleus accumbens core in rats that underwent extended-access cocaine self-administration and prolonged withdrawal (>35 days), Ca^{2+}-permeable AMPA receptor levels significantly increased, and the activation of these receptors was required for the expression of incubated cocaine seeking [389,418]. Indeed, systemic or intra-nucleus accumbens administration of mGluR1-positive allosteric modulators that decrease synaptic levels of AMPA receptors can reduce incubated cue induced cocaine seeking ([416]; Fig. 39). As noted by Wolf, the anatomical circuitry through which early neuroadaptations (e.g., in the ventral tegmental area) enable later adaptations (e.g., in the nucleus accumbens, medial prefrontal cortex, and central nucleus of the amygdala) is currently under investigation and should also incorporate the circuitry of the dorsal striatum into the network. Further studies of the mechanism of the slowly developing plasticity that occurs over weeks to months of drug exposure and abstinence is argued to be critical for advancing our understanding of the rich repertoire of experience-dependent plasticity mechanisms that can be engaged in addiction.

This hypothesis of the overexpression of CA^{2+}*-permeable AMPA receptors has had a profound influence on elucidating the neurobiology of addiction, in part because of the robustness of the observation across models and laboratories and also because it has become a marker, to some extent, of the neuronal plasticity of addiction. Reversing these synaptic effects has the potential of effecting a molecular "stop" to the evolution of neuroadaptations in the mesolimbic dopamine system and consequent neuroadaptations in the nucleus accumbens. These neuroadaptations in turn potentially drive the shift to dorsal striatal habit formation and excessive drug taking that ultimately moves the subject to the dark side when the drug is unavailable (withdrawal). The challenge that remains is finding small molecules that can access this system without significant side effects and finding the appropriate timing and treatment. Additionally, the ways in which these changes generalize to other drugs of abuse will remain an exciting area for future research.*

5.12 Molecular basis of long-term plasticity underlying addiction

Nestler EJ. Molecular basis of long-term plasticity underlying addiction. Nature Reviews Neuroscience 2001;2:119–28 [253].

Nestler EJ. Historical review: molecular and cellular mechanisms of opiate and cocaine addiction. Trends in Pharmacological Sciences 2004;25:210–8 [254].

Acknowledging that all drugs of abuse share some common neurocircuitry actions, namely the inhibition of medium spiny neurons in the nucleus accumbens either through dopamine or other G_i-coupled receptors, Nestler has focused on identifying the molecular and cellular mechanisms [253]. The conceptual framework at this molecular level is

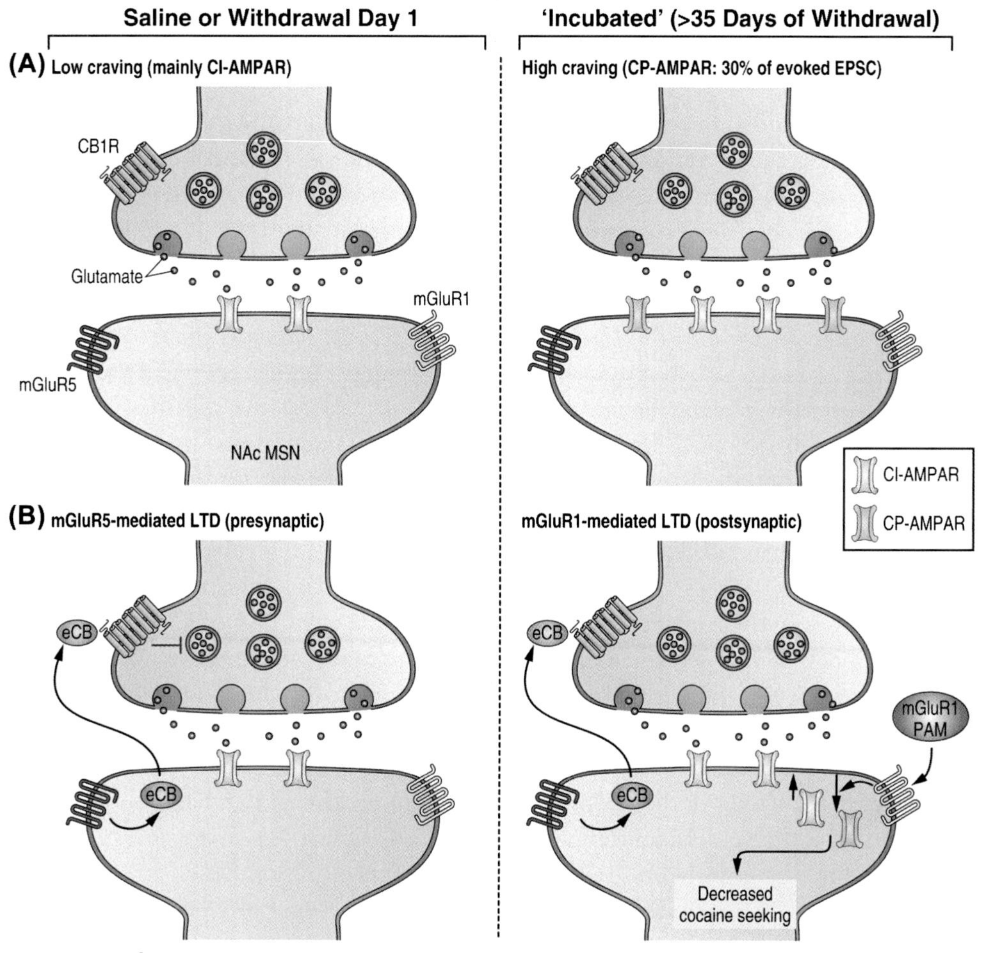

Figure 39 ***Ca^{2+}-permeable AMPA receptors in nucleus accumbens core synapses are required for the expression of incubated cocaine craving [416].*** (A) In the nucleus accumbens (NAc) core in rats that undergo extended-access cocaine self-administration and prolonged withdrawal (>35 days), Ca^{2+}-permeable AMPA receptor (CP-AMPAR) levels increase markedly, and activation of these receptors is required for the expression of incubated cocaine seeking [389,418]. This is proposes to reflect the ability of high-conductance Ca^{2+}-permeable AMPA receptors (CP-AMPAR) to augment the responsiveness of medium spiny neurons (MSNs) to glutamate drive [499], enabling MSNs to respond more robustly when a cocaine cue is presented. Some CP-AMPARs are inserted into previously silent synapses [500,501]. (B) Group I metabotropic glutamate receptor (mGluR)-mediated synaptic depression is altered in the NAc core in rats that exhibit the incubation of cocaine craving. In control rats, this synaptic depression depends on mGluR5 stimulation, endocannabinoid (eCB) formation, and the activation of presynaptic cannabinoid receptor 1 (CB1R). After prolonged withdrawal and the incubation of craving, this mechanism is disabled [502], and mGluR1-mediated synaptic depression is observed that is expressed postsynaptically by CP-AMPAR removal coupled to their replacement with lower-conductance CI-AMPARs [389,502]. By eliciting this mGluR1-mediated synaptic depression, systemic or intra-NAc administration of mGluR1 positive allosteric modulators (PAMs) can reduce incubated cue-induced cocaine seeking [389]. Similarly, mGluR1 activation removes CP-AMPARs from synapses in the ventral tegmental area in cocaine-exposed animals [503–505]. Although mGluR1-mediated synaptic depression is most obvious when CP-AMPAR levels are high, we speculate that it operates at a low level under normal conditions and that the loss of mGluR1 tone during abstinence helps to account for CP-AMPAR accumulation. Supporting this notion, decreased mGluR1 surface expression in the NAc core precedes and enables CP-AMPAR accumulation and the incubation of craving [389]. CI-AMPAR, Ca^{2+}-impermeable AMPA receptor; EPSC, excitatory postsynaptic current; LTD, long-term depression. *(Taken with permission from Wolf ME. Synaptic mechanisms underlying persistent cocaine craving. Nature Reviews Neuroscience 2016;17:351–65 [416].)*

that repeated perturbations of intracellular signal transduction pathways lead to changes in nuclear function and alterations of rates of the transcription of particular target genes. Alterations of the expression of such genes would lead to alterations of activity of the neurons where such changes occur and ultimately to changes in neural circuits in which those neurons operate ([253]; Fig. 40).

Two transcription factors in particular have been implicated in plasticity that is associated with the addiction cycle: CREB and ΔFosB. CREB regulates the transcription of genes that contain a cAMP response element site within the regulatory regions and can be found ubiquitously in genes that are expressed in the central nervous system, such as those that encode neuropeptides, synthetic enzymes for neurotransmitters, signaling proteins, and other transcription factors. CREB can be phosphorylated by protein kinase A and other protein kinases that are regulated by growth factors, putting it at a point of convergence for several intracellular messenger pathways that can regulate gene expression [253].

Much work in the addiction field has shown that the activation of CREB in the nucleus accumbens is a consequence of chronic exposure to opioids and cocaine [419–422], but note the differences observed with CREB in the amygdala with alcohol addiction (see Section *4.8. Gene transcription factor CREB: role in positive and negative affective states of alcohol addiction)*. Such activation was first recognized because CREB is activated as a consequence of upregulation of the cAMP pathway, an upregulation that has long been associated with chronic opioid administration ([254,423,424]; Fig. 41). *In vitro* in a NG108-15 neuroblastoma × glioma hybrid cell line, acute and chronic alcohol also increased cAMP levels through the activation of adenosine A_2 receptors, leading to CREB phosphorylation [425,426]. Opioids acutely inhibit adenylyl cyclase via G_i-coupled receptors, and upregulation of the cAMP pathway results from a compensatory homeostatic response of cells at various sites of the central nervous system to persistent opioid inhibition, including the nucleus accumbens [421]. Mice with mutations of the CREB gene exhibit a decrease in the development of opioid dependence, measured by physical signs of opioid withdrawal [427]. Upregulation of the cAMP pathway has been observed in various brain regions and has been hypothesized to play a role in the physical signs of opioid withdrawal and the motivational aspects of opioid withdrawal ([253]; Table 6).

Similar upregulation of the cAMP pathway in the nucleus accumbens has been observed with other drugs of abuse [428]. Nestler argued, "There is now compelling evidence that upregulation of the cAMP pathway and CREB in this brain region (nucleus accumbens) represents a mechanism of 'motivational tolerance and dependence.'" These molecular adaptations decrease an individual's sensitivity to the rewarding effects of subsequent drug exposures (tolerance) and impair the reward pathway (dependence) so that after removal of the drug the individual is left in an amotivational, depressed-like state [255,429–433]. The effects of upregulation of the cAMP pathway and CREB also contributed to an increase in the expression of dynorphin that was induced by exposure to drugs of abuse [255]. Dynorphin is a κ opioid receptor agonist that causes dysphoric-like responses. It is hypothesized that CREB increases in the dynorphin-mediated negative feedback circuit in the nucleus accumbens and contributes to aversive states that are associated with acute withdrawal ([253]; Fig. 42). However, regulation of the cAMP pathway, CREB, and dynorphin is relatively short-lived and may not explain long-term changes that are associated with the propensity to relapse [254].

The molecular changes that are associated with long-term changes in brain function as a result of chronic exposure to drugs of abuse have also been linked to changes in other transcription factors, factors that can alter gene expression and produce long-term changes in protein expression and as a result presumably neuronal function. The acute administration of drugs of abuse can cause a rapid (on the order of hours) activation of members of the Fos family, such as Fos, FosB, Fra-1, and Fra-2, in the nucleus accumbens [434,435]. Other transcription factors, isoforms of ΔFosB, accumulate over longer periods of time (days) with repeated drug administration [253]. Data suggest that an increase in ΔFosB may mediate increases in locomotor and rewarding responses to drugs of abuse [436]. The inducible expression of ΔFosB increased cocaine-seeking behavior [437]. Knockouts of ΔFosB have not supported a role for ΔFosB in the greater sensitivity to drugs of abuse [438], but such knockouts lack both products of the gene: ΔFosB and full-length FosB. Interestingly, transgenic mice that expressed a dominant-negative mutant form of c-jun, which antagonizes the transcriptional effects of ΔFosB itself, exhibited reductions of responses to drugs of abuse [439,440]. Nestler argued that ΔFosB may be a sustained molecular "switch" that helps initiate and maintain a state of addiction [441]. The ways in which changes in ΔFosB that can last for days translate into structural changes that are linked to chronic drug administration, such as a reduction of the density of dendritic spines of medium spiny neurons in the nucleus accumbens [442,443], and whether such molecular and structural changes actually underlie the vulnerability to relapse remain challenges for future work.

In summary, chronic drug administration leads to short-term upregulation of the cAMP pathway and CREB that may represent a mechanism of tolerance and dependence, but long-term changes in transcription factors, such as ΔFosB, may maintain the state of addiction and be the basis for the vulnerability to relapse.

5.13 Neuroepigenetics and addiction

Feng J, Nestler EJ. Epigenetic mechanisms of drug addiction. Current Opinion in Neurobiology 2013;23:521–8 [444].

Walker DM, Nestler EJ. Neuroepigenetics and addiction. Handbook of Clinical Neurology 2018; 148:747–65 [445].

There is a genetic loading to the vulnerability to addiction of approximately 50–60% in the human population, based on family-based, adoption, and twin studies [445]. This observation and the multifactorial nature of addiction have led to the investigation of both genetic and epigenetic mechanisms that underlie the phenotype. Walker and Nestler argue that addiction involves the dysregulation of motivational (reward) systems

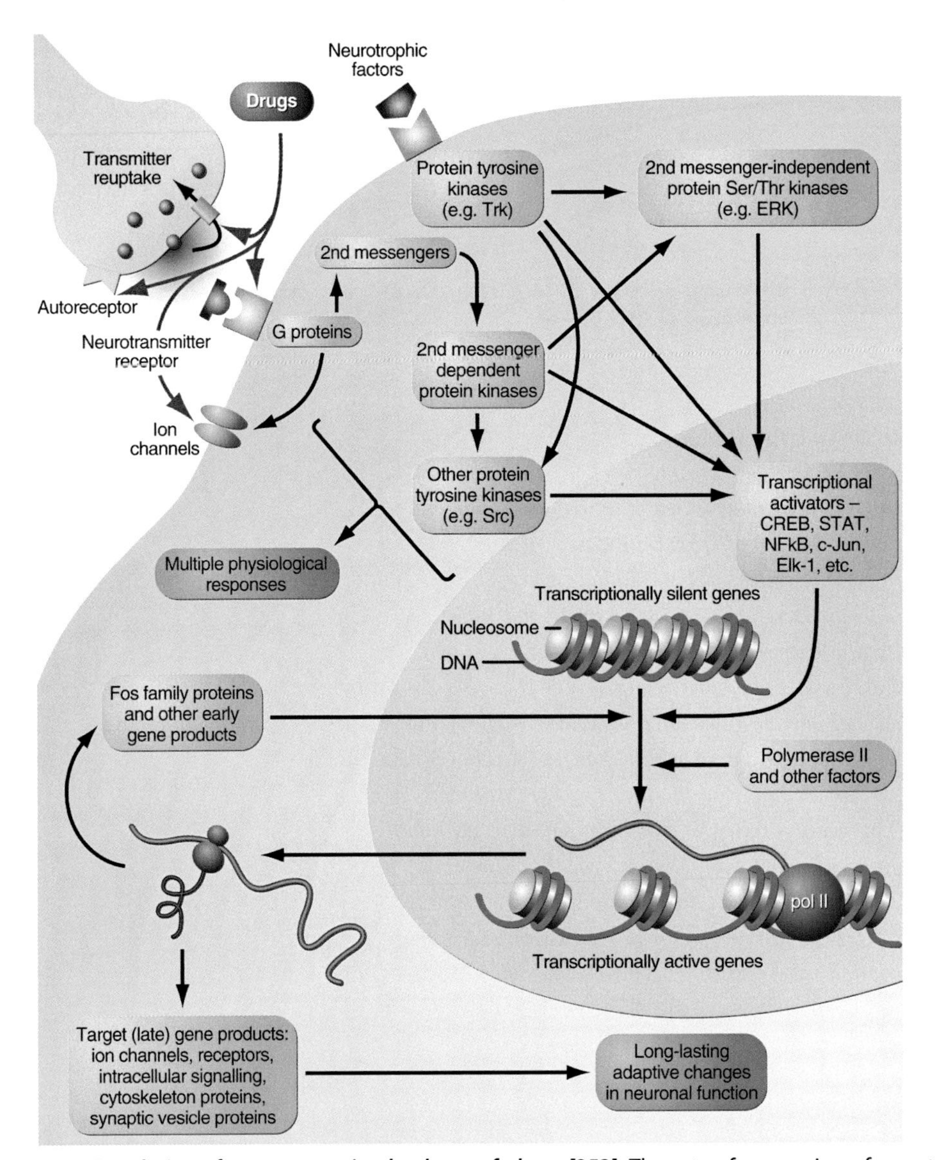

Figure 40 ***Regulation of gene expression by drugs of abuse [253].*** The rate of expression of a particular gene is controlled by its location within nucleosomes and by the activity of the transcriptional machinery [506]. A nucleosome is a tightly wound span of DNA that is bound to histones and other nuclear proteins. Transcription requires the unwinding of a nucleosome, which makes the gene accessible to a basal transcription complex. This complex consists of RNA polymerase (pol II, which transcribes the new RNA strand) and numerous regulatory proteins (some of which unwind nucleosomes through histone acetylation). Transcription factors bind to specific sites (response elements; also called promoter or enhancer elements) that are present within the regulatory regions of certain genes, and thereby increase or decrease the rate at which they are transcribed. Transcription factors act by enhancing or inhibiting the activity of the basal transcription complex, in some

in the brain and hypothesize that both genetic and epigenetic regulatory events underlie the changes throughout reward circuitry in humans and will be reflected in similar changes in animal models of addiction. They further note that a majority of the work to date on the epigenetic mechanisms of addiction has focused on the effects of drugs on the mesolimbic dopamine system (both the ventral tegmental area and nucleus accumbens, key brain regions that are involved in drug reward and neuroadaptations to chronic administration of drugs; Fig. 43).

Histone acetylation is generally associated with a permissive transcriptional state, and acute or repeated exposure to drugs of abuse, including cocaine, alcohol, morphine, and methamphetamine, increases global acetylation at the histone proteins H3 and/or H4 throughout reward circuitry [445]. Genes with enriched histone acetylation commonly consist of immediate early genes and those that are involved in neuroplasticity [444]. Although drugs of abuse can have overall effects on histone acetylation, often specific genes will change with chronic drug administration. For example, chronic exposure to both cocaine and morphine increases H3ac in the *Bdnf* and *Cdk5* promoters, genes that are associated with cell signaling [446,447].

One means of altering histone acetylation is by altering the activity or expression levels of histone acetyltransferase (HAT) enzymes that add the acetylation of lysine residues and HDACs. Walker and Nestler summarize that, in general, decreasing histone acetylation in the nucleus accumbens, either through the activation/overexpression of HDACs or inhibition/knockdown of HATs, blunts behavioral responses to drugs of abuse. Increasing histone acetylation in the nucleus accumbens through the pharmacological inhibition or knockout of HDACs or viral-mediated overexpression of HATs initially facilitates the behavioral effects of many drugs of abuse. However, there are significant selective effects of different HDAC inhibitors in drug-induced behaviors. Walker and Nestler state, "Further studies are necessary to identify how these alterations work in concert to produce a temporally specific histone landscape that is indicative of the various stages of addiction" ([445]; p. 754).

cases by altering nucleosomal structure through changes in the histone acetylation of the complex. Regulation of transcription factors is the best-understood mechanism by which changes in gene expression occur in the adult brain [507–509]. Most transcription factors are regulated by phosphorylation. Accordingly, by causing repeated perturbation of synaptic transmission and hence of protein kinases or protein phosphatases, repeated exposure to a drug of abuse would lead eventually to changes in the phosphorylation state of particular transcription factors, such as CREB, that are expressed under basal conditions. This would lead to the altered expression of their target genes. Among such target genes are those for additional transcriptional factors (such as c-Fos), which—through alterations in their levels—would alter the expression of additional target genes. Drugs of abuse could conceivably produce stable changes in gene expression through the regulation of many other types of nuclear proteins, but such actions have not yet been shown. *(Taken with permission from Nestler EJ. Molecular basis of long-term plasticity underlying addiction. Nature Reviews Neuroscience 2001;2:119–28 [253].)*

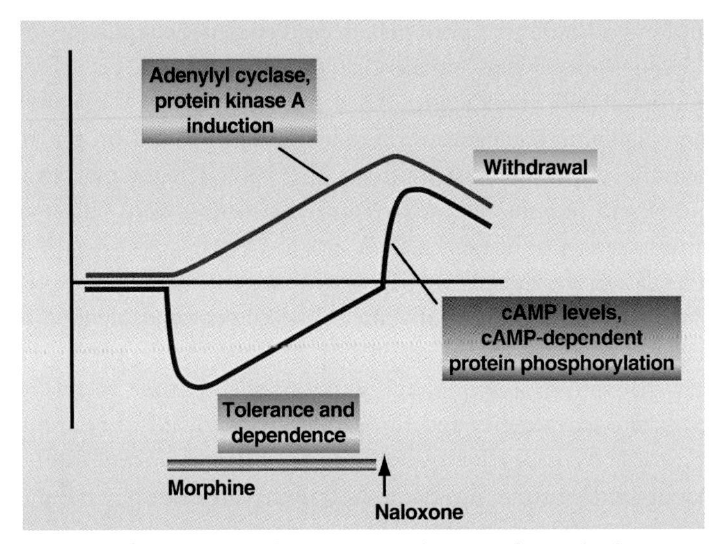

Figure 41 ***Upregulation of the cAMP pathway as a mechanism of opioid tolerance and dependence [254].*** Opiates acutely inhibit functional activity of the cAMP pathway (indicated by cellular levels of cAMP and cAMP-dependent protein phosphorylation). With continued opiate exposure, functional activity of the cAMP pathway gradually recovers and increases far above control levels following removal of the opioid (e.g., by administration of the opioid receptor antagonist naloxone). These changes in the functional state of the cAMP pathway are mediated by the induction of adenylyl cyclase and protein kinase A (PKA) in response to chronic opioid administration. The induction of these enzymes accounts for the gradual recovery of functional activity of the cAMP pathway that occurs during chronic opioid exposure (tolerance and dependence) and activation of the cAMP pathway that is observed upon the removal of opioid (withdrawal). *(Taken with permission from Nestler EJ. Historical review: molecular and cellular mechanisms of opiate and cocaine addiction. Trends in Pharmacological Sciences 2004;25:210–8 [254].)*

Table 6 Upregulation of the cAMP pathway and opiate addiction.

Site of upregulation	Functional consequence
Locus coeruleus[a]	Physical dependence and withdrawal
Ventral tegmental area[b]	Dysphoria during early withdrawal periods
Periaqueductal gray	Dysphoria during early withdrawal periods Physical dependence and withdrawal
Nucleus accumbens	Dysphoria during early withdrawal periods
Amygdala	Conditioned aspects of addiction?
Dorsal horn of spinal cord	Tolerance to opiate-induced analgesia
Myenteric plexus of gut	Tolerance to opiate-induced reductions of intestinal motility Increases in motility during withdrawal

[a]The cyclic adenosine monophosphate (cAMP) pathway is upregulated within the principal noradrenaline neurons that are located in this region.
[b]Indirect evidence indicates that the cAMP pathway may be upregulated within γ-aminobutyric acid (GABA) neurons that innervate dopamine and serotonin cells in the ventral tegmental area and periaqueductal gray, respectively. During withdrawal, the upregulated cAMP pathway would become fully functional and could contribute to a state of dysphoria by increasing the activity of GABA neurons, which would then inhibit the dopamine and serotonin neurons [170,540].
Taken with permission from Nestler EJ. Molecular basis of long-term plasticity underlying addiction. Nature Reviews Neuroscience 2001;2:119–28 [253].

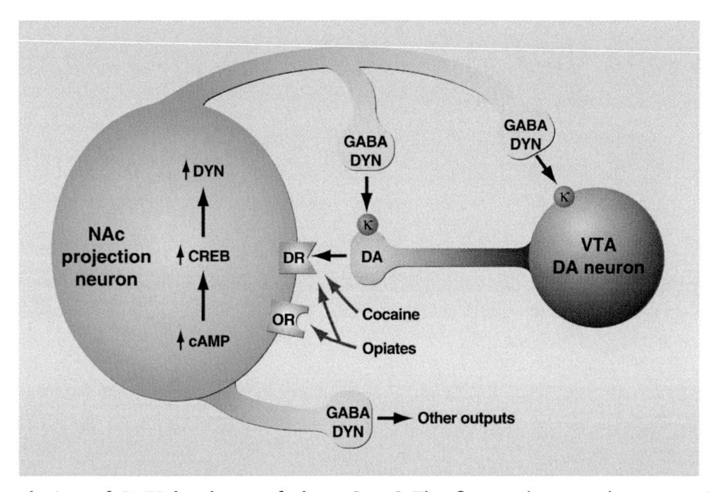

Figure 42 ***Regulation of CREB by drugs of abuse [253].*** The figure shows a dopamine (DA) neuron of the ventral tegmental area (VTA) innervating a class of γ-aminobutyric acid (GABA) projection neurons from the nucleus accumbens (NAc) that express dynorphin (DYN). Dynorphin constitutes a negative feedback mechanism in this circuit: dynorphin, released from terminals of NAc neurons, acts on κ-opioid receptors on nerve terminals and cell bodies of DA neurons to inhibit their function. Chronic exposure to cocaine or opiates upregulates the activity of this negative feedback loop through upregulation of the cAMP pathway, the activation of CREB, and the induction of dynorphin. cAMP, cyclic adenosine monophosphate; CREB, cyclic adenosine monophosphate response element binding protein; DR, dopamine receptor; OR, opioid receptor. *(Modified with permission from Carlezon WA Jr, Nestler EJ, Neve RL. Herpes simplex virus-mediated gene transfer as a tool for neuropsychiatric research. Critical Reviews in Neurobiology, 2000, 14: 47–67 [544].)*

Another means of modifying histone expression is via histone lysine methylation, which is associated with either the activation or repression of gene expression, depending on which residues are methylated and the number of methyl groups that are added. The inhibition of repressive epigenetic modifying enzymes via histone methyltransferases that mediate gene repression in the nucleus accumbens enhance drug-associated behaviors [445].

DNA methylation is also an important epigenetic regulator and occurs with the addition of a methyl group. DNA methylation at gene promoters is generally associated with repression, whereas methylation in gene bodies has been associated with active transcription. One can alter DNA methylation to study drug-elicited behaviors by inhibiting the expression of DNA methyltransferases, the enzymes that catalyze the addition of methyl groups to DNA. For example, the pharmaceutical inhibition or local knockout of DNMT3a in the nucleus accumbens enhanced cocaine reward, whereas overexpression decreased it [448].

However, as is now well known, a large number of RNA transcripts that are not translated into proteins have been identified from the complete sequencing of multiple mammalian genomes, and many of these noncoding RNAs are now known to play

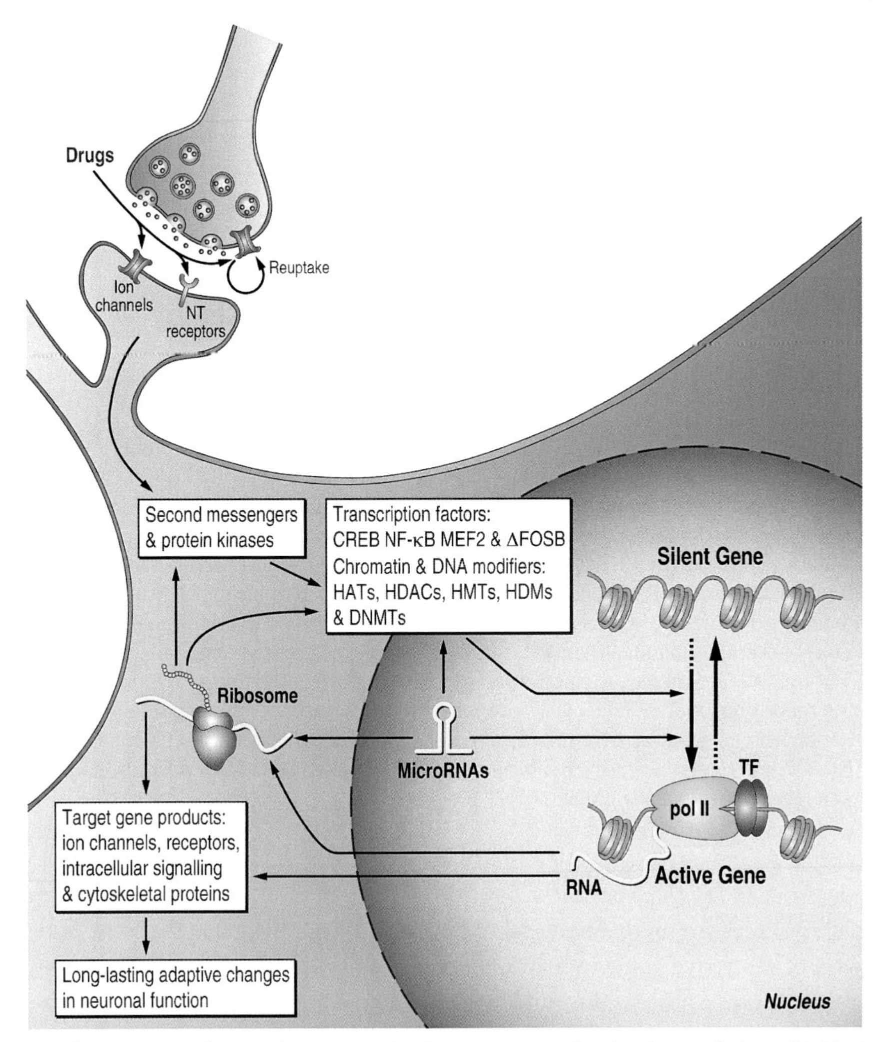

Figure 43 ***Mechanisms of transcriptional and epigenetic regulation by drugs of abuse [510].*** Drugs of abuse act through synaptic targets (reuptake mechanisms, ion channels, and neurotransmitter [NT] receptors) to alter intracellular signaling cascades. This leads to the activation or inhibition of transcription factors and of many other nuclear targets, including chromatin-regulatory proteins (shown by *thick arrows*). These processes result in the induction or repression of particular genes, which can in turn further regulate gene transcription. It is proposed that some of these drug-induced changes at the chromatin level are extremely stable and thereby underlie the long-lasting behaviors that define addiction. CREB, cAMP-response element binding protein; DNMTs, DNA methyltransferases; HATs, histone acetyltransferases; HDACs, histone deacetylases; HDMs, histone demethylases; HMTs, histone methyltransferases; MEF2, myocyte-specific enhancer factor 2; NF-κB, nuclear factor-κB; pol II, RNA polymerase II. *(Taken with permission from Robison AJ, Nestler EJ. Transcriptional and epigenic mechanisms of addiction. Nature Reviews Neuroscience 2011;12:623–37 [510].)*

an important role in cell function. For example, miRNAs are short sequences of RNA (<20 bp) that play a repressive role in gene expression by binding a target sequence on specific mRNAs and blocking translation or inducing degradation. As an example in the addiction field, miR-206 is upregulated in the prefrontal cortex but not in the ventral tegmental area, amygdala, or nucleus accumbens in rats that self-administer alcohol. The overexpression of miR-206 in the prefrontal cortex in nondependent animals increased alcohol consumption and may be related to the lower expression of BDNF [449].

Thus, Walker and Nestler conclude that drugs of abuse generally increase the likelihood of transcription in the nucleus accumbens by either increasing histone modifications that are associated with an "active" chromatin state or by reducing histone modifications that are associated with a repressive chromatin state. They further argue that "these activating marks" are associated with genes that have been implicated generally in neuroplasticity and immediate early genes. They also make a case that drugs of abuse likely affect the expression or activation of enzymes that are "writers" and "erasers" of histone modifications; if one alters these enzymes in brain regions that are associated with reward, then the behavioral responses to drugs of abuse are altered [445]. Finally, significant drug-induced epigenetic regulation presumably occurs in nongenic regions of the genome, which is also likely important in mediating the addiction phenotype. Such epigenetic transcriptional changes can be permissive, and hundreds of genes may show regulation in the opposite direction.

Although their focus was appropriately on the mesolimbic dopamine system and its terminal connections, changes in brain stress systems in addiction also have a genetic, genetic-environment, even possibly epigenetic overlay, suggesting a mechanism for genetic or epigenetic interactions in the dark side of addiction and reward/incentive salience pathways. For example, several single-nucleotide polymorphisms (SNPs) of the CRF_1 receptor gene (Crhr1) and CRF binding protein gene (CRHBP) have been associated with binge drinking in adolescents, excessive drinking, and alcohol use disorder in adult humans [450,451]. Several polymorphisms of human CRF system molecules have been associated with excessive alcohol use phenotypes, often in interactions with a history of stress [450,452]. A SNP might also alter gene regulation through epigenetic processes that are related to differences in genetic sequence. For example, specific SNP-containing intergenic transcript alleles regulate a number of chromatin modifier genes, thus directly linking SNPs and epigenetic regulation [453].

6. Synthesis: common elements of most neurobiological theories of addiction

Drug addiction can be defined as a compulsion to seek and take a drug, loss of control in limiting intake, and the emergence of a negative emotional state when access to the drug is prevented (see Chapter 1, What Is Addiction?). A heuristic framework for addiction consists of a three-stage cycle—*binge/intoxication*, *withdrawal/negative affect*, and *preoccupation/anticipation*—that represents dysregulation in three functional domains (incentive salience/habits, negative emotional states, and executive function, respectively) and is mediated by three major neurocircuitry elements (basal ganglia, extended amygdala, and prefrontal cortex, respectively). Excessive drug taking in the *binge/intoxication* stage drives the allostatic process, with the three stages feeding into each other, becoming more intense, and ultimately leading to the pathological state known as addiction ([40]; Fig. 44). Subsequently, the termination of drug taking inevitably leads to negative emotional states of acute and protracted withdrawal in the *withdrawal/negative affect* stage, which generates a second motivational drive from negative reinforcement. Protracted abstinence incorporates residual elements of negative emotional states and cue and contextual craving that form the *preoccupation/anticipation* stage.

In the *binge/intoxication* stage, most theories and models include a key role for some component of reward that usually involves the ventral tegmental area and nucleus accumbens, sometimes the central nucleus of the amygdala (Fig. 45). Drugs and alcohol at intoxicating doses have a wide but selective action on neurotransmitter systems in the brain reward systems, based on animal models of the acute reinforcing effects of drugs and alcohol, animal studies that use selective receptor antagonists for specific neurochemical systems, and human PET imaging studies. Multiple neurochemical systems are implicated in the acute reinforcing effects of alcohol, including GABA, opioid peptides, dopamine, serotonin, and glutamate. Ventral striatal-pallidal-thalamic loops have long been hypothesized to play a key role in translating motivation to action and now are argued to form the basis of "pathological" habits. New in this volume are hypotheses regarding new means of categorizing the reinforcing effects of drugs with an emphasis on choice ("Choice and the delayed reward hypothesis"; [107]) and regarding what constitutes an "addicted" rat ("Multistep general theory of the transition to addiction; [134]). Also included in this chapter under the *binge/intoxication* framework is a new hypothesis regarding the formation of pathological habits ("Drug addiction: maladaptive incentive habits"; [115]). Additionally, a new theory consolidates the role of dopamine prediction errors in addiction ("Dopamine prediction errors in reward learning and addiction: from theory to neural circuitry"; [126]).

In the *withdrawal/negative affect* stage, two neuroadaptational processes are hypothesized to be key: loss of function in the reward systems (within-system neuroadaptation) and recruitment of the brain stress systems (between-system neuroadaptation; [40]). The argument is that a decrease in dopamine and opioid peptide function leads to lower hedonic tone. Supporting this hypothesis, decreases in activity of the mesolimbic dopamine system [454] and decreases in serotonergic neurotransmission in the nucleus accumbens have been observed during drug and alcohol withdrawal in animal studies [455]. Imaging studies of humans with addiction have consistently shown long-lasting decreases in the numbers of dopamine D_2 receptors in drug- and alcohol-dependent subjects compared with controls [456]. Additionally, drug- and alcohol-dependent subjects had dramatically lower dopamine release in the striatum in response to a pharmacological challenge with

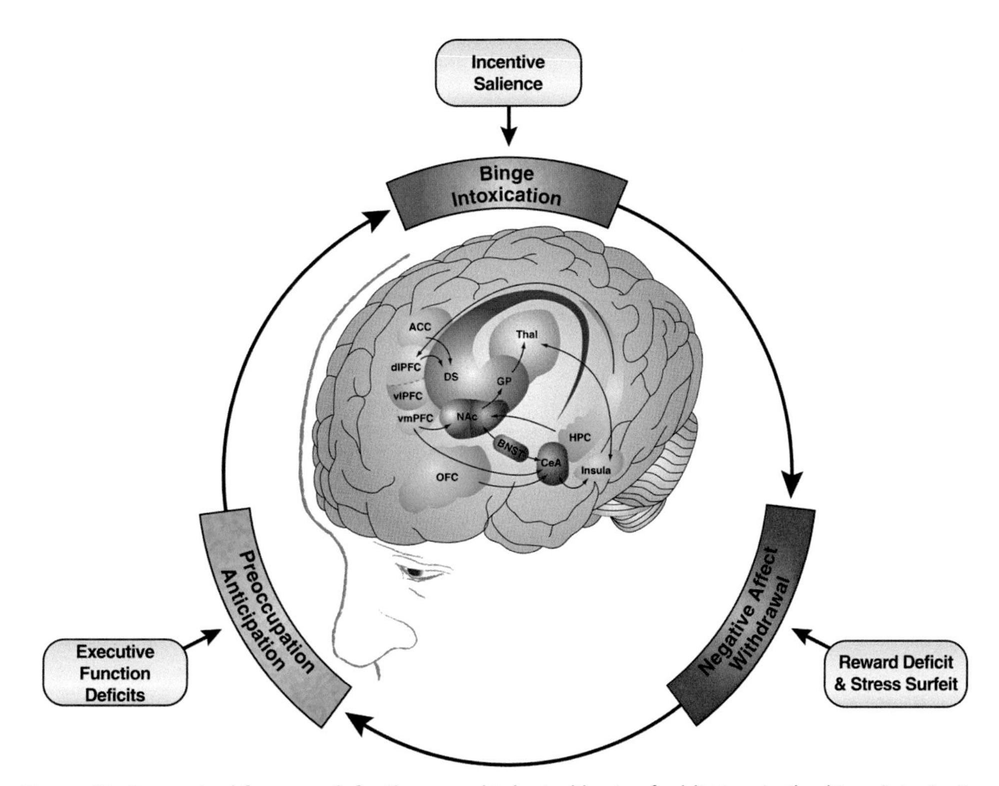

Figure 44 Conceptual framework for the neurobiological basis of addiction. In the *binge/intoxication* stage, reinforcing effects of drugs may engage reward neurotransmitters and associative mechanisms in the nucleus accumbens shell and core and then engage stimulus-response habits that depend on the dorsal striatum. Two major neurotransmitters that mediate the rewarding effects of drugs of abuse are dopamine and opioid peptides. In the *withdrawal/negative affect* stage, the negative emotional state of withdrawal may engage activation of the extended amygdala. The extended amygdala is composed of several basal forebrain structures, including the bed nucleus of the stria terminalis, central nucleus of the amygdala, and possibly a transition zone in the medial portion (or shell) of the nucleus accumbens. Major neurotransmitters in the extended amygdala that are hypothesized to function in negative reinforcement are corticotropin-releasing factor, norepinephrine, and dynorphin. There are major projections from the extended amygdala to the hypothalamus and brainstem. The *preoccupation/anticipation* (craving) stage involves the processing of conditioned reinforcement in the basolateral amygdala and the processing of contextual information by the hippocampus. Executive control depends on the prefrontal cortex and includes the representation of contingencies, the representation of outcomes, and their value and subjective states (i.e., craving and, presumably, feelings) associated with drugs. The subjective effects, termed "drug craving" in humans, involve activation of a "Go" system (more dorsolateral PFC) and inactivation of a "Stop" system (more ventromedial PFC). A major neurotransmitter that is involved in the craving stage is glutamate that is localized in pathways from frontal regions and the basolateral amygdala that project to the ventral striatum. ACC, anterior cingulate cortex; BNST, bed nucleus of the stria terminalis; CeA, central nucleus of the amygdala; DS, dorsal striatum; dlPFC, dorsolateral prefrontal cortex; GP, globus pallidus; HPC, hippocampus; NAC, nucleus accumbens; OFC, orbitofrontal cortex; Thal, thalamus; vlPFC, ventrolateral prefrontal cortex; vmPFC, ventromedial prefrontal cortex. *(Modified from Koob GF, Volkow ND. Neurocircuitry of addiction. Neuropsychopharmacology Reviews 2010;35:217–38 [erratum: 35:1051] [38].)*

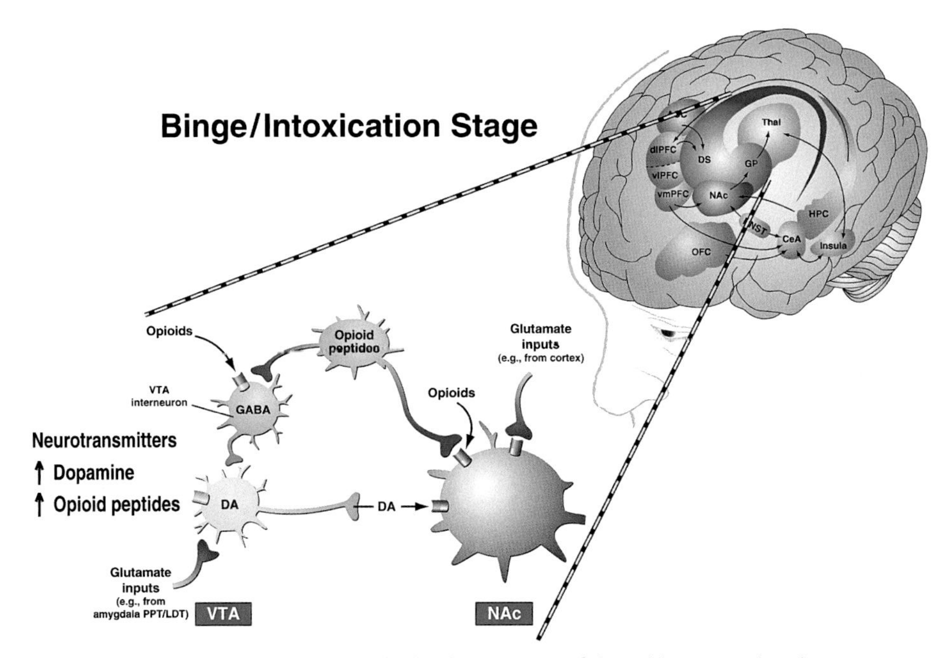

Figure 45 Neural circuitry associated with the three stages of the addiction cycle—*binge/intoxication* stage. Reinforcing effects of drugs may engage associative mechanisms and reward neurotransmitters in the nucleus accumbens shell and core and then engage stimulus-response habits that depend on the dorsal striatum. Two major neurotransmitters that mediate the rewarding effects of drugs of abuse are dopamine and opioid peptides. (Inset) Simplified schematic of converging acute actions of drugs of abuse on the ventral tegmental area (VTA) and nucleus accumbens (NAc). Drugs of abuse, despite diverse initial actions, produce some common effects on the VTA and NAc. Opioids activate the nucleus accumbens by inhibiting γ-aminobutyric acid (GABA) interneurons in the VTA, which disinhibit VTA dopamine neurons. Opioids also directly act on opioid receptors on NAc neurons. *(Modified with permission from Nestler EJ. Is there a common molecular pathway for addiction? Nature Neuroscience 2005;8:1445–9 [469].)*

the stimulant drug methylphenidate [456]. These findings suggest an overall reduction of the sensitivity of the dopamine component of reward circuitry to natural reinforcers and other drugs in individuals with addiction [457].

Another major neuroadaptation that can contribute to the negative emotional state that drives negative reinforcement in the *withdrawal/negative affect* stage is termed a between-system neuroadaptation. Here, brain neurocircuits and neurochemical systems that are involved in arousal-stress modulation are engaged within neurocircuitry of the brain stress systems in an attempt to overcome the chronic presence of the perturbing drug or alcohol and to restore normal function despite the presence of drug. The neuroanatomical entity that is termed the extended amygdala [77] may represent a common anatomical substrate that integrates brain arousal-stress systems with hedonic processing systems to produce the between-system opponent process (Fig. 46).

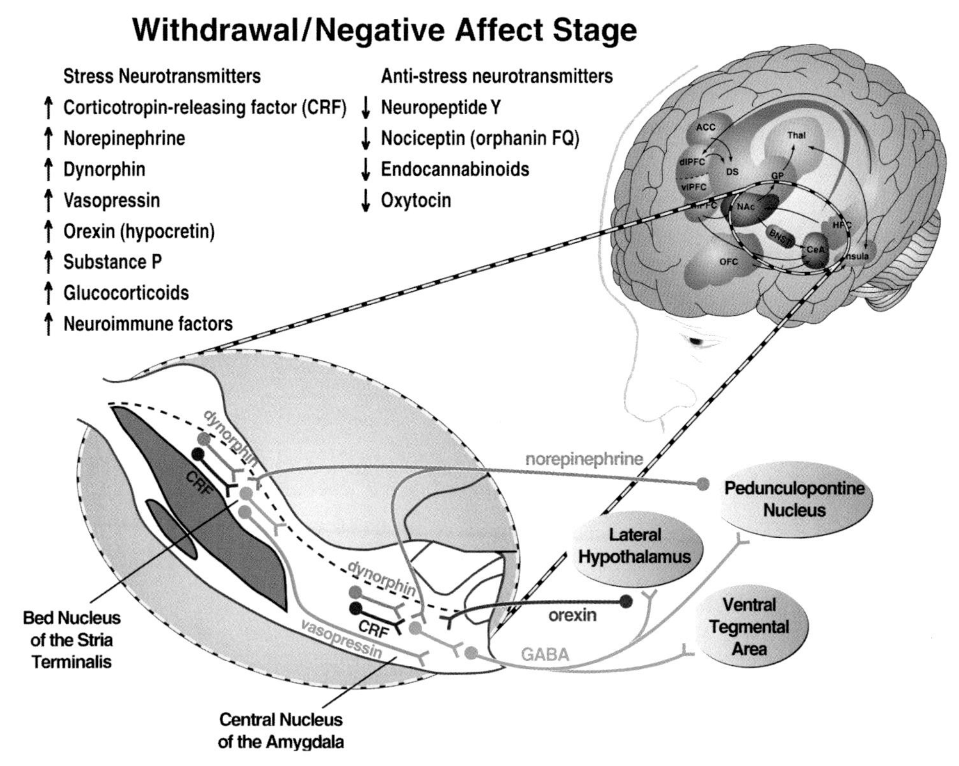

Figure 46 Neural circuitry associated with the three stages of the addiction cycle—*withdrawal/negative affect* stage. The negative emotional state of withdrawal may engage activation of the extended amygdala. (Inset) The extended amygdala is composed of several basal forebrain structures (see text), including the bed nucleus of the stria terminalis, central nucleus of the amygdala, and possibly a transition area in the medial portion (shell) of the nucleus accumbens. Major neurotransmitters in the extended amygdala that are hypothesized to play a role in negative reinforcement are corticotropin-releasing factor, norepinephrine, and dynorphin. The extended amygdala has major projections to the hypothalamus and brainstem. *(Adapted from Koob GF. A role for brain stress systems in addiction. Neuron 2008;59:11–34 [458]; George O, Koob GF. Control of craving by the prefrontal cortex. Proceedings of the National Academy of Sciences USA 2013;110:4165–6 [541].)*

A key brain stress system that is mediated by CRF systems in both the extended amygdala and HPA axis is dysregulated by the chronic administration of all major drugs with dependence or abuse potential, with a common response of elevated adrenocorticotropic hormone, corticosterone, and amygdala CRF during acute withdrawal from chronic drug administration [458]. In animal models of alcohol dependence, extrahypothalamic CRF systems become hyperactive during alcohol withdrawal, with an increase in extracellular CRF in the central nucleus of the amygdala and bed nucleus of the stria terminalis in dependent rats [458]. Other brain neurotransmitter or neuromodulatory systems that have pro-stress actions also converge on the extended amygdala and include norepinephrine, vasopressin, substance P, hypocretin (orexin), and dynorphin, all of which may contribute to stress-like states and negative emotional states that are associated with drug withdrawal or protracted abstinence [458].

Neurotransmitter systems that are implicated in anti-stress actions include NPY, nociceptin, and endocannabinoids. The combination of decreases in reward neurotransmitter function, the recruitment of brain stress systems, and/or the suppression of anti-stress systems provides a powerful motivation for reengaging in drug taking and drug seeking. Thus, the vulnerability to substance use disorder may involve not only a sensitized stress system but also a hypoactive stress buffer, and behavioral and pharmacological interventions that block pro-stress systems and stimulate anti-stress systems may be particularly interesting targets for the treatment of substance use disorder.

New in this chapter are neurobiological hypotheses regarding the role of negative reinforcement in addiction and the ways in which neuroplasticity in brain negative emotional systems can drive negative reinforcement ("Negative reinforcement in drug addiction: the darkness within"; [133]) and the ways in which "psychological" symptoms of opioid withdrawal, including learned associations of opioid withdrawal, can persist in an allostatic state for months, even years ("Neurobiology of opioid dependence in creating addiction vulnerability"; [217]). Additionally, an argument is made that the sensitized negative emotional state (hyperkatifeia) that drives negative reinforcement in opioid addiction has parallels and overlaps with hyperalgesia in the pain domain ("Opioids, pain, the brain, and hyperkatifeia: a framework for the rational use of opioids for pain and relevance to addiction"; [213]). Finally, a hypothesis is presented that proposes that epigenetic modifications within the amygdala provide a molecular basis for the negative emotional state that is associated with alcohol withdrawal and that such epigenetic modifications are engaged by repeated alcohol use, genetic vulnerability, and alcohol exposure during crucial developmental periods ("Epigenetic basis of the dark side of alcohol addiction"; [256]).

The *preoccupation/anticipation* or "craving" stage of the addiction cycle has long been hypothesized to be a key part of the neurocircuitry that mediates relapse in humans (Fig. 47). Dysregulation of the frontal cortex mediates elements of both impulsivity and compulsivity, in addition to elements of protracted abstinence and craving. Human imaging studies reveal neurocircuitry dysregulation during the *preoccupation/anticipation* stage in substance use disorder that includes compromised frontal cortical executive function and dysregulated substrates that mediate craving. Lower frontal cortex activity was shown to parallel deficits in executive function in neuropsychologically challenging tasks in individuals with alcohol use disorder with and without Wernicke-Korsakoff's syndrome [459,460]. Such frontal cortex-derived executive function disorders in addiction have been linked to deficits in the ability of behavioral treatments to effect recovery from alcohol use disorder post-detoxification ([460]; see Volume 3, *Alcohol*).

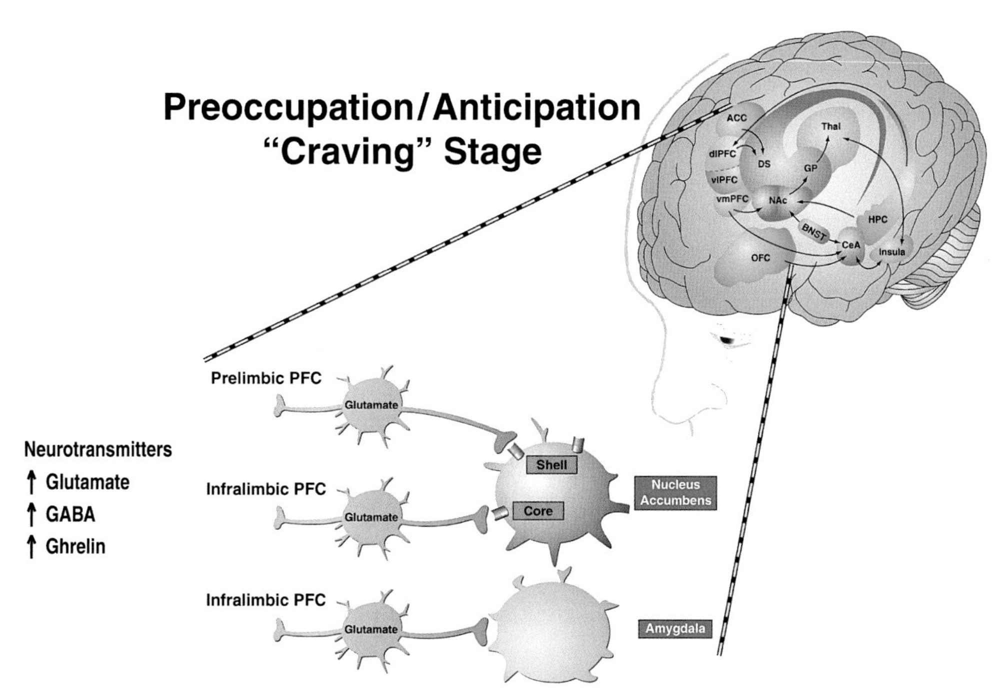

Figure 47 Neural circuitry associated with the three stages of the addiction cycle—*preoccupation/anticipation* (craving) stage. This stage involves the processing of conditioned reinforcement in the basolateral amygdala and processing of contextual information in the hippocampus. Executive control depends on the prefrontal cortex and includes the representation of contingencies, the representation of outcomes, their value, and subjective states (i.e., craving and, presumably, feelings) associated with drugs. A major neurotransmitter that is involved in the craving stage is glutamate that is localized in pathways from frontal regions and the basolateral amygdala that project to the ventral striatum. ACC, anterior cingulate cortex; dlPFC, dorsolateral prefrontal cortex; vlPFC, ventrolateral prefrontal cortex; vmPFC, ventromedial prefrontal cortex; OFC, orbitofrontal cortex; DS, dorsal striatum; NAc, nucleus accumbens; GP, globus pallidus; Thal, thalamus; BNST, bed nucleus of the stria terminalis; HPC, hippocampus; CeA, central nucleus of the amygdala. (Inset) Increased activity in the prelimbic and infralimbic cortices in the prefrontal cortex (PFC) in initiating and inhibiting the reinstatement of drug-seeking behavior. Prelimbic glutamate projections are hypothesized to contribute to incentive salience and habit formation, whereas infralimbic glutamate projections are hypothesized to contribute to the inhibition of drug seeking (NAc) and inhibition of brain stress systems. A combination of high prelimbic and low infralimbic glutamate activity may drive drug seeking by driving craving and disinhibiting restraints on impulsivity and compulsivity. *(Modified with permission from Koob GF, Everitt BJ, Robbins TW. Reward, motivation, and addiction. In: Squire LG, Berg D, Bloom FE, Du Lac S, Ghosh A, Spitzer N, editors. Fundamental neuroscience, 3rd ed. Amsterdam: Academic Press; 2008. p. 987–1016 [542]. (Inset) Modified from Kalivas PW. The glutamate homeostasis hypothesis of addiction. Nature Reviews Neuroscience 2009;10:561–72 [384].)*

Craving responses to cues in human imaging studies activate an overarching cognitive control network in the brain, involving dorsolateral prefrontal, anterior cingulate, and parietal cortices, all of which support a broad range of executive functions [461]. These authors hypothesized that a frontal—cingulate—parietal—subcortical cognitive control network is consistently recruited across a range of traditional executive function tasks, many of which show deficits in humans with alcohol use disorder. Indeed, the most prominent activation by alcohol-related cues involves the dorsolateral prefrontal cortex, cingulate cortex, and orbitofrontal cortex [462]. Such drug cue-evoked responses are hypothesized to reflect the neural representations of reward and incentive salience constructs [463]. One parsimonious view of the human imaging data that is consistent with animal model data is that there is a "Go" system in the dorsal prefrontal/cingulate cortex that drives impulsivity and craving and a "Stop" system in the ventromedial prefrontal cortex that inhibits impulsivity and craving [39].

In humans, stress and stressors have long been associated with relapse and the vulnerability to relapse [464]. Individuals with addiction are hypersensitive to pain during withdrawal, particularly in the face of negative affect [465]. Indeed, the leading precipitant of relapse is negative emotion/affect, including elements of anger, frustration, sadness, anxiety, and guilt [466].

Studies of "craving" in animal models can be divided into three domains: drug seeking that is induced by the drug itself (drug priming-induced reinstatement), drug seeking that is induced by stimuli that are paired with drug taking (cue- and context-induced reinstatement), and drug seeking that is induced by an acute stressor or a state of stress (stress-induced reinstatement; [467]). Neurotransmitter systems that are involved in drug- and cue-induced reinstatement involve a glutamatergic projection from the frontal cortex to the nucleus accumbens that is modulated by dopamine activity in the frontal cortex, basolateral amygdala, and ventral subiculum [97]. In contrast, the stress-induced reinstatement of drug-related responding in animal models appears to depend on the activation of both CRF and norepinephrine in elements of the extended amygdala (central nucleus of the amygdala and bed nucleus of the stria terminalis; [138]; see Chapter 1, What Is Addiction?). New in this chapter are hypotheses regarding the CNDS theory ("competing neurobehavioral decision systems") based on competition between impulsive and executive function systems to account for self-control failure and to integrate the neuroscience of addiction with developmental processes, socioeconomic status, and comorbidities [333]. Others argue from a parallel neurocircuitry perspective that addiction involves hyperactivity in the amygdala or impulsive system that drives emotional (somatic) responses that exaggerate the rewarding impact of available incentives and hypoactivity in the prefrontal cortex or reflective system ("Somatic marker theory of addiction"; [347]). A new synthesis of molecular-cellular changes in the glutamatergic system that are associated with an animal model of craving is presented in "Neurobiology of cue-induced incubation ("craving"): synaptic mechanism hypothesis" [416].

At a molecular-cellular level, drugs have been reported to alter the expression of certain transcription factors (nuclear proteins that bind to regulatory regions of genes, thereby regulating their transcription into mRNA), as well as a wide variety of proteins that are involved in neurotransmission in several key brain regions. Other molecular changes within the incentive salience-reward circuitry that may be common to all drugs of abuse include the upregulation of a postsynaptic G_s/cAMP/protein kinase A signaling pathway in the nucleus accumbens that is central to establishing and maintaining the addicted state [468]. Transcription factors, such as CREB and downstream ΔFosB, nuclear factor κB, and CDK5, can also alter gene expression and produce long-term changes in protein expression and thus neuronal function [469] and even structural changes in the cytoskeleton of neurons via actions on actin [470]. Indeed, chronic drug exposure can alter the morphology of neurons in dopamine-regulated circuits. For example, in rodents, chronic cocaine, alcohol, or amphetamine administration alters neuronal dendritic branching and spine density in the nucleus accumbens and prefrontal cortex—an adaptation that is thought to play a role in the enhanced incentive motivational value of the drug in addiction [471–473].

There is growing evidence that epigenetic mechanisms mediate many of the drug-induced changes in gene expression patterns that lead to structural, synaptic, and behavioral plasticity in the brain [444,474,475]. The dynamic and often long-lasting changes that occur in the transcription factors ΔFosB, CREB, and nuclear factor κB after chronic drug administration are particularly interesting because they appear to modulate the synthesis of proteins that are involved in key aspects of the addiction phenotype, such as synaptic plasticity [476]. These molecular changes that are described above can influence all three stages of the addiction cycle, in effect loading the circuits that contribute to neuroadaptations in the reward-habit, stress-emotion, executive function, and interoceptive networks in the brain whose dysfunctions coalesce to drive compulsive alcohol and drug intake.

We must mention one note of caution concerning a point we made in the first edition of this book in 2006 [477]. Despite accumulating evidence that suggests that the role of the midbrain dopamine system is more to facilitate responding to salient incentives than to mediate primary reward, the simplistic view continues to be *drug reward = activation of the mesolimbic dopamine system*. As a result, the focus of most molecular-cellular studies to date has been on neuroplasticity of the mesolimbic dopamine system and its projection areas. Such studies have identified numerous potential cellular targets that could mediate long-term changes that are associated with development of the channeling of the behavioral repertoire toward drug seeking that is the hallmark of compulsive drug taking and the loss of control over drug taking in addiction. We argued in 2006, "A challenge for future research will be to move out from under the constraints of the mesocorticolimbic dopamine system to other neurocircuits and targets." Although there has been a shift in focus to both the *withdrawal/negative affect* and *preoccupation/anticipation* stages of the addiction cycle, a large proportion of the preclinical neurobiology of addiction remains to be studies that seek to modulate the mesolimbic dopamine system. In the present book, we provide a more catholic view of the field, with an emphasis on work with alcohol and opioids and mechanisms that are related to the dark side of addiction.

Note also that the neurocircuitry models that are presented in Figs. 44–47 above are a minimalistic model that has, we hope, heuristic value. Other structures can easily be added that obviously contribute to some aspects of addiction, such as the habenula, periaqueductal gray, and hypothalamus. However, the structures that are included in these models are those that are most represented in the reviews cited above. Emphasis obviously varies from investigator to investigator, with some who emphasize a critical role for aspects of incentive salience and habits (basal ganglia: striatal-pallidal-thalamic-cortical loops), negative emotional states of withdrawal or hyperkatifeia (extended amygdala), and impulsivity and executive function (prefrontal cortex). Critical for the future of the neuroscience of addiction will be to identify individual differences in the vulnerability and resilience to addiction by understanding the weight of the combination of environmental loading (history of stress and drug exposure) and genetic loading and their interactions via molecular-cellular changes to the function of specific neurocircuits. This challenge remains largely unexplored.

References

[1] Wise RA. Action of drugs of abuse on brain reward systems. Pharmacology Biochemistry and Behavior 1980;13(Suppl. 1):213–23.

[2] de Wit H, Wise RA. Blockade of cocaine reinforcement in rats with the dopamine receptor blocker pimozide, but not with the noradrenergic blockers phentolamine and phenoxybenzamine. Canadian Journal of Psychology 1977;31:195–203.

[3] Yokel RA, Wise RA. Increased lever pressing for amphetamine after pimozide in rats: implications for a dopamine theory of reward. Science 1975;187:547–9.

[4] Yokel RA, Wise RA. Amphetamine- type reinforcement by dopaminergic agonists in the rat. Psychopharmacology 1978;58:289–96.

[5] Wise RA, Spindler J, deWit H, Gerberg GJ. Neuroleptic-induced "anhedonia" in rats: pimozide blocks reward quality of food. Science 1978a;201:262–4.

[6] Wise RA, Spindler J, Legault L. Major attenuation of food reward with performance-sparing doses of pimozide in the rat. Canadian Journal of Psychology 1978b;32:77–85.

[7] Wise RA, Bozarth MA. A psychomotor stimulant theory of addiction. Psychological Review 1987; 94:469–92.

[8] Glickman SE, Schiff BB. A biological theory of reinforcement. Psychological Review 1967;74: 81–109.

[9] Esposito RU, Faulkner W, Kornetsky C. Specific modulation of brain stimulation reward by haloperidol. Pharmacology Biochemistry and Behavior 1979;10:937–40.

[10] Fouriezos G, Hansson P, Wise RA. Neuroleptic-induced attenuation of brain stimulation reward in rats. Journal of Comparative and Physiological Psychology 1978;92:661–71.

[11] Bozarth MA, Wise RA. Heroin reward is dependent on a dopaminergic substrate. Life Sciences 1981; 29:1881–6.

[12] Yokel RA, Wise RA. Attenuation of intravenous amphetamine reinforcement by central dopamine blockade in rats. Psychopharmacology 1976;48:311–8.

[13] Amalric M, Koob GF. Low doses of methylnaloxonium in the nucleus accumbens antagonize hyperactivity induced by heroin in the rat. Pharmacology Biochemistry and Behavior 1985;23:411–5.
[14] Carlezon Jr WA, Wise RA. Rewarding actions of phencyclidine and related drugs in nucleus accumbens shell and frontal cortex. Journal of Neuroscience 1996;16:3112–22.
[15] Cunningham CL, Malott DH, Dickinson SD, Risinger FO. Haloperidol does not alter expression of ethanol-induced conditioned place preference. Behavioural Brain Research 1992;50:1–5.
[16] Pettit HO, Ettenberg A, Bloom FE, Koob GF. Destruction of dopamine in the nucleus accumbens selectively attenuates cocaine but not heroin self-administration in rats. Psychopharmacology 1984; 84:167–73.
[17] Rassnick S, Stinus L, Koob GF. The effects of 6-hydroxydopamine lesions of the nucleus accumbens and the mesolimbic dopamine system on oral self-administration of ethanol in the rat. Brain Research 1993b;623:16–24.
[18] Vaccarino FJ, Amalric M, Swerdlow NR, Koob GF. Blockade of amphetamine but not opiate-induced locomotion following antagonism of dopamine function in the rat. Pharmacology Biochemistry and Behavior 1986;24:61–5.
[19] Wise RA. Brain reward circuitry: insights from unsensed incentives. Neuron 2002;36:229–40.
[20] Ljungberg T, Apicella P, Schultz W. Responses of monkey dopamine neurons during learning of behavioral reactions. Journal of Neurophysiology 1992;67:145–63.
[21] Mirenowicz J, Schultz W. Importance of unpredictability for reward responses in primate dopamine neurons. Journal of Neurophysiology 1994;72:1024–7.
[22] Schultz W. Getting formal with dopamine and reward. Neuron 2002;36:241–63.
[23] Schultz W, Apicella P, Ljungberg T. Responses of monkey dopamine neurons to reward and conditioned stimuli during successive steps of learning a delayed response task. Journal of Neuroscience 1993;13:900–13.
[24] Schultz W, Dayan P, Montague PR. A neural substrate of prediction and reward. Science 1997;275: 1593–9.
[25] Koob GF. Dopamine, addiction and reward. Seminars in Neuroscience 1992a;4:139–48.
[26] Salamone JD. Complex motor and sensorimotor functions of striatal and accumbens dopamine: involvement in instrumental behavior processes. Psychopharmacology 1992;107:160–74.
[27] Salamone JD. The involvement of nucleus accumbens dopamine in appetitive and aversive motivation. Behavioural Brain Research 1994;61:117–33.
[28] Ranje C, Ungerstedt U. Lack of acquisition in dopamine denervated animals tested in an underwater Y-maze. Brain Research 1977;134:95–111.
[29] Koob GF. Drugs of abuse: anatomy, pharmacology, and function of reward pathways. Trends in Pharmacological Sciences 1992b;13:177–84.
[30] Lyness WH, Friedle NM, Moore KE. Destruction of dopaminergic nerve terminals in nucleus accumbens: effect on d-amphetamine self-administration. Pharmacology Biochemistry and Behavior 1979;11:553–6.
[31] Roberts DCS, Koob GF, Klonoff P, Fibiger HC. Extinction and recovery of cocaine self-administration following 6-hydroxydopamine lesions of the nucleus accumbens. Pharmacology Biochemistry and Behavior 1980;12:781–7.
[32] Hyytia P, Koob GF. GABA-A receptor antagonism in the extended amygdala decreases ethanol self-administration in rats. European Journal of Pharmacology 1995;283:151–9.
[33] Nauta JH, Haymaker W. Hypothalamic nuclei and fiber connections. In: Haymaker W, Anderson E, Nauta WJH, editors. The hypothalamus. Springfield, IL: Charles C. Thomas; 1969. p. 136–209.
[34] Alheid GF, Heimer L. New perspectives in basal forebrain organization of special relevance for neuropsychiatric disorders: the striatopallidal, amygdaloid, and corticopetal components of substantia innominata. Neuroscience 1988;27:1–39.
[35] Groenewegen HJ, Berendse HW, Wolters JG, Lohman AH. The anatomical relationship of the prefrontal cortex with the striatopallidal system, the thalamus and the amygdala: evidence for a parallel organization. In: Uylings HBM, van Eden CG, de Bruin JPC, Corner MA, Feenstra MGP, editors. The prefrontal cortex: its structure, function, and pathology. Progress in brain research, vol. 85. New York: Elsevier; 1990. p. 95–116.

[36] Wise RA, Koob GF. The development and maintenance of drug addiction. Neuropsychopharmacology 2014;39:254–62.
[37] Wise RA. Dopamine, learning and motivation. Nature Reviews Neuroscience 2004;5:483–94.
[38] Koob GF, Volkow ND. Neurocircuitry of addiction. Neuropsychopharmacology Reviews 2010;35: 217–38 [erratum: 35:1051].
[39] Koob GF, Volkow ND. Neurobiology of addiction: a neurocircuitry analysis. Lancet Psychiatry 2016;3:760–73.
[40] Koob GF, Le Moal M. Drug abuse: hedonic homeostatic dysregulation. Science 1997;278:52–8.
[41] Baumeister RF, Heatherton TF, Tice DM, editors. Losing control: how and why people fail at self-regulation. San Diego: Academic Press; 1994.
[42] Goldstein RZ, Volkow ND. Drug addiction and its underlying neurobiological basis: neuroimaging evidence for the involvement of the frontal cortex. American Journal of Psychiatry 2002;159: 1642–52.
[43] Koob GF, editor. Impulse control disorders. San Diego: Cognella; 2013b, ISBN 978-1-62131-212-3.
[44] McBride WJ, Li TK. Animal models of alcoholism: neurobiology of high alcohol-drinking behavior in rodents. Critical Reviews in Neurobiology 1998;12:339–69.
[45] Gatto GJ, McBride WJ, Murphy JM, Lumeng L, Li TK. Ethanol self-infusion into the ventral tegmental area by alcohol-preferring rats. Alcohol 1994;11:557–64.
[46] Rodd-Henricks ZA, McKinzie DL, Crile RS, Murphy JM, McBride WJ. Regional heterogeneity for the intracranial self-administration of ethanol within the ventral tegmental area of female Wistar rats. Psychopharmacology 2000;149:217–24.
[47] Rodd ZA, Bell RL, Zhang Y, Murphy JM, Goldstein A, Zaffaroni A, Li TK, McBride WJ. Regional heterogeneity for the intracranial self-administration of ethanol and acetaldehyde within the ventral tegmental area of alcohol-preferring (P) rats: involvement of dopamine and serotonin. Neuropsychopharmacology 2005;30:330–8.
[48] Rodd ZA, Melendez RI, Bell RL, Kuc KA, Zhang Y, Murphy JM, McBride WJ. Intracranial self-administration of ethanol within the ventral tegmental area of male Wistar rats: evidence for involvement of dopamine neurons. Journal of Neuroscience 2004;24:1050–7.
[49] Rodd-Henricks ZA, Melendez RI, Zaffaroni A, Goldstein A, McBride WJ, Li TK. The reinforcing effects of acetaldehyde in the posterior ventral tegmental area of alcohol-preferring rats. Pharmacology Biochemistry and Behavior 2002;72:55–64.
[50] Rodd ZA, Bell RL, Zhang Y, Goldstein A, Zaffaroni A, McBride WJ, Li TK. Salsolinol produces reinforcing effects in the nucleus accumbens shell of alcohol-preferring (P) rats. Alcoholism: Clinical and Experimental Research 2003;27:440–9.
[51] Nowak KL, McBride WJ, Lumeng L, Li TK, Murphy JM. Blocking GABA(A) receptors in the anterior ventral tegmental area attenuates ethanol intake of the alcohol-preferring P rat. Psychopharmacology 1998;139:108–16.
[52] June HL, Eggers MW, Warren-Reese C, DeLong J, Ricks-Cord A, Durr LF, Cason CR. The effects of the novel benzodiazepine receptor inverse agonist Ru 34000 on ethanol-maintained behaviors. European Journal of Pharmacology 1998a;350:151–8.
[53] June HL, Torres L, Cason CR, Hwang BH, Braun MR, Murphy JM. The novel benzodiazepine inverse agonist RO19-4603 antagonizes ethanol motivated behaviors: neuropharmacological studies. Brain Research 1998b;784:256–75.
[54] Katner SN, McBride WJ, Lumeng L, Li TK, Murphy JM. Alcohol intake of P rats is regulated by muscarinic receptors in the pedunculopontine nucleus and VTA. Pharmacology Biochemistry and Behavior 1997;58:497–504.
[55] Hodge CW, Chappelle AM, Samson HH. GABAergic transmission in the nucleus accumbens is involved in the termination of ethanol self-administration in rats. Alcoholism: Clinical and Experimental Research 1995;19:1486–93.
[56] Tomkins DM, Sellers EM, Fletcher PJ. Median and dorsal raphe injections of the 5-HT1A agonist, 8-OH-DPAT, and the GABAA agonist, muscimol, increase voluntary ethanol intake in Wistar rats. Neuropharmacology 1994;33:349–58.

[57] Di Chiara G. Nucleus accumbens shell and core dopamine: differential role in behavior and addiction. Behavioural Brain Research 2002;137:75–114.
[58] Ito R, Robbins TW, Everitt BJ. Differential control over cocaine-seeking behavior by nucleus accumbens core and shell. Nature Neuroscience 2004;7:389–97.
[59] Di Chiara G, North RA. Neurobiology of opiate abuse. Trends in Pharmacological Sciences 1992; 13:185–93.
[60] Groenewegen HJ, Berendse HW, Haber SN. Organization of the output of the ventral striatopallidal system in the rat: ventral pallidal efferents. Neuroscience 1993;57:113–42.
[61] Haber SN, Fudge JL, McFarland NR. Striatonigrostriatal pathways in primates form an ascending spiral from the shell to the dorsolateral striatum. Journal of Neuroscience 2000;20:2369–82.
[62] Cadoni C, Di Chiara G. Reciprocal changes in dopamine responsiveness in the nucleus accumbens shell and core and in the dorsal caudate-putamen in rats sensitized to morphine. Neuroscience 1999; 90:447–55.
[63] Pontieri FE, Tanda G, Di Chiara G. Intravenous cocaine, morphine, and amphetamine preferentially increase extracellular dopamine in the "shell" as compared with the "core" of the rat nucleus accumbens. Proceedings of the National Academy of Sciences of the United States of America 1995;92:12304–8.
[64] Pontieri FE, Tanda G, Orzi F, Di Chiara G. Effects of nicotine on the nucleus accumbens and similarity to those of addictive drugs. Nature 1996;382:255–7.
[65] Tanda G, Pontieri FE, Frau R, Di Chiara G. Contribution of blockade of the noradrenaline carrier to the increase of extracellular dopamine in the rat prefrontal cortex by amphetamine and cocaine. European Journal of Neuroscience 1997;9:2077–85.
[66] Bassareo V, De Luca MA, Di Chiara G. Differential Expression of Motivational Stimulus Properties by Dopamine in nucleus accumbens shell versus core and prefrontal cortex. Journal of Neuroscience 2002;22:4709–19.
[67] Bassareo V, Di Chiara G. Differential influence of associative and nonassociative learning mechanisms on the responsiveness of prefrontal and accumbal dopamine transmission to food stimuli in rats fed ad libitum. Journal of Neuroscience 1997;17:851–61.
[68] Bassareo V, Di Chiara G. Modulation of feeding-induced activation of mesolimbic dopamine transmission by appetitive stimuli and its relation to motivational state. European Journal of Neuroscience 1999a;11:4389–97.
[69] Bassareo V, Di Chiara G. Differential responsiveness of dopamine transmission to food-stimuli in nucleus accumbens shell/core compartments. Neuroscience 1999b;89:637–41.
[70] Tanda G, Di Chiara G. A dopamine-mu1 opioid link in the rat ventral tegmentum shared by palatable food (Fonzies) and non-psychostimulant drugs of abuse. European Journal of Neuroscience 1998;10:1179–87.
[71] Di Chiara G. Drug addiction as dopamine-dependent associative learning disorder. European Journal of Pharmacology 1999;375:13–30.
[72] Parkinson JA, Olmstead MC, Burns LH, Robbins TW, Everitt BJ. Dissociation in effects of lesions of the nucleus accumbens core and shell on appetitive pavlovian approach behavior and the potentiation of conditioned reinforcement and locomotor activity by D-amphetamine. Journal of Neuroscience 1999;19:2401–11.
[73] Hall J, Parkinson JA, Connor TM, Dickinson A, Everitt BJ. Involvement of the central nucleus of the amygdala and nucleus accumbens core in mediating Pavlovian influences on instrumental behaviour. European Journal of Neuroscience 2001;13:1984–92.
[74] Whitelaw RB, Markou A, Robbins TW, Everitt BJ. Excitotoxic lesions of the basolateral amygdala impair the acquisition of cocaine-seeking behaviour under a second-order schedule of reinforcement. Psychopharmacology 1996;127:213–24.
[75] Robledo P, Robbins TW, Everitt BJ. Effects of excitotoxic lesions of the central amygdaloid nucleus on the potentiation of reward-related stimuli by intra-accumbens amphetamine. Behavioral Neuroscience 1996;110:981–90.

[76] Burns LH, Robbins TW, Everitt BJ. Differential effects of excitotoxic lesions of the basolateral amygdala, ventral subiculum and medial prefrontal cortex on responding with conditioned reinforcement and locomotor activity potentiated by intra-accumbens infusions of D-amphetamine. Behavioural Brain Research 1993;55:167−83.
[77] Heimer L, Alheid G. Piecing together the puzzle of basal forebrain anatomy. In: Napier TC, Kalivas PW, Hanin I, editors. The basal forebrain: anatomy to function. Advances in experimental medicine and biology, vol. 295. New York: Plenum Press; 1991. p. 1−42.
[78] Peoples LL, Cavanaugh D. Differential changes in signal and background firing of accumbal neurons during cocaine self-administration. Journal of Neurophysiology 2003;90:993−1010.
[79] Peoples LL, Lynch KG, Lesnock J, Gangadhar N. Accumbal neural responses during the initiation and maintenance of intravenous cocaine self-administration. Journal of Neurophysiology 2004;91:314−23.
[80] Carelli RM, King VC, Hampson RE, Deadwyler SA. Firing patterns of nucleus accumbens neurons during cocaine self-administration in rats. Brain Research 1993;626:14−22.
[81] Chang JY, Sawyer SF, Lee R-S, Woodward DJ. Electrophysiological and pharmacological evidence for the role of the nucleus accumbens in cocaine self-administration in freely moving rats. Journal of Neuroscience 1994;14:1224−44.
[82] Janak PH, Chang JY, Woodward DJ. Neuronal spike activity in the nucleus accumbens of behaving rats during ethanol self-administration. Brain Research 1999;817:172−84.
[83] Peoples LL, West MO. Phasic firing of single neurons in the rat nucleus accumbens correlated with the timing of intravenous cocaine self-administration. Journal of Neuroscience 1996;16:3459−73.
[84] Uzwiak AJ, Guyette FX, West MO, Peoples LL. Neurons in accumbens subterritories of the rat: phasic firing time-locked within seconds of intravenous cocaine self-infusion. Brain Research 1997;767:363−9.
[85] Peoples LL, Uzwiak AJ, Gee F, Fabbricatore AT, Muccino KJ, Mohta BD, West MO. Phasic accumbal firing may contribute to the regulation of drug taking during intravenous cocaine self-administration sessions. In: McGinty JF, editor. Advancing from the ventral striatum to the extended amygdala: implications for neuropsychiatry and drug abuse. Annals of the New York Academy of Sciences, vol. 877. New York: New York Academy of Sciences; 1999. p. 781−7.
[86] Peoples LL, Uzwiak AJ, Guyette FX, West MO. Tonic inhibition of single nucleus accumbens neurons in the rat: a predominant but not exclusive firing pattern induced by cocaine self-administration sessions. Neuroscience 1998;86:13−22.
[87] Willuhn I, Burgeno LM, Everitt BJ, Phillips PE. Hierarchical recruitment of phasic dopamine signaling in the striatum during the progression of cocaine use. Proceedings of the National Academy of Sciences of the United States of America 2012;109:20703−8.
[88] Pennartz CM, Groenewegen HJ, Lopes da Silva FH. The nucleus accumbens as a complex of functionally distinct neuronal ensembles: an integration of behavioural, electrophysiological and anatomical data. Progress in Neurobiology 1994;42:719−61.
[89] Grant S, Contoreggi C, London ED. Drug abusers show impaired performance in a laboratory test of decision making. Neuropsychologia 2000;38:1180−7.
[90] Koob GF, Le Moal M. Drug addiction, dysregulation of reward, and allostasis. Neuropsychopharmacology 2001;24:97−129.
[91] Rogers RD, Everitt BJ, Baldacchino A, Blackshaw AJ, Swainson R, Wynne K, Baker NB, Hunter J, Carthy T, Booker E, London M, Deakin JF, Sahakian BJ, Robbins TW. Dissociable deficits in the decision-making cognition of chronic amphetamine abusers, opiate abusers, patients with focal damage to prefrontal cortex, and tryptophan-depleted normal volunteers: evidence for monoaminergic mechanisms. Neuropsychopharmacology 1999;20:322−39.
[92] Volkow ND, Fowler JS, Wang GJ. The addicted human brain: insights from imaging studies. Journal of Clinical Investigation 2003;111:1444−51.
[93] Hyman SE, Malenka RC. Addiction and the brain: the neurobiology of compulsion and its persistence. Nature Reviews Neuroscience 2001;2:695−703.
[94] Thomas MJ, Malenka RC. Synaptic plasticity in the mesolimbic dopamine system. Transactions of the Royal Society of London B Biological Sciences 2003;358:815−9.

[95] Robinson TE, Berridge KC. The neural basis of drug craving: an incentive-sensitization theory of addiction. Brain Research Reviews 1993;18:247–91.
[96] Carlezon Jr WA, Nestler EJ. Elevated levels of GluR1 in the midbrain: a trigger for sensitization to drugs of abuse? Trends in Neurosciences 2002;25:610–5.
[97] Everitt BJ, Wolf ME. Psychomotor stimulant addiction: a neural systems perspective. Journal of Neuroscience 2002;22:3312–20 [erratum: 22(16):1a].
[98] Kalivas PW. Interactions between dopamine and excitatory amino acids in behavioral sensitization to psychostimulants. Drug and Alcohol Dependence 1995;37:95–100.
[99] Kombian SB, Malenka RC. Simultaneous LTP of non-NMDA- and LTD of NMDA-receptor-mediated responses in the nucleus accumbens. Nature 1994;368:242–6.
[100] Bonci A, Malenka RC. Properties and plasticity of excitatory synapses on dopaminergic and GABAergic cells in the ventral tegmental area. Journal of Neuroscience 1999;19:3723–30.
[101] Pennartz CM, Ameerun RF, Groenewegen HJ, Lopes da Silva FH. Synaptic plasticity in an in vitro slice preparation of the rat nucleus accumbens. European Journal of Neuroscience 1993;5:107–17.
[102] Thomas MJ, Malenka RC, Bonci A. Modulation of long-term depression by dopamine in the mesolimbic system. Journal of Neuroscience 2000;20:5581–6.
[103] Jones S, Kornblum JL, Kauer JA. Amphetamine blocks long-term synaptic depression in the ventral tegmental area. Journal of Neuroscience 2000;20:5575–80.
[104] Ungless MA, Whistler JL, Malenka RC, Bonci A. Single cocaine exposure in vivo induces long-term potentiation in dopamine neurons. Nature 2001;411:583–7.
[105] Saal D, Dong Y, Bonci A, Malenka RC. Drugs of abuse and stress trigger a common synaptic adaptation in dopamine neurons. Neuron 2003;37:577–82 [erratum: 38:359].
[106] Thomas MJ, Beurrier C, Bonci A, Malenka RC. Long-term depression in the nucleus accumbens: a neural correlate of behavioral sensitization to cocaine. Nature Neuroscience 2001;4:1217–23.
[107] Ahmed SH. Individual decision-making in the causal pathway to addiction: contributions and limitations of rodent models. Pharmacology Biochemistry and Behavior 2018;164:22–31.
[108] Allain F, Minogianis EA, Roberts DC, Samaha AN. How fast and how often: the pharmacokinetics of drug use are decisive in addiction. Neuroscience & Biobehavioral Reviews 2015;56:166–79.
[109] Roberts WA. Are animals stuck in time? Psychological Bulletin 2002;128:473–89.
[110] Tobin H, Logue AW. Self-control across species (*Columba livia*, *Homo sapiens*, and *Rattus norvegicus*). Journal of Comparative Psychology 1994;108:126–33.
[111] Lenoir M, Serre F, Cantin L, Ahmed SH. Intense sweetness surpasses cocaine reward. PLoS One 2007;2:e698.
[112] Ainslie G. Specious reward: a behavioral theory of impulsiveness and impulse control. Psychological Bulletin 1975;82:463–96.
[113] Lenoir M, Cantin L, Vanhille N, Serre F, Ahmed SH. Extended heroin access increases heroin choices over a potent nondrug alternative. Neuropsychopharmacology 2013;38:1209–20.
[114] Kenny PJ, Polis I, Koob GF, Markou A. Low dose cocaine self-administration transiently increases but high dose cocaine persistently decreases brain reward function in rats. European Journal of Neuroscience 2003;17:191–5.
[115] Belin D, Belin-Rauscent A, Murray JE, Everitt BJ. Addiction: failure of control over maladaptive incentive habits. Current Opinion in Neurobiology 2013;23:564–72.
[116] Belin-Rauscent A, Fouyssac M, Bonci A, Belin D. How preclinical models evolved to resemble the diagnostic criteria of drug addiction. Biological Psychiatry 2016;79:39–46.
[117] Dickinson A. Actions and habits: the development of behavioural autonomy. Philosophical Transactions of the Royal Society of London B Biological Sciences 1985;308:67–78.
[118] Yin HH, Knowlton BJ. The role of the basal ganglia in habit formation. Nature Reviews Neuroscience June 2006;7(6):464–76.
[119] Coutureau E, Marchand AR, Di Scala G. Goal-directed responding is sensitive to lesions to the prelimbic cortex or basolateral nucleus of the amygdala but not to their disconnection. Behavioral Neuroscience 2009;123:443–8.
[120] Balleine B, Killcross AS, Dickinson A. The effect of lesions of the basolateral amygdala on instrumental conditioning. Journal of Neuroscience 2003;23:666–75.

[121] Yin H, Ostlund SB, Knowlton B, Balleine B. The role of the dorsomedial striatum in instrumental conditioning. European Journal of Neuroscience 2005;22:513–23.

[122] Yin H, Knowlton B, Balleine B. Lesions of dorsolateral striatum preserve outcome expectancy but disrupt habit formation in instrumental learning. European Journal of Neuroscience 2004;19:181–9.

[123] Faure A. Lesion to the nigrostriatal dopamine system disrupts stimulus-response habit formation. Journal of Neuroscience 2005;25:2771–80.

[124] Lingawi NW, Balleine BW. Amygdala central nucleus interacts with dorsolateral striatum to regulate the acquisition of habits. Journal of Neuroscience 2012;32:1073–81.

[125] Coutureau E, Killcross S. Inactivation of the infralimbic prefrontal cortex reinstates goal-directed responding in overtrained rats. Behavioural Brain Research 2003;146:167–74.

[126] Keiflin R, Janak PH. Dopamine prediction errors in reward learning and addiction: from theory to neural circuitry. Neuron 2015;88:247–63.

[127] Redish AD. Addiction as a computational process gone awry. Science 2004;306:1944–7.

[128] Robinson TE, Berridge KC. Addiction. Annual Review of Psychology 2003;54:25–53.

[129] Zernig G, Ahmed SH, Cardinal RN, Morgan D, Acquas E, Foltin RW, Vezina P, Negus SS, Crespo JA, Stockl P, et al. Explaining the escalation of drug use in substance dependence: models and appropriate animal laboratory tests. Pharmacology 2007;80:65–119.

[130] Robbins TW. Relationship between reward-enhancing and stereotypical effects of psychomotor stimulant drugs. Nature 1976;264:57–9.

[131] Willuhn I, Burgeno LM, Groblewski PA, Phillips PE. Excessive cocaine use results from decreased phasic dopamine signaling in the striatum. Nature Neuroscience 2014;17:704–9.

[132] George O, Koob GF, Vendruscolo LF. Negative reinforcement via motivational withdrawal is the driving force behind the transition to addiction. Psychopharmacology 2014;231:3911–7.

[133] Koob GF. Negative reinforcement in drug addiction: the darkness within. Current Opinion in Neurobiology 2013a;23:559–63.

[134] Piazza PV, Deroche-Gamonet V. A multistep general theory of transition to addiction. Psychopharmacology 2013;229:387–413.

[135] Nielsen DA, Kreek MJ. Common and specific liability to addiction: approaches to association studies of opioid addiction. Drug and Alcohol Dependence 2012;123:S33–41.

[136] Verweij KJ, Zietsch BP, Lynskey MT, Medland SE, Neale MC, Martin NG, Boomsma DI, Vink JM. Genetic and environmental influences on cannabis use initiation and problematic use: a meta-analysis of twin studies. Addiction 2010;105:417–30.

[137] Piazza PV, Le Moal ML. Pathophysiological basis of vulnerability to drug abuse: role of an interaction between stress, glucocorticoids, and dopaminergic neurons. Annual Review of Pharmacology and Toxicology 1996;36:359–78.

[138] Shaham Y, Shalev U, Lu L, De Wit H, Stewart J. The reinstatement model of drug relapse: history, methodology and major findings. Psychopharmacology 2003;168:3–20.

[139] Aston-Jones G, Harris GC. Brain substrates for increased drug seeking during protracted withdrawal. Neuropharmacology 2004;47(Suppl. 1):167–79.

[139a] Koob GF, Kreek MJ. Stress, dysregulation of drug reward pathways, and the transition to drug dependence. Am J Psychiatry 2007;164:1149–59.

[140] Kreek MJ, Koob GF. Drug dependence: stress and dysregulation of brain reward pathways. Drug and Alcohol Dependence 1998;51:23–47.

[141] Koob GF, Kreek MJ. Stress, dysregulation of drug reward pathways, and the transition to drug dependence. American Journal of Psychiatry 2007;164:1149–59.

[142] Carroll ME, Meisch RA. Increased drug-reinforced behavior due to food deprivation. Advances in Behavioral Pharmacology 1984;4:47–88.

[143] Piazza PV, Le Moal M. The role of stress in drug self-administration. Trends in Pharmacological Sciences 1998;19:67–74.

[144] Piazza PV, Deminiere JM, Le Moal M, Simon H. Factors that predict individual vulnerability to amphetamine self-administration. Science 1989;245:1511–3.

[145] Piazza PV, Maccari S, Deminiere JM, Le Moal M, Mormede P, Simon H. Corticosterone levels determine individual vulnerability to amphetamine self-administration. Proceedings of the National Academy of Sciences of the United States of America 1991b;88:2088–92.

[146] Mantsch JR, Saphier D, Goeders NE. Corticosterone facilitates the acquisition of cocaine self-administration in rats: opposite effects of the type II glucocorticoid receptor agonist dexamethasone. Journal of Pharmacology and Experimental Therapeutics 1998;287:72–80.
[147] Marinelli M, Piazza PV, Deroche V, Maccari S, Le Moal M, Simon H. Corticosterone circadian secretion differentially facilitates dopamine-mediated psychomotor effect of cocaine and morphine. Journal of Neuroscience 1994;14:2724–31.
[148] Marinelli M, Rouge-Pont F, Deroche V, Barrot M, De Jesus-Oliveira C, Le Moal M, Piazza PV. Glucocorticoids and behavioral effects of psychostimulants. I: locomotor response to cocaine depends on basal levels of glucocorticoids. Journal of Pharmacology and Experimental Therapeutics 1997b; 281:1392–400.
[149] Marinelli M, Rouge-Pont F, De Jesus-Oliveira C, Le Moal M, Piazza PV. Acute blockade of corticosterone secretion decreases the psychomotor stimulant effects of cocaine. Neuropsychopharmacology 1997a;16:156–61.
[150] Barrot M, Marinelli M, Abrous DN, Rouge-Pont F, Le Moal M, Piazza PV. The dopaminergic hyper-responsiveness of the shell of the nucleus accumbens is hormone-dependent. European Journal of Neuroscience 2000;12:973–9.
[151] Mantsch JR, Yuferov V, Mathieu-Kia AM, Ho A, Kreek MJ. Neuroendocrine alterations in a high-dose, extended-access rat self-administration model of escalating cocaine use. Psychoneuroendocrinology 2003;28:836–62.
[152] Zhou Y, Spangler R, Ho A, Kreek MJ. Increased CRH mRNA levels in the rat amygdala during short-term withdrawal from chronic 'binge' cocaine. Molecular Brain Research 2003;114:73–9.
[153] Zhou Y, Spangler R, LaForge KS, Maggos CE, Ho A, Kreek MJ. Corticotropin-releasing factor and type 1 corticotropin-releasing factor receptor messenger RNAs in rat brain and pituitary during "binge"-pattern cocaine administration and chronic withdrawal. Journal of Pharmacology and Experimental Therapeutics 1996;279:351–8.
[154] Koob GF. Corticotropin-releasing factor, norepinephrine and stress. Biological Psychiatry 1999;46: 1167–80.
[155] Koob GF. Neuroadaptive mechanisms of addiction: studies on the extended amygdala. European Neuropsychopharmacology 2003b;13:442–52.
[156] Richter RM, Weiss F. In vivo CRF release in rat amygdala is increased during cocaine withdrawal in self-administering rats. Synapse 1999;32:254–61.
[157] Merlo-Pich E, Lorang M, Yeganeh M, Rodriguez de Fonseca F, Raber J, Koob GF, Weiss F. Increase of extracellular corticotropin-releasing factor-like immunoreactivity levels in the amygdala of awake rats during restraint stress and ethanol withdrawal as measured by microdialysis. Journal of Neuroscience 1995;15:5439–47.
[158] Olive MF, Koenig HN, Nannini MA, Hodge CW. Elevated extracellular CRF levels in the bed nucleus of the stria terminalis during ethanol withdrawal and reduction by subsequent ethanol intake. Pharmacology Biochemistry and Behavior 2002;72:213–20.
[159] Ghozland S, Zorrilla E, Parsons LH, Koob GF. Mecamylamine increases extracellular CRF levels in the central nucleus of the amygdala or nicotine-dependent rats. Society for Neuroscience Abstracts; 2004 [abstract# 708.8].
[160] Rodriguez de Fonseca F, Carrera MRA, Navarro M, Koob GF, Weiss F. Activation of corticotropin-releasing factor in the limbic system during cannabinoid withdrawal. Science 1997;276:2050–4.
[161] Weiss F, Ciccocioppo R, Parsons LH, Katner S, Liu X, Zorrilla EP, Valdez GR, Ben-Shahar O, Angeletti S, Richter RR. Compulsive drug-seeking behavior and relapse: neuroadaptation, stress, and conditioning factors. In: Quinones-Jenab, editor. The biological basis of cocaine addiction. Annals of the New York Academy of Sciences, vol. 937. New York: New York Academy of Sciences; 2001a. p. 1–26.
[162] Baldwin HA, Rassnick S, Rivier J, Koob GF, Britton KT. CRF antagonist reverses the "anxiogenic" response to ethanol withdrawal in the rat. Psychopharmacology 1991;103:227–32.
[163] Rassnick S, Heinrichs SC, Britton KT, Koob GF. Microinjection of a corticotropin-releasing factor antagonist into the central nucleus of the amygdala reverses anxiogenic-like effects of ethanol withdrawal. Brain Research 1993a;605:25–32.

[164] Heinrichs SC, Menzaghi F, Schulteis G, Koob GF, Stinus L. Suppression of corticotropin-releasing factor in the amygdala attenuates aversive consequences of morphine withdrawal. Behavioural Pharmacology 1995;6:74–80.
[165] Stinus L, Cador M, Zorrilla EP, Koob GF. Buprenorphine and a CRF1 antagonist block the acquisition of opiate withdrawal-induced conditioned place aversion in rats. Neuropsychopharmacology 2005;30:90–8.
[166] Basso AM, Spina M, Rivier J, Vale W, Koob GF. Corticotropin-releasing factor antagonist attenuates the "anxiogenic-like" effect in the defensive burying paradigm but not in the elevated plus-maze following chronic cocaine in rats. Psychopharmacology 1999;145:21–30.
[167] Sarnyai Z, Biro E, Gardi J, Vecsernyes M, Julesz J, Telegdy G. Brain corticotropin-releasing factor mediates "anxiety-like" behavior induced by cocaine withdrawal in rats. Brain Research 1995; 675:89–97.
[168] Valdez GR, Roberts AJ, Chan K, Davis H, Brennan M, Zorrilla EP, Koob GF. Increased ethanol self-administration and anxiety-like behavior during acute withdrawal and protracted abstinence: regulation by corticotropin-releasing factor. Alcoholism: Clinical and Experimental Research 2002;26:1494–501.
[169] Aston-Jones G, Delfs JM, Druhan J, Zhu Y. The bed nucleus of the stria terminalis. A target site for noradrenergic actions in opiate withdrawal. In: McGinty JF, editor. Advancing from the ventral striatum to the extended amygdala: implications for neuropsychiatry and drug abuse. Annals of the New York Academy of Sciences, vol. 877. New York: New York Academy of Sciences; 1999. p. 486–98.
[170] Delfs JM, Zhu Y, Druhan JP, Aston-Jones G. Noradrenaline in the ventral forebrain is critical for opiate withdrawal-induced aversion. Nature 2000;403:430–4.
[171] Koob GF, Le Moal M. Drug addiction and allostasis. In: Schulkin J, editor. Allostasis, homeostasis, and the costs of physiological adaptation. New York: Cambridge University Press; 2004. p. 150–63.
[172] Heilig M, Koob GF, Ekman R, Britton KT. Corticotropin-releasing factor and neuropeptide Y: role in emotional integration. Trends in Neurosciences 1994;17:80–5.
[173] Valdez GR, Zorrilla EP, Roberts AJ, Koob GF. Antagonism of corticotropin-releasing factor attenuates the enhanced responsiveness to stress observed during protracted ethanol abstinence. Alcohol 2003;29:55–60.
[174] Harris GC, Aston-Jones G. Enhanced morphine preference following prolonged abstinence: association with increased Fos expression in the extended amygdala. Neuropsychopharmacology 2003;28: 292–9.
[175] Ahmed SH, Walker JR, Koob GF. Persistent increase in the motivation to take heroin in rats with a history of drug escalation. Neuropsychopharmacology 2000;22:413–21.
[176] Erb S, Shaham Y, Stewart J. Stress reinstates cocaine-seeking behavior after prolonged extinction and a drug-free period. Psychopharmacology 1996;128:408–12.
[177] Le AD, Quan B, Juzytch W, Fletcher PJ, Joharchi N, Shaham Y. Reinstatement of alcohol-seeking by priming injections of alcohol and exposure to stress in rats. Psychopharmacology 1998;135: 169–74.
[178] Martin-Fardon R, Ciccocioppo R, Massi M, Weiss F. Nociceptin prevents stress-induced ethanol- but not cocaine-seeking behavior in rats. NeuroReport 2000;11:1939–43.
[179] Shaham Y, Stewart J. Stress reinstates heroin-seeking in drug-free animals: an effect mimicking heroin, not withdrawal. Psychopharmacology 1995;119:334–41.
[180] Shaham Y, Funk D, Erb S, Brown TJ, Walker CD, Stewart J. Corticotropin-releasing factor, but not corticosterone, is involved in stress-induced relapse to heroin-seeking in rats. Journal of Neuroscience 1997;17:2605–14.
[181] Erb S, Shaham Y, Stewart J. The role of corticotropin-releasing factor and corticosterone in stress- and cocaine-induced relapse to cocaine seeking in rats. Journal of Neuroscience 1998;18:5529–36.
[182] Lu L, Ceng X, Huang M. Corticotropin-releasing factor receptor type I mediates stress-induced relapse to opiate dependence in rats. NeuroReport 2000;11:2373–8.
[183] Lu L, Liu D, Ceng X. Corticotropin-releasing factor receptor type 1 mediates stress-induced relapse to cocaine-conditioned place preference in rats. European Journal of Pharmacology 2001;415: 203–8.

[184] Shaham Y, Erb S, Leung S, Buczek Y, Stewart J. CP-154,526, a selective, non-peptide antagonist of the corticotropin-releasing factor1 receptor attenuates stress-induced relapse to drug seeking in cocaine- and heroin-trained rats. Psychopharmacology 1998;137:184–90.
[185] Erb S, Stewart J. A role for the bed nucleus of the stria terminalis, but not the amygdala, in the effects of corticotropin-releasing factor on stress-induced reinstatement of cocaine seeking. Journal of Neuroscience 1999;19:RC35.
[186] Shaham Y, Erb S, Stewart J. Stress-induced relapse to heroin and cocaine seeking in rats: a review. Brain Research Reviews 2000a;33:13–33.
[187] Erb S, Salmaso N, Rodaros D, Stewart J. A role for the CRF-containing pathway from central nucleus of the amygdala to bed nucleus of the stria terminalis in the stress-induced reinstatement of cocaine seeking in rats. Psychopharmacology 2001;158:360–5.
[188] Erb S, Hitchcott PK, Rajabi H, Mueller D, Shaham Y, Stewart J. Alpha-2 adrenergic receptor agonists block stress-induced reinstatement of cocaine seeking. Neuropsychopharmacology 2000;23:138–50.
[189] Highfield D, Yap J, Grimm JW, Shalev U, Shaham Y. Repeated lofexidine treatment attenuates stress-induced, but not drug cues-induced reinstatement of a heroin-cocaine mixture (speedball) seeking in rats. Neuropsychopharmacology 2001;25:320–31.
[190] Shaham Y, Highfield D, Delfs J, Leung S, Stewart J. Clonidine blocks stress-induced reinstatement of heroin seeking in rats: an effect independent of locus coeruleus noradrenergic neurons. European Journal of Neuroscience 2000b;12:292–302.
[191] Leri F, Flores J, Rodaros D, Stewart J. Blockade of stress-induced but not cocaine-induced reinstatement by infusion of noradrenergic antagonists into the bed nucleus of the stria terminalis or the central nucleus of the amygdala. Journal of Neuroscience 2002;22:5713–8.
[192] Shalev U, Grimm JW, Shaham Y. Neurobiology of relapse to heroin and cocaine seeking: a review. Pharmacological Reviews 2002;54:1–42.
[193] Koob GF, Sanna PP, Bloom FE. Neuroscience of addiction. Neuron 1998;21:467–76.
[194] Johnston JB. Further contributions to the study of the evolution of the forebrain. Journal of Comparative Neurology 1923;35:337–481.
[195] Alheid GF, De Olmos JS, Beltramino CA. Amygdala and extended amygdala. In: Paxinos G, editor. The rat nervous system. San Diego: Academic Press; 1995. p. 495–578.
[196] Koob GF. Alcoholism: allostasis and beyond. Alcoholism: Clinical and Experimental Research 2003a;27:232–43.
[197] Allen YS, Roberts GW, Bloom SR, Crow TJ, Polak JM. Neuropeptide Y in the stria terminalis: evidence for an amygdalofugal projection. Brain Research 1984;321:357–62.
[198] Dong HW, Petrovich GD, Swanson LW. Topography of projections from amygdala to bed nuclei of the stria terminalis. Brain Research Reviews 2001;38:192–246.
[199] Gray TS, Magnuson DJ. Peptide immunoreactive neurons in the amygdala and the bed nucleus of the stria terminalis project to the midbrain central gray in the rat. Peptides 1992;13:451–60.
[200] Kozicz T. Axon terminals containing tyrosine hydroxylase- and dopamine-beta-hydroxylase immunoreactivity form synapses with galanin immunoreactive neurons in the lateral division of the bed nucleus of the stria terminalis in the rat. Brain Research 2001;914:23–33.
[201] McDonald AJ, Shammah-Lagnado SJ, Shi C, Davis M. Cortical afferents to the extended amygdala. In: McGinty JF, editor. Advancing from the ventral striatum to the extended amygdala: implications for neuropsychiatry and drug abuse. Annals of the New York Academy of Sciences, vol. 877. New York: New York Academy of Sciences; 1999. p. 309–38.
[202] Phelix CF, Paull WK. Demonstration of distinct corticotropin releasing factor-containing neuron populations in the bed nucleus of the stria terminalis: a light and electron microscopic immunocytochemical study in the rat. Histochemistry 1990;94:345–64.
[203] Gray TS, Piechowski RA, Yracheta JM, Rittenhouse PA, Bethea CL, Van de Kar LD. Ibotenic acid lesions in the bed nucleus of the stria terminalis attenuate conditioned stress-induced increases in prolactin, ACTH and corticosterone. Neuroendocrinology 1993;57:517–24.
[204] Lesur A, Gaspar P, Alvarez C, Berger B. Chemoanatomic compartments in the human bed nucleus of the stria terminalis. Neuroscience 1989;32:181–94.

[205] Nijsen MJ, Croiset G, Diamant M, De Wied D, Wiegant VM. CRH signalling in the bed nucleus of the stria terminalis is involved in stress-induced cardiac vagal activation in conscious rats. Neuropsychopharmacology 2001;24:1–10.
[206] Pompei P, Tayebaty SJ, De Caro G, Schulkin J, Massi M. Bed nucleus of the stria terminalis: site for the antinatriorexic action of tachykinins in the rat. Pharmacology Biochemistry and Behavior 1991; 40:977–81.
[207] Caine SB, Heinrichs SC, Coffin VL, Koob GF. Effects of the dopamine D1 antagonist SCH 23390 microinjected into the accumbens, amygdala or striatum on cocaine self-administration in the rat. Brain Research 1995;692:47–56.
[208] Epping-Jordan MP, Markou A, Koob GF. The dopamine D-1 receptor antagonist SCH 23390 injected into the dorsolateral bed nucleus of the stria terminalis decreased cocaine reinforcement in the rat. Brain Research 1998;784:105–15.
[209] Heyser CJ, Roberts AJ, Schulteis G, Koob GF. Central administration of an opiate antagonist decreases oral ethanol self-administration in rats. Alcoholism: Clinical and Experimental Research 1999;23:1468–76.
[210] Moller C, Wiklund L, Sommer W, Thorsell A, Heilig M. Decreased experimental anxiety and voluntary ethanol consumption in rats following central but not basolateral amygdala lesions. Brain Research 1997;760:94–101.
[211] Roberts AJ, Cole M, Koob GF. Intra-amygdala muscimol decreases operant ethanol self-administration in dependent rats. Alcoholism: Clinical and Experimental Research 1996;20: 1289–98.
[212] Koob GF, Schulkin J. Addiction and stress: an allostatic view. Neuroscience & Biobehavioral Reviews 2018 [in press].
[213] Shurman J, Koob GF, Gutstein HB. Opioids, pain, the brain, and hyperkatifeia: a framework for the rational use of opioids for pain. Pain Medicine 2010;11:1092–8.
[214] Solomon RL, Corbit JD. An opponent-process theory of motivation. I. Temporal dynamics of affect. Psychological Review 1974;81(2):119–45.
[215] Baker TB, Piper ME, McCarthy DE, Majeskie MR, Fiore MC. Addiction motivation reformulated: an affective processing model of negative reinforcement. Psychological Review 2004;111(1):33–51.
[216] Case A, Deaton A. Rising morbidity and mortality in midlife among white non-Hispanic Americans in the 21st century. Proceedings of the National Academy of Sciences of the United States of America 2015;112:15078–83.
[217] Evans CJ, Cahill CM. Neurobiology of opioid dependence in creating addiction vulnerability. F1000Research 2016;5. F1000 Faculty Rev-1748.
[218] O'Brien CP, Testa T, O'Brien TJ, Brady JP, Wells B. Conditioned narcotic withdrawal in humans. Science 1977;195:1000–2.
[219] Kenny PJ, Chen SA, Kitamura O, Markou A, Koob GF. Conditioned withdrawal drives heroin consumption and decreases reward sensitivity. Journal of Neuroscience 2006;26:5894–900.
[220] Gracy KN, Dankiewicz LA, Koob GF. Opiate withdrawal-induced Fos immunoreactivity in the rat extended amygdala parallels the development of conditioned place aversion. Neuropsychopharmacology 2001;24:152–60.
[221] Ren ZY, Shi J, Epstein DH, Wang J, Lu L. Abnormal pain response in pain-sensitive opiate addicts after prolonged abstinence predicts increased drug craving. Psychopharmacology 2009;204:423–9.
[222] Compton P, Athanasos P, Elashoff D. Withdrawal hyperalgesia after acute opioid physical dependence in nonaddicted humans: a preliminary study. The Journal of Pain 2003;4:511–9.
[223] Egli M, Koob GF, Edwards S. Alcohol dependence as a chronic pain disorder. Neuroscience & Biobehavioral Reviews 2012;36:2179–92.
[224] Neugebauer V. The amygdala: different pains, different mechanisms. Pain 2007;127:1–2.
[225] Bester H, Menendez L, Besson JM, Bernard JF. Spino (trigemino) parabrachiohypothalamic pathway: electrophysiological evidence for an involvement in pain processes. Journal of Neurophysiology 1995;73:568–85.
[226] Price DD. Psychological and neural mechanisms of the affective dimension of pain. Science 2000; 288:1769–72.

[227] Koob GF. The dark side of emotion: the addiction perspective. European Journal of Pharmacology 2015;753:73–87.
[228] Neugebauer V, Li W. Processing of nociceptive mechanical and thermal information in central amygdala neurons with knee-joint input. Journal of Neurophysiology 2002;87:103–12.
[229] Ji G, Fu Y, Ruppert KA, Neugebauer V. Pain-related anxiety-like behavior requires CRF1 receptors in the amygdala. Molecular Pain 2007;3:13.
[230] Edwards S, Vendruscolo LF, Schlosburg JE, Misra KK, Wee S, Park PE, Schulteis G, Koob GF. Development of mechanical hypersensitivity in rats during heroin and ethanol dependence: alleviation by CRF1 receptor antagonism. Neuropharmacology 2012;62:1142–51.
[231] Park PE, Schlosburg JE, Vendruscolo LF, Schulteis G, Edwards S, Koob GF. Chronic CRF_1 receptor blockade reduces heroin intake escalation and dependence-induced hyperalgesia. Addiction Biology 2015;20:275–84.
[232] Carelli RM, West EA. When a good taste turns bad: neural mechanisms underlying the emergence of negative affect and associated natural reward devaluation by cocaine. Neuropharmacology 2014;76: 360–9.
[233] Wheeler RA, Twining RC, Jones JL, Slater JM, Grigson PS, Carelli RM. Behavioral and electrophysiological indices of negative affect predict cocaine self-administration. Neuron 2008;57:774–85.
[234] Wheeler RA, Aragona BJ, Fuhrmann KA, Jones JL, Day JJ, Cacciapaglia F, Wightman RM, Carelli RM. Cocaine cues drive opposing context dependent shifts in reward processing and emotional state. Biological Psychiatry 2011;69:1067–74.
[235] Wheeler RA, Carelli RM. Dissecting motivational circuitry to understand substance abuse. Neuropharmacology 2009;56(Suppl. 1):149–59.
[236] Nyland JE, Grigson PS. A drug-paired taste cue elicits withdrawal and predicts cocaine self-administration. Behavioural Brain Research 2013;240:87–90.
[237] Pandey SC. The gene transcription factor cyclic AMP-responsive element binding protein: role in positive and negative affective states of alcohol addiction. Pharmacology & Therapeutics 2004; 104:47–58.
[238] Diamond I, Gordon AS. Cellular and molecular neuroscience of alcoholism. Physiological Reviews 1997;77:1–20.
[239] Tabakoff B, Hoffman PL. Alcohol: neurobiology. In: Lowinson JH, Ruiz P, Millman RB, editors. Substance abuse: a comprehensive textbook. 2nd ed. Baltimore: Williams and Wilkins; 1992. p. 152–85.
[240] Gordon AS, Mochly-Rosen D, Diamond I. Alcoholism: a possible G protein disorder. In: Milligan G, Wakelam MJO, editors. G proteins, signal transduction and disease. London: Academic Press; 1992. p. 191–216.
[241] Gordon AS, Collier K, Diamond I. Ethanol regulation of adenosine receptor-stimulated cAMP levels in a clonal neural cell line: an in vitro model of cellular tolerance to ethanol. Proceedings of the National Academy of Sciences of the United States of America 1986;83:2105–8.
[242] Pandey SC, Roy A, Mittal N. Effects of chronic ethanol intake and its withdrawal on the expression and phosphorylation of the creb gene transcription factor in rat cortex. Journal of Pharmacology and Experimental Therapeutics 2001;296:857–68.
[243] Pandey SC, Zhang D, Mittal N, Nayyar D. Potential role of the gene transcription factor cyclic AMP-responsive element binding protein in ethanol withdrawal-related anxiety. Journal of Pharmacology and Experimental Therapeutics 1999;288:866–78.
[244] Pandey SC, Roy A, Zhang H. The decreased phosphorylation of cyclic adenosine monophosphate (cAMP) response element binding (CREB) protein in the central amygdala acts as a molecular substrate for anxiety related to ethanol withdrawal in rats. Alcoholism: Clinical and Experimental Research 2003;27:396–409.
[245] Li J, Li YH, Yuan XR. Changes of phosphorylation of cAMP response element binding protein in rat nucleus accumbens after chronic ethanol intake: naloxone reversal. Acta Pharmacologica Sinica 2003; 24:930–6.
[246] Zhang H, Pandey SC. Effects of PKA modulation on the expression of neuropeptide Y in rat amygdaloid structures during ethanol withdrawal. Peptides 2003;24:1397–402.

[247] Graves L, Dalvi A, Lucki I, Blendy JA, Abel T. Behavioral analysis of CREB alphadelta mutation on a B6/129 F1 hybrid background. Hippocampus 2002;12:18–26.
[248] Pandey SC, Roy A, Zhang H, Xu T. Partial deletion of the cAMP response element-binding protein gene promotes alcohol-drinking behaviors. Journal of Neuroscience 2004;24:5022–30.
[249] Valverde O, Mantamadiotis T, Torrecilla M, Ugedo L, Pineda J, Bleckmann S, Gass P, Kretz O, Mitchell JM, Schutz G, Maldonado R. Modulation of anxiety-like behavior and morphine dependence in CREB-deficient mice. Neuropsychopharmacology 2004;29:1122–33.
[250] Chance WT, Sheriff S, Peng F, Balasubramaniam A. Antagonism of NPY-induced feeding by pretreatment with cyclic AMP response element binding protein antisense oligonucleotide. Neuropeptides 2000;34:167–72.
[251] Pandey SC. Anxiety and alcohol abuse disorders: a common role for CREB and its target, the neuropeptide Y gene. Trends in Pharmacological Sciences 2003;24:456–60.
[252] Pluzarev O, Pandey SC. Modulation of CREB expression and phosphorylation in the rat nucleus accumbens during nicotine exposure and withdrawal. Journal of Neuroscience Research 2004;77: 884–91.
[253] Nestler EJ. Molecular basis of long-term plasticity underlying addiction. Nature Reviews Neuroscience 2001;2:119–28.
[254] Nestler EJ. Historical review: molecular and cellular mechanisms of opiate and cocaine addiction. Trends in Pharmacological Sciences 2004;25:210–8.
[255] Carlezon Jr WA, Thome J, Olson VG, Lane-Ladd SB, Brodkin ES, Hiroi N, Duman RS, Neve RL, Nestler EJ. Regulation of cocaine reward by CREB. Science 1998;282:2272–5.
[256] Pandey SC, Kyzar EJ, Zhang H. Epigenetic basis of the dark side of alcohol addiction. Neuropharmacology 2017;122:74–84.
[257] Pandey SC, Ugale R, Zhang H, Tang L, Prakash A. Brain chromatin remodeling: a novel mechanism of alcoholism. Journal of Neuroscience 2008a;28:3729–37.
[258] Pandey SC, Zhang H, Ugale R, Prakash A, Xu T, Misra K. Effector immediate-early gene arc in the amygdala plays a critical role in alcoholism. Journal of Neuroscience 2008b;28:2589–600.
[259] Pandey SC, Sakharkar AJ, Tang L, Zhang H. Potential role of adolescent alcohol exposure-induced amygdaloid histone modifications in anxiety and alcohol intake during adulthood. Neurobiology of Disease 2015;82:607–19.
[260] You C, Zhang H, Sakharkar AJ, Teppen T, Pandey SC. Reversal of deficits in dendritic spines, BDNF and Arc expression in the amygdala during alcohol dependence by HDAC inhibitor treatment. The International Journal of Neuropsychopharmacology 2014;17:313–22.
[261] Tunstall BJ, Carmack SA, Koob GF, Vendruscolo LF. Dysregulation of brain stress systems mediates compulsive alcohol drinking. Current Opinion in Behavioral Sciences 2017;13:85–90.
[262] Jentsch JD, Taylor JR. Impulsivity resulting from frontostriatal dysfunction in drug abuse: implications for the control of behavior by reward-related stimuli. Psychopharmacology 1999;146:373–90.
[263] Killcross S, Robbins TW, Everitt BJ. Different types of fear-conditioned behaviour mediated by separate nuclei within amygdala. Nature 1997;388:377–80.
[264] Gentile CG, Jarrell TW, Teich A, McCabe PM, Schneiderman N. The role of amygdaloid central nucleus in the retention of differential Pavlovian conditioning of bradycardia in rabbits. Behavioural Brain Research 1986;20:263–73.
[265] Iwata J, LeDoux JE, Meeley MP, Arneric S, Reis DJ. Intrinsic neurons in the amygdaloid field projected to by the medial geniculate body mediate emotional responses conditioned to acoustic stimuli. Brain Research 1986;383:195–214.
[266] Kapp BS, Frysinger RC, Gallagher M, Haselton JR. Amygdala central nucleus lesions: effect on heart rate conditioning in the rabbit. Physiology & Behavior 1979;23:1109–17.
[267] LeDoux JE, Iwata J, Cicchetti P, Reis DJ. Different projections of the central amygdaloid nucleus mediate autonomic and behavioral correlates of conditioned fear. Journal of Neuroscience 1988;8: 2517–29.
[268] Meil WM, See RE. Lesions of the basolateral amygdala abolish the ability of drug associated cues to reinstate responding during withdrawal from self-administered cocaine. Behavioural Brain Research 1997;87:139–48.

[269] Alderson HL, Robbins TW, Everitt BJ. The effects of excitotoxic lesions of the basolateral amygdala on the acquisition of heroin-seeking behaviour in rats. Psychopharmacology 2000;153:111–9.
[270] Hitchcott PK, Harmer CJ, Phillips GD. Enhanced acquisition of discriminative approach following intra-amygdala d-amphetamine. Psychopharmacology 1997;132:237–46.
[271] Cestari V, Mele A, Oliverio A, Castellano C. Amygdala lesions block the effect of cocaine on memory in mice. Brain Research 1996;713:286–9.
[272] Milner B. Some cognitive effects of frontal-lobe lesions in man. Philosophical Transactions of the Royal Society of London B Biological Sciences 1982;298:211–26.
[273] Damasio AR. The somatic marker hypothesis and the possible functions of the prefrontal cortex. Philosophical Transactions: Series B. Biological Sciences 1996;351:1413–20.
[274] Iversen SD, Mishkin M. Perseverative interference in monkeys following selective lesions of the inferior prefrontal convexity. Experimental Brain Research 1970;11:376–86.
[275] Ridley RM, Clark BA, Durnford LJ, Baker HF. Stimulus-bound perseveration after frontal ablations in marmosets. Neuroscience 1993;52:595–604.
[276] Blanc G, Herve D, Simon H, Lisoprawski A, Glowinski J, Tassin JP. Response to stress of mesocortico-frontal dopaminergic neurones in rats after long-term isolation. Nature 1980;284: 265–7.
[277] Herman JP, Guillonneau D, Dantzer R, Scatton B, Semerdjian-Rouquier L, Le Moal M. Differential effects of inescapable footshocks and of stimuli previously paired with inescapable footshocks on dopamine turnover in cortical and limbic areas of the rat. Life Sciences 1982;30:2207–14.
[278] Sawaguchi T, Goldman-Rakic PS. D1 dopamine receptors in prefrontal cortex: involvement in working memory. Science 1991;251:947–50.
[279] Simon H, Scatton B, Moal ML. Dopaminergic A10 neurones are involved in cognitive functions. Nature 1980;286:150–1.
[280] Jentsch JD, Verrico CD, Le D, Roth RH. Repeated exposure to delta 9-tetrahydrocannabinol reduces prefrontal cortical dopamine metabolism in the rat. Neuroscience Letters 1998;246:169–72.
[281] Karoum F, Suddath RL, Wyatt RJ. Chronic cocaine and rat brain catecholamines: long-term reduction in hypothalamic and frontal cortex dopamine metabolism. European Journal of Pharmacology 1990;186:1–8.
[282] Ricaurte GA, Schuster CR, Seiden LS. Long-term effects of repeated methylamphetamine administration on dopamine and serotonin neurons in the rat brain: a regional study. Brain Research 1980; 193:153–63.
[283] Robinson TE, Becker JB. Enduring changes in brain and behavior produced by chronic amphetamine administration: a review and evaluation of animal models of amphetamine psychosis. Brain Research 1986;396:157–98.
[284] Jentsch JD, Taylor JR, Elsworth JD, Redmond Jr DE, Roth RH. Altered frontal cortical dopaminergic transmission in monkeys after subchronic phencyclidine exposure: involvement in frontostriatal cognitive deficits. Neuroscience 1999;90:823–32.
[285] Sorg BA, Kalivas PW. Effects of cocaine and footshock stress on extracellular dopamine levels in the medial prefrontal cortex. Neuroscience 1993;53:695–703.
[286] Robinson TE, Becker JB, Moore CJ, Castaneda E, Mittleman G. Enduring enhancement in frontal cortex dopamine utilization in an animal model of amphetamine psychosis. Brain Research 1985; 343:374–7.
[287] Hamamura T, Fibiger HC. Enhanced stress-induced dopamine release in the prefrontal cortex of amphetamine-sensitized rats. European Journal of Pharmacology 1993;237:65–71.
[288] Louilot A, Le Moal M, Simon H. Opposite influences of dopaminergic pathways to the prefrontal cortex or the septum on the dopaminergic transmission in the nucleus accumbens: an in vivo voltammetric study. Neuroscience 1989;29:45–56.
[289] Piazza PV, Deminiere JM, Maccari S, Le Moal M, Mormede P, Simon H. Individual vulnerability to drug self-administration: action of corticosterone on dopaminergic systems as a possible pathophysiological mechanism. In: Willner P, Scheel-Kruger J, editors. The mesolimbic dopamine system: from motivation to action. Chichester: John Wiley and Sons; 1991a. p. 473–95.

[290] Le Moal M, Simon H. Mesocorticolimbic dopaminergic network: functional and regulatory roles. Physiological Reviews 1991;71:155–234.

[291] Tassin JP, Stinus L, Simon H, Blanc G, Thierry AM, Le Moal M, Cardo B, Glowinski J. Relationship between the locomotor hyperactivity induced by A10 lesions and the destruction of the fronto-cortical dopaminergic innervation in the rat. Brain Research 1978;141:267–81.

[292] Banks KE, Gratton A. Possible involvement of medial prefrontal cortex in amphetamine-induced sensitization of mesolimbic dopamine function. European Journal of Pharmacology 1995;282: 157–67.

[293] Pfefferbaum A, Sullivan EV, Mathalon DH, Lim KO. Frontal lobe volume loss observed with magnetic resonance imaging in older chronic alcoholics. Alcoholism: Clinical and Experimental Research 1997;21:521–9.

[294] Sullivan EV, Harding AJ, Pentney R, Dlugos C, Martin PR, Parks MH, Desmond JE, Chen SH, Pryor MR, De Rosa E, Pfefferbaum A. Disruption of frontocerebellar circuitry and function in alcoholism. Alcoholism: Clinical and Experimental Research 2003;27:301–9.

[295] Oscar-Berman M, Hutner N. Frontal lobe changes after chronic alcohol ingestion. In: Hunt WA, Nixon SJ, editors. Alcohol-induced brain damage. NIAAA research monograph, vol. 22. Bethesda, MD: National Institute on Alcohol Abuse and Alcoholism; 1993. p. 121–56.

[296] Parsons OA, Butters N, Nathan PE, editors. Neuropsychology of alcoholism: implications for diagnosis and treatment. New York: Guilford Press; 1987.

[297] Sullivan EV. Human brain vulnerability to alcoholism: evidence from neuroimaging studies. In: Noronha A, Eckardt M, Warren K, editors. Review of NIAAA's neuroscience and behavioral research portfolio. NIAAA research monograph, vol. 34. Bethesda, MD: National Institute on Alcohol Abuse and Alcoholism; 2000. p. 473–508.

[298] Harper C, Kril J. Patterns of neuronal loss in the cerebral cortex in chronic alcoholic patients. Journal of the Neurological Sciences 1989;92:81–9.

[299] Kubota M, Nakazaki S, Hirai S, Saeki N, Yamaura A, Kusaka T. Alcohol consumption and frontal lobe shrinkage: study of 1432 non-alcoholic subjects. Journal of Neurology Neurosurgery and Psychiatry 2001;71:104–6.

[300] Parsons O. Impaired neuropsychological cognitive functioning in sober alcohols. In: Hunt WA, Nixon SJ, editors. Alcohol-induced brain damage. NIAAA research monograph, vol. 22. Bethesda, MD: National Institute on Alcohol Abuse and Alcoholism; 1993. p. 173–94.

[301] Sullivan EV, Mathalon DH, Zipursky RB, Kersteen-Tucker Z, Knight RT, Pfefferbaum A. Factors of the Wisconsin Card Sorting Test as measures of frontal-lobe function in schizophrenia and in chronic alcoholism. Psychiatry Research 1993;46:175–99.

[302] Sullivan EV, Desmond JE, Lim KO, Pfefferbaum A. Speed and efficiency but not accuracy or timing deficits of limb movements in alcoholic men and women. Alcoholism: Clinical and Experimental Research 2002;26:705–13.

[303] Sullivan EV, Rosenbloom MJ, Pfefferbaum A. Pattern of motor and cognitive deficits in detoxified alcoholic men. Alcoholism: Clinical and Experimental Research 2000b;24:611–21.

[304] Sullivan EV, Rosenbloom MJ, Lim KO, Pfefferbaum A. Longitudinal changes in cognition, gait, and balance in abstinent and relapsed alcoholic men: relationships to changes in brain structure. Neuropsychology 2000a;14:178–88.

[305] Noel X, Paternot J, Van der Linden M, Sferrazza R, Verhas M, Hanak C, Kornreich C, Martin P, De Mol J, Pelc I, Verbanck P. Correlation between inhibition, working memory and delimited frontal area blood flow measure by 99mTc-Bicisate SPECT in alcohol-dependent patients. Alcohol and Alcoholism 2001;36:556–63.

[306] Fama R, Pfefferbaum A, Sullivan EV. Perceptual learning in detoxified alcoholic men: contributions from explicit memory, executive function, and age. Alcoholism: Clinical and Experimental Research 2004;28:1657–65.

[307] Drevets WC, Gautier C, Price JC, Kupfer DJ, Kinahan PE, Grace AA, Price JL, Mathis CA. Amphetamine-induced dopamine release in human ventral striatum correlates with euphoria. Biological Psychiatry 2001;49:81–96.

[308] Laruelle M, Abi-Dargham A, van Dyck CH, Rosenblatt W, Zea-Ponce Y, Zoghbi SS, Baldwin RM, Charney DS, Hoffer PB, Kung HF. SPECT imaging of striatal dopamine release after amphetamine challenge. Journal of Nuclear Medicine 1995;36:1182–90.
[309] Volkow ND, Wang GJ, Fowler JS, Logan J, Gatley SJ, Wong C, Hitzemann R, Pappas NR. Reinforcing effects of psychostimulants in humans are associated with increases in brain dopamine and occupancy of D(2) receptors. Journal of Pharmacology and Experimental Therapeutics 1999b;291: 409–15.
[310] Volkow ND, Wang G, Fowler JS, Logan J, Gerasimov M, Maynard L, Ding Y, Gatley SJ, Gifford A, Franceschi D. Therapeutic doses of oral methylphenidate significantly increase extracellular dopamine in the human brain. Journal of Neuroscience 2001;21:RC121.
[311] Volkow ND, Fowler JS, Wang GJ. Role of dopamine in drug reinforcement and addiction in humans: results from imaging studies. Behavioural Pharmacology 2002;13:355–66.
[312] Volkow ND, Wang GJ, Fowler JS, Logan J, Gatley SJ, Hitzemann R, Chen AD, Dewey SL, Pappas N. Decreased striatal dopaminergic responsiveness in detoxified cocaine-dependent subjects. Nature 1997;386:830–3.
[313] Martin-Solch C, Magyar S, Kunig G, Missimer J, Schultz W, Leenders KL. Changes in brain activation associated with reward processing in smokers and nonsmokers. A positron emission tomography study. Experimental Brain Research 2001;139:278–86.
[314] Volkow ND, Wang GJ, Fowler JS, Franceschi D, Thanos PK, Wong C, Gatley SJ, Ding YS, Molina P, Schlyer D, Alexoff D, Hitzemann R, Pappas N. Cocaine abusers show a blunted response to alcohol intoxication in limbic brain regions. Life Sciences 2000;66:PL161–7.
[315] Volkow ND, Fowler JS, Wolf AP, Hitzemann R, Dewey S, Bendriem B, Alpert R, Hoff A. Changes in brain glucose metabolism in cocaine dependence and withdrawal. American Journal of Psychiatry 1991;148:621–6.
[316] Volkow ND, Wang GJ, Fowler JS, Hitzemann R, Angrist B, Gatley SJ, Logan J, Ding YS, Pappas N. Association of methylphenidate-induced craving with changes in right striato-orbitofrontal metabolism in cocaine abusers: implications in addiction. American Journal of Psychiatry 1999a;156: 19–26.
[317] Grant S, London ED, Newlin DB, Villemagne VL, Liu X, Contoreggi C, Phillips RL, Kimes AS, Margolin A. Activation of memory circuits during cue-elicited cocaine craving. Proceedings of the National Academy of Sciences of the United States of America 1996;93:12040–5.
[318] Wang GJ, Volkow ND, Fowler JS, Cervany P, Hitzemann RJ, Pappas NR, Wong CT, Felder C. Regional brain metabolic activation during craving elicited by recall of previous drug experiences. Life Sciences 1999;64:775–84.
[319] Brody AL, Mandelkern MA, London ED, Childress AR, Lee GS, Bota RG, Ho ML, Saxena S, Baxter Jr LR, Madsen D, Jarvik ME. Brain metabolic changes during cigarette craving. Archives of General Psychiatry 2002;59:1162–72.
[320] Adinoff B, Devous Sr MD, Best SM, George MS, Alexander D, Payne K. Limbic responsiveness to procaine in cocaine-addicted subjects. American Journal of Psychiatry 2001;158:390–8.
[321] Volkow ND, Hitzemann R, Wang GJ, Fowler JS, Wolf AP, Dewey SL, Handlesman L. Long-term frontal brain metabolic changes in cocaine abusers. Synapse 1992b;11:184–90 [erratum: 12:86].
[322] Childress AR, Mozley PD, McElgin W, Fitzgerald J, Reivich M, O'Brien CP. Limbic activation during cue-induced cocaine craving. American Journal of Psychiatry 1999;156:11–8.
[323] Kilts CD, Schweitzer JB, Quinn CK, Gross RE, Faber TL, Muhammad F, Ely TD, Hoffman JM, Drexler KP. Neural activity related to drug craving in cocaine addiction. Archives of General Psychiatry 2001;58:334–41.
[324] Ingvar M, Ghatan PH, Wirsen-Meurling A, Risberg J, Von Heijne G, Stone-Elander S, Ingvar DH. Alcohol activates the cerebral reward system in man. Journal of Studies on Alcohol 1998;59:258–69.
[325] Nakamura H, Tanaka A, Nomoto Y, Ueno Y, Nakayama Y. Activation of fronto-limbic system in the human brain by cigarette smoking: evaluated by a CBF measurement. Keio Journal of Medicine 2000;49(Suppl. 1):A122–4.

[326] Volkow ND, Gillespie H, Mullani N, Tancredi L, Grant C, Valentine A, Hollister L. Brain glucose metabolism in chronic marijuana users at baseline and during marijuana intoxication. Psychiatry Research 1996;67:29–38.
[327] Volkow ND, Mullani N, Gould KL, Adler S, Krajewski K. Cerebral blood flow in chronic cocaine users: a study with positron emission tomography. British Journal of Psychiatry 1988;152:641–8.
[328] Garavan H, Pankiewicz J, Bloom A, Cho JK, Sperry L, Ross TJ, Salmeron BJ, Risinger R, Kelley D, Stein EA. Cue-induced cocaine craving: neuroanatomical specificity for drug users and drug stimuli. American Journal of Psychiatry 2000;157:1789–98.
[329] Maas LC, Lukas SE, Kaufman MJ, Weiss RD, Daniels SL, Rogers VW, Kukes TJ, Renshaw PF. Functional magnetic resonance imaging of human brain activation during cue-induced cocaine craving. American Journal of Psychiatry 1998;155:124–6.
[330] Wexler BE, Gottschalk CH, Fulbright RK, Prohovnik I, Lacadie CM, Rounsaville BJ, Gore JC. Functional magnetic resonance imaging of cocaine craving. American Journal of Psychiatry 2001; 158:86–95.
[331] Volkow ND, Hitzemann R, Wang GJ, Fowler JS, Burr G, Pascani K, Dewey SL, Wolf AP. Decreased brain metabolism in neurologically intact healthy alcoholics. American Journal of Psychiatry 1992a;149:1016–22.
[332] Volkow ND, Wang GJ, Hitzemann R, Fowler JS, Overall JE, Burr G, Wolf AP. Recovery of brain glucose metabolism in detoxified alcoholics. American Journal of Psychiatry 1994;151:178–83.
[333] Bickel WK, Jarmolowicz DP, Mueller ET, Gatchalian KM, McClure SM. Are executive function and impulsivity antipodes? A conceptual reconstruction with special reference to addiction. Psychopharmacology 2012a;221(3):361–87.
[334] Bickel WK, Jarmolowicz DP, Mueller ET, Gatchalian KM. The behavioral economics and neuroeconomics of reinforcer pathologies: implications for etiology and treatment of addiction. Current Psychiatry Reports 2011;13:406–15.
[335] Bickel WK, Jarmolowicz DP, Mueller ET, Koffarnus MN, Gatchalian KM. Excessive discounting of delayed reinforcers as a trans-disease process contributing to addiction and other disease-related vulnerabilities: emerging evidence. Pharmacology & Therapeutics 2012b;134:287–97.
[336] Bickel WK, Miller ML, Yi R, Kowal BP, Lindquist DM, Pitcock JA. Behavioral and neuroeconomics of drug addiction: competing neural systems and temporal dis- counting processes. Drug and Alcohol Dependence 2007;90S:S85–91.
[337] Bickel WK, Yi R. Temporal discounting as a measure of executive function: insights from the competing neuro-behavioral decision system hypothesis of addiction. Advances in Health Economics and Health Services Research 2008;20:289–309.
[338] Bickel WK, Snider SE, Quisenberry AJ, Stein JS, Hanlon CA. Competing neurobehavioral decision systems theory of cocaine addiction: from mechanisms to therapeutic opportunities. Progress in Brain Research 2016;223:269–93.
[339] Bickel WK, Quisenberry AJ, Moody L, Wilson AG. Therapeutic opportunities for self-control repair in addiction and related disorders: change and the limits of change in trans-disease processes. Clinical Psychological Science 2015;3:140–53.
[340] Koob GF. Antireward, compulsivity, and addiction: seminal contributions of Dr. Athina Markou to motivational dysregulation in addiction. Psychopharmacology 2017;234:1315–32.
[341] Koob GF, Le Moal M. Plasticity of reward neurocircuitry and the "dark side" of drug addiction. Nature Neuroscience 2005;8:1442–4.
[342] Koob GF, Zorrilla EP. Impulsivity derived from the dark side: neurocircuits that contribute to negative urgency. Frontiers in Behavioral Neuroscience 2018 [in press].
[343] Cyders MA, Smith GT. Emotion-based dispositions to rash action: positive and negative urgency. Psychological Bulletin 2008;134(6):807–28.
[344] Billieux J, Van der Linden M, Ceschi G. Which dimensions of impulsivity are related to cigarette craving? Addictive Behaviors 2007;32(6):1189–99.
[345] Stautz K, Cooper A. Impulsivity-related personality traits and adolescent alcohol use: a meta-analytic review. Clinical Psychology Review 2013;33(4):574–92.

[346] Zorrilla EP, Koob GF. The dark side of compulsive eating and food addiction: affective dysregulation, negative reinforcement, and negative urgency. In: Cottone P, Koob GF, Sabino V, Moore S, editors. Food addiction and compulsive eating behavior: research perspectives. New York: Elsevier; 2018 [in press].

[347] Verdejo-Garcıa A, Bechara A. A somatic marker theory of addiction. Neuropharmacology 2009;56:48–62.

[348] Damasio AR. Descartes' error: emotion, reason, and the human brain. New York: Grosset/Putnam; 1994.

[349] Bechara A. Risky business: emotion, decision-making and addiction. Journal of Gambling Studies 2003;19:23–51.

[350] Paulus MP, Hozack N, Frank L, Brown GG, Schuckit MA. Decision making by methamphetamine-dependent subjects is associated with error-rate-independent decrease in prefrontal and parietal activation. Biological Psychiatry 2003;53:65–74.

[351] Naqvi NH, Rudrauf D, Damasio H, Bechara A. Damage to the insula disrupts addiction to cigarette smoking. Science 2007;315:531–4.

[352] Townshend JM, Duka T. Mixed emotions: alcoholics' impairment in the recognition of specific facial expressions. Neuropsychologia 2003;41:773–82.

[353] Kornreich C, Blairy S, Philippot P, Dan B, Foisy M, Hess U, Le Bon O, Pelc I, Verbank P. Impaired emotional facial expression recognition in alcoholism compared with obsessive-compulsive disorders and normal controls. Psychiatry Research 2001;102:235–48.

[354] Kalivas PW, McFarland K. Brain circuitry and the reinstatement of cocaine-seeking behavior. Psychopharmacology 2003;168:44–56.

[355] McFarland K, Davidge SB, Lapish CC, Kalivas PW. Limbic and motor circuitry underlying footshock-induced reinstatement of cocaine-seeking behavior. Journal of Neuroscience 2004;24:1551–60.

[356] Grimm JW, See RE. Dissociation of primary and secondary reward-relevant limbic nuclei in an animal model of relapse. Neuropsychopharmacology 2000;22:473–9.

[357] McFarland K, Kalivas PW. The circuitry mediating cocaine-induced reinstatement of drug-seeking behavior. Journal of Neuroscience 2001;21:8655–63.

[358] Anderson SM, Bari AA, Pierce RC. Administration of the D1-like dopamine receptor antagonist SCH-23390 into the medial nucleus accumbens shell attenuates cocaine priming-induced reinstatement of drug-seeking behavior in rats. Psychopharmacology 2003;168:132–8.

[359] Cornish JL, Kalivas PW. Glutamate transmission in the nucleus accumbens mediates relapse in cocaine addiction. Journal of Neuroscience 2000;20:RC89.

[360] Park WK, Bari AA, Jey AR, Anderson SM, Spealman RD, Rowlett JK, Pierce RC. Cocaine administered into the medial prefrontal cortex reinstates cocaine-seeking behavior by increasing AMPA receptor-mediated glutamate transmission in the nucleus accumbens. Journal of Neuroscience 2002;22:2916–25.

[361] Cornish JL, Duffy P, Kalivas PW. A role for nucleus accumbens glutamate transmission in the relapse to cocaine-seeking behavior. Neuroscience 1999;93:1359–67.

[362] McFarland K, Lapish CC, Kalivas PW. Prefrontal glutamate release into the core of the nucleus accumbens mediates cocaine-induced reinstatement of drug-seeking behavior. Journal of Neuroscience 2003;23:3531–7.

[363] Kantak KM, Black Y, Valencia E, Green-Jordan K, Eichenbaum HB. Dissociable effects of lidocaine inactivation of the rostral and caudal basolateral amygdala on the maintenance and reinstatement of cocaine-seeking behavior in rats. Journal of Neuroscience 2002;22:1126–36.

[364] Di Ciano P, Everitt BJ. Contribution of the ventral tegmental area to cocaine-seeking maintained by a drug-paired conditioned stimulus in rats. European Journal of Neuroscience 2004a;19:1661–7.

[365] Fuchs RA, Evans KA, Parker MP, See RE. Differential involvement of orbitofrontal cortex subregions in conditioned cue-induced and cocaine-primed reinstatement of cocaine seeking in rats. Journal of Neuroscience 2004;24:6600–10.

[366] See RE, Kruzich PJ, Grimm JW. Dopamine, but not glutamate, receptor blockade in the basolateral amygdala attenuates conditioned reward in a rat model of relapse to cocaine-seeking behavior. Psychopharmacology 2001;154:301–10.
[367] Di Ciano P, Everitt BJ. Direct interactions between the basolateral amygdala and nucleus accumbens core underlie cocaine-seeking behavior by rats. Journal of Neuroscience 2004b;24:7167–73.
[368] Weiss F, Maldonado-Vlaar CS, Parsons LH, Kerr TM, Smith DL, Ben-Shahar O. Control of cocaine-seeking behavior by drug-associated stimuli in rats: effects on recovery of extinguished operant-responding and extracellular dopamine levels in amygdala and nucleus accumbens. Proceedings of the National Academy of Sciences of the United States of America 2000;97:4321–6.
[369] Ciccocioppo R, Sanna PP, Weiss F. Cocaine-predictive stimulus induces drug-seeking behavior and neural activation in limbic brain regions after multiple months of abstinence: reversal by D(1) antagonists. Proceedings of the National Academy of Sciences of the United States of America 2001;98:1976–81.
[370] Di Ciano P, Everitt BJ. Dissociable effects of antagonism of NMDA and AMPA/KA receptors in the nucleus accumbens core and shell on cocaine-seeking behavior. Neuropsychopharmacology 2001; 25:341–60.
[371] Capriles N, Rodaros D, Sorge RE, Stewart J. A role for the prefrontal cortex in stress- and cocaine-induced reinstatement of cocaine seeking in rats. Psychopharmacology 2003;168:66–74.
[372] Kalivas PW, McFarland K, Bowers S, Szumlinski K, Xi ZX, Baker D. Glutamate transmission and addiction to cocaine. In: Moghaddam B, Wolf ME, editors. Glutamate and disorders of cognition and motivation. Annals of the New York Academy of Sciences, vol. 1003. New York: New York Academy of Sciences; 2003. p. 169–75.
[373] Kalivas PW. Glutamate systems in cocaine addiction. Current Opinion in Pharmacology 2004;4: 23–9.
[374] Xi ZX, Ramamoorthy S, Baker DA, Shen H, Samuvel DJ, Kalivas PW. Modulation of group II metabotropic glutamate receptor signaling by chronic cocaine. Journal of Pharmacology and Experimental Therapeutics 2002;303:608–15.
[375] Bowers MS, McFarland K, Lake RW, Peterson YK, Lapish CC, Gregory ML, Lanier SM, Kalivas PW. Activator of G protein signaling 3: a gatekeeper of cocaine sensitization and drug seeking. Neuron 2004;42:269–81.
[376] Baker DA, McFarland K, Lake RW, Shen H, Tang XC, Toda S, Kalivas PW. Neuroadaptations in cystine-glutamate exchange underlie cocaine relapse. Nature Neuroscience 2003;6:743–9.
[377] Ghasemzadeh MB, Permenter LK, Lake R, Worley PF, Kalivas PW. Homer1 proteins and AMPA receptors modulate cocaine-induced behavioural plasticity. European Journal of Neuroscience 2003; 18:1645–51.
[378] Szumlinski KK, Dehoff MH, Kang SH, Frys KA, Lominac KD, Klugmann M, Rohrer J, Griffin 3rd W, Toda S, Champtiaux NP, Berry T, Tu JC, Shealy SE, During MJ, Middaugh LD, Worley PF, Kalivas PW. Homer proteins regulate sensitivity to cocaine. Neuron 2004;43:401–13.
[379] Chiamulera C, Epping-Jordan MP, Zocchi A, Marcon C, Cottiny C, Tacconi S, Corsi M, Orzi F, Conquet F. Reinforcing and locomotor stimulant effects of cocaine are absent in mGluR5 null mutant mice. Nature Neuroscience 2001;4:873–4.
[380] McGeehan AJ, Olive MF. The mGluR5 antagonist MPEP reduces the conditioned rewarding effects of cocaine but not other drugs of abuse. Synapse 2003;47:240–2.
[381] Scofield MD, Heinsbroek JA, Gipson CD, Kupchik YM, Spencer S, Smith AC, Roberts-Wolfe D, Kalivas PW. The nucleus accumbens: mechanisms of addiction across drug classes reflect the importance of glutamate homeostasis. Pharmacological Reviews 2016;68:816–71.
[382] Moussawi K, Pacchioni A, Moran M, Olive MF, Gass JT, Lavin A, Kalivas PW. N-Acetylcysteine reverses cocaine-induced metaplasticity. Nature Neuroscience 2009;12:182–9.
[383] Moussawi K, Zhou W, Shen H, Reichel CM, See RE, Carr DB, Kalivas PW. Reversing cocaine-induced synaptic potentiation provides enduring protection from relapse. Proceedings of the National Academy of Sciences of the United States of America 2011;108:385–90.
[384] Kalivas PW. The glutamate homeostasis hypothesis of addiction. Nature Reviews Neuroscience 2009;10:561–72.

[385] Zhao S, Studer D, Chai X, Graber W, Brose N, Nestel S, Young C, Rodriguez EP, Saetzler K, Frotscher M. Structural plasticity of spines at giant mossy fiber synapses. Frontiers in Neural Circuits 2012;6:103.

[386] Kopec C, Malinow R. Neuroscience: matters of size. Science 2006;314:1554–5.

[387] Toda S, Shen HW, Peters J, Cagle S, Kalivas PW. Cocaine increases actin cycling: effects in the reinstatement model of drug seeking. Journal of Neuroscience 2006;26:1579–87.

[388] Wolf ME, Ferrario CR. AMPA receptor plasticity in the nucleus accumbens after repeated exposure to cocaine. Neuroscience & Biobehavioral Reviews 2010;35:185–211.

[389] Loweth JA, Scheyer AF, Milovanovic M, LaCrosse AL, Flores-Barrera E, Werner CT, Li X, Ford KA, Le T, Olive MF, Szumlinski KK, Tseng KY, Wolf ME. Synaptic depression via mGluR1 positive allosteric modulation suppresses cue-induced cocaine craving. Nature Neuroscience 2014; 17:73–80.

[390] Li X, Xi ZX, Markou A. Metabotropic glutamate 7 (mGlu7) receptor: a target for medication development for the treatment of cocaine dependence. Neuropharmacology 2013;66:12–23.

[391] Scofield MD, Kalivas PW. Astrocytic dysfunction and addiction: consequences of impaired glutamate homeostasis. The Neuroscientist 2014;20:610–22.

[392] Koob GF, Lloyd GK, Mason BJ. Development of pharmacotherapies for drug addiction: a Rosetta Stone approach. Nature Reviews Drug Discovery 2009;8:500–15.

[393] See RE, Fuchs RA, Ledford CC, McLaughlin J. Drug addiction, relapse, and the amygdala. In: Shinnick-Gallagher P, Pitkanen A, Shekhar A, Cahill L, editors. The amygdala in brain function: basic and clinical approaches. Annals of the New York Academy of Sciences, vol. 985. New York: New York Academy of Sciences; 2003. p. 294–307.

[394] Tiffany ST, Carter BL. Is craving the source of compulsive drug use? Journal of Psychopharmacology 1998;12:23–30.

[395] Carter BL, Tiffany ST. Meta-analysis of cue-reactivity in addiction research. Addiction 1999;94: 327–40.

[396] Kruzich PJ, See RE. Differential contributions of the basolateral and central amygdala in the acquisition and expression of conditioned relapse to cocaine-seeking behavior. Journal of Neuroscience 2001;21:RC155.

[397] McLaughlin J, See RE. Selective inactivation of the dorsomedial prefrontal cortex and the basolateral amygdala attenuates conditioned-cued reinstatement of extinguished cocaine-seeking behavior in rats. Psychopharmacology 2003;168:57–65.

[398] Weiss F, Martin-Fardon R, Ciccocioppo R, Kerr TM, Smith DL, Ben-Shahar O. Enduring resistance to extinction of cocaine-seeking behavior induced by drug-related cues. Neuropsychopharmacology 2001b;25:361–72.

[399] Bonson KR, Grant SJ, Contoreggi CS, Links JM, Metcalfe J, Weyl HL, Kurian V, Ernst M, London ED. Neural systems and cue-induced cocaine craving. Neuropsychopharmacology 2002; 26:376–86.

[400] Weissenborn R, Robbins TW, Everitt BJ. Effects of medial prefrontal or anterior cingulate cortex lesions on responding for cocaine under fixed-ratio and second-order schedules of reinforcement in rats. Psychopharmacology 1997;134:242–57.

[401] Kaddis FG, Uretsky NJ, Wallace LJ. DNQX in the nucleus accumbens inhibits cocaine-induced conditioned place preference. Brain Research 1995;697:76–82.

[402] Layer RT, Uretsky NJ, Wallace LJ. Effects of the AMPA/kainate receptor antagonist DNQX in the nucleus accumbens on drug-induced conditioned place preference. Brain Research 1993;617: 267–73.

[403] Hotsenpiller G, Giorgetti M, Wolf ME. Alterations in behaviour and glutamate transmission following presentation of stimuli previously associated with cocaine exposure. European Journal of Neuroscience 2001;14:1843–55.

[404] Everitt BJ, Cardinal RN, Hall J, Parkinson JA, Robbins TW. Differential involvement of amygdala subsystems in appetitive conditioning and drug addiction. In: Aggleton JP, editor. The amygdala: a functional analysis. New York: Oxford University Press; 2000. p. 353–90.

[405] Vorel SR, Liu X, Hayes RJ, Spector JA, Gardner EL. Relapse to cocaine-seeking after hippocampal theta burst stimulation. Science 2001;292:1175–8.
[406] Gallagher M, Graham PW, Holland PC. The amygdala central nucleus and appetitive Pavlovian conditioning: lesions impair one class of conditioned behavior. Journal of Neuroscience 1990;10: 1906–11.
[407] Parkinson JA, Robbins TW, Everitt BJ. Dissociable roles of the central and basolateral amygdala in appetitive emotional learning. European Journal of Neuroscience 2000a;12:405–13.
[408] Parkinson JA, Willoughby PJ, Robbins TW, Everitt BJ. Disconnection of the anterior cingulate cortex and nucleus accumbens core impairs Pavlovian approach behavior: further evidence for limbic cortical-ventral striatopallidal systems. Behavioral Neuroscience 2000b;114:42–63.
[409] Li Y, Hu XT, Berney TG, Vartanian AJ, Stine CD, Wolf ME, White FJ. Both glutamate receptor antagonists and prefrontal cortex lesions prevent induction of cocaine sensitization and associated neuroadaptations. Synapse 1999;34:169–80.
[410] Pierce RC, Reeder DC, Hicks J, Morgan ZR, Kalivas PW. Ibotenic acid lesions of the dorsal prefrontal cortex disrupt the expression of behavioral sensitization to cocaine. Neuroscience 1998;82: 1103–14.
[411] Wolf ME, Dahlin SL, Hu XT, Xue CJ, White K. Effects of lesions of prefrontal cortex, amygdala, or fornix on behavioral sensitization to amphetamine: comparison with N-methyl-D-aspartate antagonists. Neuroscience 1995;69:417–39.
[412] Wolf ME. The role of excitatory amino acids in behavioral sensitization to psychomotor stimulants. Progress in Neurobiology 1998;54:679–720.
[413] Robbins TW, Everitt BJ. Drug addiction: bad habits add up. Nature 1999;398:567–70.
[414] Everitt BJ, Dickinson A, Robbins TW. The neuropsychological basis of addictive behaviour. Brain Research Reviews 2001;36:129–38.
[415] Markou A, Koob GF. Post-cocaine anhedonia: an animal model of cocaine withdrawal. Neuropsychopharmacology 1991;4:17–26.
[416] Wolf ME. Synaptic mechanisms underlying persistent cocaine craving. Nature Reviews Neuroscience 2016;17:351–65.
[417] Cruz FC, Koya E, Guez-Barber DH, Bossert JM, Lupica CR, Shaham Y, Hope BT. New technologies for examining the role of neuronal ensembles in drug addiction and fear. Nature Reviews Neuroscience 2013;14:743–54.
[418] Conrad KL, Tseng KY, Uejima JL, Reimers JM, Heng LJ, Shaham Y, Marinelli M, Wolf ME. Formation of accumbens GluR2-lacking AMPA receptors mediates incubation of cocaine craving. Nature 2008;454:118–21.
[419] Shaw-Lutchman TZ, Barrot M, Wallace T, Gilden L, Zachariou V, Impey S, Duman RS, Storm D, Nestler EJ. Regional and cellular mapping of cAMP response element-mediated transcription during naltrexone-precipitated morphine withdrawal. Journal of Neuroscience 2002;22:3663–72.
[420] Shaw-Lutchman TZ, Impey S, Storm D, Nestler EJ. Regulation of CRE-mediated transcription in mouse brain by amphetamine. Synapse 2003;48:10–7.
[421] Terwilliger RZ, Beitner-Johnson D, Sevarino KA, Crain SM, Nestler EJ. A general role for adaptations in G-proteins and the cyclic AMP system in mediating the chronic actions of morphine and cocaine on neuronal function. Brain Research 1991;548:100–10.
[422] Turgeon SM, Pollack AE, Fink JS. Enhanced CREB phosphorylation and changes in c-Fos and FRA expression in striatum accompany amphetamine sensitization. Brain Research 1997;749:120–6.
[423] Collier HO, Francis DL. Morphine abstinence is associated with increased brain cyclic AMP. Nature 1975;255:159–62.
[424] Sharma SK, Klee WA, Nirenberg M. Dual regulation of adenylate cyclase accounts for narcotic dependence and tolerance. Proceedings of the National Academy of Sciences of the United States of America 1975;72:3092–6.
[425] Constantinescu A, Diamond I, Gordon AS. Ethanol-induced translocation of cAMP-dependent protein kinase to the nucleus: mechanism and functional consequences. Journal of Biological Chemistry 1999;274:26985–91.

[426] Sapru MK, Diamond I, Gordon AS. Adenosine receptors mediate cellular adaptation to ethanol in NG108-15 cells. Journal of Pharmacology and Experimental Therapeutics 1994;271:542–8.
[427] Maldonado R, Blendy JA, Tzavara E, Gass P, Roques BP, Hanoune J, Schutz G. Reduction of morphine abstinence in mice with a mutation in the gene encoding CREB. Science 1996;273: 657–9.
[428] Ortiz J, Fitzgerald LW, Charlton M, Lane S, Trevisan L, Guitart X, Shoemaker W, Duman RS, Nestler EJ. Biochemical actions of chronic ethanol exposure in the mesolimbic dopamine system. Synapse 1995;21:289–98.
[429] Barrot M, Olivier JD, Perrotti LI, DiLeone RJ, Berton O, Eisch AJ, Impey S, Storm DR, Neve RL, Yin JC, Zachariou V, Nestler EJ. CREB activity in the nucleus accumbens shell controls gating of behavioral responses to emotional stimuli. Proceedings of the National Academy of Sciences of the United States of America 2002;99:11435–40.
[430] Newton SS, Thome J, Wallace TL, Shirayama Y, Schlesinger L, Sakai N, Chen J, Neve R, Nestler EJ, Duman RS. Inhibition of cAMP response element-binding protein or dynorphin in the nucleus accumbens produces an antidepressant-like effect. Journal of Neuroscience 2002;22: 10883–90.
[431] Pliakas AM, Carlson RR, Neve RL, Konradi C, Nestler EJ, Carlezon Jr WA. Altered responsiveness to cocaine and increased immobility in the forced swim test associated with elevated cAMP response element-binding protein expression in nucleus accumbens. Journal of Neuroscience 2001;21: 7397–403.
[432] Self DW, Genova LM, Hope BT, Barnhart WJ, Spencer JJ, Nestler EJ. Involvement of cAMP-dependent protein kinase in the nucleus accumbens in cocaine self-administration and relapse of cocaine-seeking behavior. Journal of Neuroscience 1998;18:1848–59.
[433] Walters CL, Blendy JA. Different requirements for cAMP response element binding protein in positive and negative reinforcing properties of drugs of abuse. Journal of Neuroscience 2001;21: 9438–44.
[434] Hope BT, Nye HE, Kelz MB, Self DW, Iadarola MJ, Nakabeppu Y, Duman RS, Nestler EJ. Induction of a long-lasting AP-1 complex composed of altered Fos-like proteins in brain by chronic cocaine and other chronic treatments. Neuron 1994;13:1235–44.
[435] Moratalla R, Elibol B, Vallejo M, Graybiel AM. Network-level changes in expression of inducible Fos-Jun proteins in the striatum during chronic cocaine treatment and withdrawal. Neuron 1996;17: 147–56.
[436] Kelz MB, Chen J, Carlezon Jr WA, Whisler K, Gilden L, Beckmann AM, Steffen C, Zhang YJ, Marotti L, Self DW, Tkatch T, Baranauskas G, Surmeier DJ, Neve RL, Duman RS, Picciotto MR, Nestler EJ. Expression of the transcription factor deltaFosB in the brain controls sensitivity to cocaine. Nature 1999;401:272–6.
[437] Colby CR, Whisler K, Steffen C, Nestler EJ, Self DW. Striatal cell type-specific overexpression of DeltaFosB enhances incentive for cocaine. Journal of Neuroscience 2003;23:2488–93.
[438] Hiroi N, Brown JR, Haile CN, Ye H, Greenberg ME, Nestler EJ. FosB mutant mice: loss of chronic cocaine induction of Fos-related proteins and heightened sensitivity to cocaine's pyschomotor and rewarding effects. Proceedings of the National Academy of Sciences of the United States of America 1997;94:10397–402.
[439] McClung CA, Ulery PG, Perrotti LI, Zachariou V, Berton O, Nestler EJ. DeltaFosB: a molecular switch for long-term adaptation in the brain. Molecular Brain Research 2004;132:146–54.
[440] Peakman MC, Colby C, Perrotti LI, Tekumalla P, Carle T, Ulery P, Chao J, Duman C, Steffen C, Monteggia L, Allen MR, Stock JL, Duman RS, McNeish JD, Barrot M, Self DW, Nestler EJ, Schaeffer E. Inducible, brain region-specific expression of a dominant negative mutant of c-Jun in transgenic mice decreases sensitivity to cocaine. Brain Research 2003;970:73–86.
[441] Nestler EJ, Barrot M, Self DW. DeltaFosB: a sustained molecular switch for addiction. Proceedings of the National Academy of Sciences of the United States of America 2001;98:11042–6.
[442] Robinson TE, Kolb B. Persistent structural modifications in nucleus accumbens and prefrontal cortex neurons produced by previous experience with amphetamine. Journal of Neuroscience 1997;17: 8491–7.

[443] Robinson TE, Kolb B. Morphine alters the structure of neurons in the nucleus accumbens and neocortex of rats. Synapse 1999;33:160–2.
[444] Feng J, Nestler EJ. Epigenetic mechanisms of drug addiction. Current Opinion in Neurobiology 2013;23:521–8.
[445] Walker DM, Nestler EJ. Neuroepigenetics and addiction. Handbook of Clinical Neurology 2018; 148:747–65.
[446] Mashayekhi FJ, Rasti M, Rahvar M, et al. Expression levels of the BDNF gene and histone modifications around its promoters in the ventral tegmental area and locus ceruleus of rats during forced abstinence from morphine. Neurochemical Research 2012;37:1517–23.
[447] Schmidt HD, Sangrey GR, Darnell SB, et al. Increased brain-derived neurotrophic factor (BDNF) expression in the ventral tegmental area during cocaine abstinence is associated with increased histone acetylation at BDNF exon I-containing promoters. Journal of Neurochemistry 2012;120:202–9.
[448] LaPlant Q, Vialou V, Covington 3rd HE, et al. Dnmt3a regulates emotional behavior and spine plasticity in the nucleus accumbens. Nature Neuroscience 2010;13:1137–43.
[449] Tapocik JD, Barbier E, Flanigan M, Solomon M, Pincus A, Pilling A, Sun H, Schank JR, King C, Heilig M. microRNA-206 in rat medial prefrontal cortex regulates BDNF expression and alcohol drinking. Journal of Neuroscience 2014;34:4581–8.
[450] Blomeyer D, Treutlein J, Esser G, Schmidt MH, Schumann G, Laucht M. Interaction between CRHR1 gene and stressful life events predicts adolescent heavy alcohol use. Biological Psychiatry 2008;63:146–51.
[451] Treutlein J, Kissling C, Frank J, Wiemann S, Dong L, Depner M, Saam C, Lascorz J, Soyka M, Preuss UW, Rujescu D, Skowronek MH, Rietschel M, Spanagel R, Heinz A, Laucht M, Mann K, Schumann G. Genetic association of the human corticotropin releasing hormone receptor 1 (*CRHR1*) with binge drinking and alcohol intake patterns in two independent samples. Molecular Psychiatry 2006;11:594–602.
[452] Schmid B, Blomeyer D, Treutlein J, Zimmermann US, Buchmann AF, Schmidt MH, Esser G, Rietschel M, Banaschewski T, Schumann G, Laucht M. Interacting effects of CRHR1 gene and stressful life events on drinking initiation and progression among 19-year-olds. The International Journal of Neuropsychopharmacology 2010;13:703–14.
[453] Zaina S, Pérez-Luque EL, Lund G. Genetics talks to epigenetics? The interplay between sequence variants and chromatin structure. Current Genomics 2010;11:359–67.
[454] Diana M, Pistis M, Carboni S, Gessa GL, Rossetti ZL. Profound decrement of mesolimbic dopaminergic neuronal activity during ethanol withdrawal syndrome in rats: electrophysiological and biochemical evidence. Proceedings of the National Academy of Sciences of the United States of America 1993;90:7966–9.
[455] Koob GF. Neurocircuitry of alcohol addiction: synthesis from animal models. In: Sullivan EV, Pfefferbaum A, editors. Alcohol and the nervous system. Handbook of clinical neurology, vol. 125. Amsterdam: Elsevier; 2014. p. 33–54.
[456] Volkow ND, Fowler JS, Wang GJ, Baler R, Telang F. Imaging dopamine's role in drug abuse and addiction. Neuropharmacology 2009;56(Suppl. 1):3–8.
[457] Ashok AH, Mizuno Y, Volkow ND, Howes OD. JAMA Psychiatry 2017;74(5):511–9.
[458] Koob GF. A role for brain stress systems in addiction. Neuron 2008;59:11–34.
[459] Oscar-Berman M. Function and dysfunction of prefrontal brain circuitry in alcoholic Korsakoff's syndrome. Neuropsychology Review 2012;22:154–69.
[460] Sullivan EV, Pfefferbaum A. Neurocircuitry in alcoholism: a substrate of disruption and repair. Psychopharmacology 2005;180:583–94.
[461] Niendam TA, Laird AR, Ray KL, Dean YM, Glahn DC, Carter CS. Meta-analytic evidence for a superordinate cognitive control network subserving diverse executive functions. Cognitive, Affective, & Behavioral Neuroscience 2012;12:241–68.
[462] Olbrich HM, Valerius G, Paris C, Hagenbuch F, Ebert D, Juengling FD. Brain activation during craving for alcohol measured by positron emission tomography. Australian and New Zealand Journal of Psychiatry February 2006;40(2):171–8.

[463] Jasinska AJ, Stein EA, Kaiser J, Naumer MJ, Yalachkov Y. Factors modulating neural reactivity to drug cues in addiction: a survey of human neuroimaging studies. Neuroscience & Biobehavioral Reviews 2014;38:1–16.

[464] Marlatt G, Gordon J. Determinants of relapse: implications for the maintenance of behavioral change. In: Davidson P, Davidson S, editors. Behavioral medicine: changing health lifestyles. New York: Brunner/Mazel; 1980. p. 410–52.

[465] Jochum T, Boettger MK, Burkhardt C, Juckel G, Bär KJ. Increased pain sensitivity in alcohol withdrawal syndrome. European Journal of Pain 2010;14:713–8.

[466] Zywiak WH, Connors GJ, Maisto SA, Westerberg VS. Relapse research and the Reasons for Drinking Questionnaire: a factor analysis of Marlatt's relapse taxonomy. Addiction 1996;91(Suppl. l): s121–30.

[467] Martin-Fardon R, Weiss F. Modeling relapse in animals. Current Topics in Behavioral Neuroscience 2013;13:403–32.

[468] Edwards S, Koob GF. Neurobiology of dysregulated motivational systems in drug addiction. Future Neurology 2010;5:393–410.

[469] Nestler EJ. Is there a common molecular pathway for addiction? Nature Neuroscience 2005;8: 1445–9.

[470] Russo SJ, Dietz DM, Dumitriu D, Morrison JH, Malenka RC, Nestler EJ. The addicted synapse: mechanisms of synaptic and structural plasticity in nucleus accumbens. Trends in Neurosciences 2010;33:267–76.

[471] Kroener S, Mulholland PJ, New NN, Gass JT, Becker HC, Chandler J. Chronic alcohol exposure alters behavioral and synaptic plasticity of the rodent prefrontal cortex. PLoS One 2012;7(5):e37541. https://doi.org/10.1371/journal.pone.0037541.g001.

[472] Robinson TE, Gorny G, Mitton E, Kolb B. Cocaine self-administration alters the morphology of dendrites and dendritic spines in the nucleus accumbens and neocortex. Synapse 2001;39:257–66.

[473] Zhou FC, Anthony B, Dunn KW, Lindquist WB, Xu ZC, Deng P. Chronic alcohol drinking alters neuronal dendritic spines in the brain reward center nucleus accumbens. Brain Research 2007;1134: 148–61.

[474] Cadet JL, McCoy MT, Jayanthi S. Epigenetics and addiction. Clinical Pharmacology & Therapeutics 2016;99:502–11.

[475] Kyzar EJ, Floreani C, Teppen TL, Pandey SC. Adolescent alcohol exposure: burden of epigenetic reprogramming, synaptic remodeling, and adult psychopathology. Frontiers in Neuroscience 2016;10:222.

[476] Nestler EJ. Transcriptional mechanisms of drug addiction. Clinical Psychopharmacological Neuroscience 2012;10:136–43.

[477] Koob GF, Le Moal M. Neurobiology of addiction. London: Academic Press; 2006.

[478] Olsen RW, McCabe RT, Wamsley JK. GABAA receptor subtypes: autoradiographic comparison of GABA, benzodiazepine, and convulsant binding sites in the rat central nervous system. Journal of Chemical Neuroanatomy 1990;3:59–76.

[479] Sequier JM, Richards JG, Malherbe P, Price GW, Mathews S, Mohler H. Mapping of brain areas containing RNA homologous to cDNAs encoding the alpha and beta subunits of the rat GABAA gamma-aminobutyrate receptor. Proceedings of the National Academy of Sciences of the United States of America 1988;85:7815–9.

[480] Shivers BD, Killisch I, Sprengel R, Sontheimer H, Kohler M, Schofield PR, Seeburg PH. Two novel GABAA receptor subunits exist in distinct neuronal subpopulations. Neuron 1989;3:327–37.

[481] Corbit LH, Nie H, Janak PH. Habitual alcohol seeking: time course and the contribution of subregions of the dorsal striatum. Biological Psychiatry 2012;72:389–95.

[482] Olmstead M, Lafond M, Everitt B, Dickinson A. Cocaine seeking by rats is a goal-directed action. Behavioral Neuroscience 2001;115:394–402.

[483] Zapata A, Minney VL, Shippenberg TS. Shift from goal-directed to habitual cocaine seeking after prolonged experience in rats. Journal of Neuroscience 2010;30:15457–63.

[484] Belin D, Jonkman S, Dickinson A, Robbins T, Everitt B. Parallel and interactive learning processes within the basal ganglia: relevance for the understanding of addiction. Behavioural Brain Research 2009;199:89–102.

[485] Belin-Rauscent A, Everitt BJ, Belin D. Intrastriatal shifts mediate the transition from drug-seeking actions to habits. Biological Psychiatry 2012;72:343–5.

[486] Murray JE, Belin D, Everitt BJ. Double dissociation of the dorsomedial and dorsolateral striatal control over the acquisition and performance of cocaine seeking. Neuropsychopharmacology 2012a;37:2456–66.

[487] Porrino L. Cocaine self-administration produces a progressive involvement of limbic, association, and sensorimotor striatal domains. Journal of Neuroscience 2004a;24:3554–62.

[488] Porrino LJ, Daunais JB, Smith HR, Nader MA. The expanding effects of cocaine: studies in a nonhuman primate model of cocaine self-administration. Neuroscience & Biobehavioral Reviews 2004b;27:813–20.

[489] Belin D, Everitt BJ. Cocaine seeking habits depend upon dopamine-dependent serial connectivity linking the ventral with the dorsal striatum. Neuron 2008;57:432–41.

[490] Belin D, Everitt BJ. Drug addiction: the neural and psychological basis of a compulsive incentive habit. In: Steiner H, Tseng KY, editors. Handbook of basal ganglia structure and function. Handbook of behavioral neuroscience, vol. 20. New York: Elsevier; 2010. p. 571–92.

[491] Vanderschuren LJ, Di Ciano P, Everitt BJ. Involvement of the dorsal striatum in cue-controlled cocaine seeking. Journal of Neuroscience 2005;25:8665–70.

[492] Murray JE, Belin D, Everitt BJ. Basolateral and central nuclei of the amygdala required for the transition to dorsolateral. dopamine control over habitual cocaine seeking. Neuropsychopharmacology 2012b;38:S79–197.

[493] Finch DM. Neurophysiology of converging synaptic inputs from the rat prefrontal cortex, amygdala, midline thalamus, and hippocampal formation onto single neurons of the caudate/putamen and nucleus accumbens. Hippocampus 1996;6:495–512.

[494] O'Donnell P, Grace AA. Synaptic interactions among excitatory afferents to nucleus accumbens neurons: hippocampal gating of prefrontal cortical input. Journal of Neuroscience 1995;15:3622–39.

[495] Feltenstein MW, See RE. Systems level neuroplasticity in drug addiction. Cold Spring Harbor Perspectives in Medicine 2013;3:a011916.

[496] Marchant NJ, Kaganovsky K, Shaham Y, Bossert JM. Role of corticostriatal circuits in context-induced reinstatement of drug seeking. Brain Research 2015;1628:219–32.

[497] Li X, Caprioli D, Marchant NJ. Recent updates on incubation of drug craving: a mini-review. Addiction Biology 2015;20:872–6.

[498] Pickens CL, Airavaara M, Theberge F, Fanous S, Hope BT, Shaham Y. Neurobiology of the incubation of drug craving. Trends in Neurosciences 2011;34:411–20.

[499] Purgianto A, Scheyer AF, Loweth JA, Ford KA, Tseng KY, Wolf ME. Different adaptations in AMPA receptor transmission in the nucleus accumbens after short vs long access cocaine self-administration regimens. Neuropsychopharmacology 2013;38:1789–97.

[500] Lee BR, Ma YY, Huang YH, Wang X, Otaka M, Ishikawa M, Neumann PA, Graziane NM, Brown TE, Suska A, Guo C, Lobo MK, Sesack SR, Wolf ME, Nestler EJ, Shaham Y, Schlüter OM, Dong Y. Maturation of silent synapses in amygdala-accumbens projection contributes to incubation of cocaine craving. Nature Neuroscience 2013;16:1644–51.

[501] Ma YY, Lee BR, Wang X, Guo C, Liu L, Cui R, Lan Y, Balcita-Pedicino JJ, Wolf ME, Sesack SR, Shaham Y, Schlüter OM, Huang YH, Dong Y. Bidirectional modulation of incubation of cocaine craving by silent synapse-based remodeling of prefrontal cortex to accumbens projections. Neuron 2014;83:1453–67.

[502] McCutcheon JE, Loweth JA, Ford KA, Marinelli M, Wolf ME, Tseng KY. Group I mGluR activation reverses cocaine-induced accumulation of calcium-permeable AMPA receptors in nucleus accumbens synapses via a protein kinase C-dependent mechanism. Journal of Neuroscience 2011;31:14536–41.

[503] Bellone C, Luscher C. Cocaine triggered AMPA receptor redistribution is reversed in vivo by mGluR-dependent long-term depression. Nature Neuroscience 2006;9:636–41.

[504] Mameli M, Balland B, Lujan R, Luscher C. Rapid synthesis and synaptic insertion of GluR2 for mGluR-LTD in the ventral tegmental area. Science 2007;317:530–3.
[505] Mameli M, Halbout B, Creton C, Engblom D, Parkitna JR, Spanagel R, Lüscher C. Cocaine-evoked synaptic plasticity: persistence in the VTA triggers adaptations in the NAc. Nature Neuroscience 2009;12:1036–41.
[506] Carey M, Smale ST. Transcriptional regulation in eukaryotes: concepts, strategies, and techniques. Cold Spring Harbor, NY: Cold Spring Harbor Laboratory Press; 2000.
[507] Berke JD, Hyman SE. Addiction, dopamine, and the molecular mechanisms of memory. Neuron 2000;25:515–32.
[508] Nestler EJ, Aghajanian GK. Molecular and cellular basis of addiction. Science 1997;278:58–63.
[509] Nestler EJ, Hope BT, Widnell KL. Drug addiction: a model for the molecular basis of neural plasticity. Neuron 1993;11:995–1006.
[510] Robison AJ, Nestler EJ. Transcriptional and epigenetic mechanisms of addiction. Nature Reviews Neuroscience 2011;12:623–37.
[511] Shaham Y, Stewart J. Exposure to mild stress enhances the reinforcing efficacy of intravenous heroin self-administration in rats. Psychopharmacology 1994;114:523–7.
[512] Piazza PV, Deminiere JM, Le Moal M, Simon H. Stress- and pharmacologically-induced behavioral sensitization increases vulnerability to acquisition of amphetamine self-administration. Brain Research 1990;514:22–6.
[513] Goeders NE, Guerin GF. Non-contingent electric footshock facilitates the acquisition of intravenous cocaine self-administration in rats. Psychopharmacology 1994;114:63–70.
[514] Shaham Y. Immobilization stress-induced oral opioid self-administration and withdrawal in rats: role of conditioning factors and the effect of stress on "relapse" to opioid drugs. Psychopharmacology 1993;111:477–85.
[515] Haney M, Maccari S, Le Moal M, Simon H, Piazza PV. Social stress increases the acquisition of cocaine self-administration in male and female rats. Brain Research 1995;698:46–52.
[516] Miczek KA, Mutschler NH. Activational effects of social stress on IV cocaine self-administration in rats. Psychopharmacology 1996;128:256–64.
[517] Piazza PV, Rouge-Pont F, Deminiere JM, Kharoubi M, Le Moal M, Simon H. Dopaminergic activity is reduced in the prefrontal cortex and increased in the nucleus accumbens of rats predisposed to develop amphetamine self-administration. Brain Research 1991c;567:169–74.
[518] Maccari S, Piazza PV, Deminiere JM, Lemaire V, Mormede P, Simon H, Angelucci L, Le Moal M. Life events-induced decrease of corticosteroid type I receptors is associated with reduced corticosterone feedback and enhanced vulnerability to amphetamine self-administration. Brain Research 1991; 547:7–12.
[519] Hadaway PF, Alexander BK, Coambs RB, Beyerstein B. The effect of housing and gender on preference for morphine-sucrose solutions in rats. Psychopharmacology 1979;66:87–91.
[520] Marks-Kaufman R, Lewis MJ. Early housing experience modifies morphine self-administration and physical dependence in adult rats. Addictive Behaviors 1984;9:235–43.
[521] Schenk S, Gorman K, Amit Z. Age-dependent effects of isolation housing on the self-administration of ethanol in laboratory rats. Alcohol 1990;7:321–6.
[522] Schenk S, Lacelle G, Gorman K, Amit Z. Cocaine self-administration in rats influenced by environmental conditions: implications for the etiology of drug abuse. Neuroscience Letters 1987;81: 227–31.
[523] Wolffgramm J, Heyne A. Social behavior, dominance, and social deprivation of rats determine drug choice. Pharmacology Biochemistry and Behavior 1991;38:389–99.
[524] Bozarth MA, Murray A, Wise RA. Influence of housing conditions on the acquisition of intravenous heroin and cocaine self-administration in rats. Pharmacology Biochemistry and Behavior 1989;33: 903–7.
[525] Alexander BK, Beyerstein BL, Hadaway PF, Coambs RB. Effect of early and later colony housing on oral ingestion of morphine in rats. Pharmacology Biochemistry and Behavior 1981;15:571–6.
[526] Ramsey NF, Van Ree JM. Emotional but not physical stress enhances intravenous cocaine self-administration in drug-naive rats. Brain Research 1993;608:216–22.

[527] Deminiere JM, Piazza PV, Guegan G, Abrous N, Maccari S, Le Moal M, Simon H. Increased locomotor response to novelty and propensity to intravenous amphetamine self-administration in adult offspring of stressed mothers. Brain Research 1992;586:135–9.
[528] Boyle AE, Gill K, Smith BR, Amit Z. Differential effects of an early housing manipulation on cocaine-induced activity and self-administration in laboratory rats. Pharmacology Biochemistry and Behavior 1991;39:269–74.
[529] Phillips GD, Howes SR, Whitelaw RB, Wilkinson LS, Robbins TW, Everitt BJ. Isolation rearing enhances the locomotor response to cocaine and a novel environment, but impairs the intravenous self-administration of cocaine. Psychopharmacology 1994;115:407–18.
[530] Watanabe T, Yamamoto R, Maeda A, Nakagawa T, Minami M, Satoh M. Effects of excitotoxic lesions of the central or basolateral nucleus of the amygdala on naloxone-precipitated withdrawal-induced conditioned place aversion in morphine-dependent rats. Brain Research 2002;958:423–8.
[531] Manning BH. A lateralized deficit in morphine antinociception after unilateral inactivation of the central amygdala. Journal of Neuroscience 1998;18:9453–70.
[532] Manning BH, Mayer DJ. The central nucleus of the amygdala contributes to the production of morphine antinociception in the formalin test. Pain 1995;63:141–52.
[533] Yadin E, Thomas E, Strickland CE, Grishkat HL. Anxiolytic effects of benzodiazepines in amygdala-lesioned rats. Psychopharmacology 1991;103:473–9.
[534] Hayes RJ, Gardner EL. The basolateral complex of the amygdala mediates the modulation of intracranial self-stimulation threshold by drug-associated cues. European Journal of Neuroscience 2004; 20:273–80.
[535] Yun IA, Fields HL. Basolateral amygdala lesions impair both cue- and cocaine-induced reinstatement in animals trained on a discriminative stimulus task. Neuroscience 2003;121:747–57.
[536] Fuchs RA, Weber SM, Rice HJ, Neisewander JL. Effects of excitotoxic lesions of the basolateral amygdala on cocaine-seeking behavior and cocaine conditioned place preference in rats. Brain Research 2002;929:15–25.
[537] Schulteis G, Ahmed SH, Morse AC, Koob GF, Everitt BJ. Conditioning and opiate withdrawal: the amygdala links neutral stimuli with the agony of overcoming drug addiction. Nature 2000;405: 1013–4.
[538] Cador M, Robbins TW, Everitt BJ. Involvement of the amygdala in stimulus-reward associations: interaction with the ventral striatum. Neuroscience 1989;30:77–86.
[539] Chudasama Y, Robbins TW. Dissociable contributions of the orbitofrontal and infralimbic cortex to pavlovian autoshaping and discrimination reversal learning: further evidence for the functional heterogeneity of the rodent frontal cortex. Journal of Neuroscience 2003;23:8771–80.
[540] Rasmussen K, Beitner-Johnson DB, Krystal JH, Aghajanian GK, Nestler EJ. Opiate withdrawal and the rat locus coeruleus: behavioral, electrophysiological, and biochemical correlates. Journal of Neuroscience 1990;10:2308–17.
[541] George O, Koob GF. Control of craving by the prefrontal cortex. Proceedings of the National Academy of Sciences of the United States of America 2013;110:4165–6.
[542] Koob GF, Everitt BJ, Robbins TW. Reward, motivation, and addiction. In: Squire LG, Berg D, Bloom FE, Du Lac S, Ghosh A, Spitzer N, editors. Fundamental neuroscience. 3rd ed. Amsterdam: Academic Press; 2008. p. 987–1016.
[543] Pfefferbaum A, Sullivan EV, Mathalon DH, Shear PK, Rosenbloom MJ, Lim KO. Longitudinal changes in magnetic resonance imaging brain volumes in abstinent and relapsed alcoholics. Alcohol Clin Exp Res 1995;19:1177–91.
[544] Carlezon Jr WA, Nestler EJ, Neve RL. Herpes simplex virus-mediated gene transfer as a tool for neuropsychiatric research. Critical Reviews in Neurobiology 2000;14:47–67.

Index

'*Note:* Page numbers followed by "f" indicate figures, "t" indicate tables.'

B

C

Printed in the United States
By Bookmasters